T0275681

HUMAN MODELING FOR BIO-INSPIRED ROBOTICS

HUMAN MODELING FOR BIO-INSPIRED ROBOTICS

Mechanical Engineering in Assistive Technologies

Editors

JUN UEDA
Georgia Institute of Technology, Atlanta, GA, United States

YUICHI KURITA
Hiroshima University, Hiroshima, Japan

AMSTERDAM • BOSTON • HEIDELBERG • LONDON
NEW YORK • OXFORD • PARIS • SAN DIEGO
SAN FRANCISCO • SINGAPORE • SYDNEY • TOKYO
Academic Press is an imprint of Elsevier

Academic Press is an imprint of Elsevier
125 London Wall, London EC2Y 5AS, United Kingdom
525 B Street, Suite 1800, San Diego, CA 92101-4495, United States
50 Hampshire Street, 5th Floor, Cambridge, MA 02139, United States
The Boulevard, Langford Lane, Kidlington, Oxford OX5 1GB, United Kingdom

Notices
Knowledge and best practice in this field are constantly changing. As new research and
experience broaden our understanding, changes in research methods, professional
practices, or medical treatment may become necessary.

Practitioners and researchers must always rely on their own experience and knowledge
in evaluating and using any information, methods, compounds, or experiments described
herein. In using such information or methods they should be mindful of their own safety
and the safety of others, including parties for whom they have a professional responsibility.

To the fullest extent of the law, neither the Publisher nor the authors, contributors, or
editors, assume any liability for any injury and/or damage to persons or property as a
matter of products liability, negligence or otherwise, or from any use or operation of any
methods, products, instructions, or ideas contained in the material herein.

Library of Congress Cataloging-in-Publication Data
A catalog record for this book is available from the Library of Congress

British Library Cataloguing-in-Publication Data
A catalogue record for this book is available from the British Library

ISBN: 978-0-12-803137-7

For information on all Academic Press publications
visit our website at https://www.elsevier.com/

Working together
to grow libraries in
developing countries

www.elsevier.com • www.bookaid.org

Publisher: Joe Hayton
Acquisition Editor: Fiona Geraghty
Editorial Project Manager: Maria Convey
Production Project Manager: Nicky Carter
Designer: Matthew Limbert

Typeset by SPi Global, India

CONTENTS

11. Adaptive Human-Robot Physical Interaction for Robot Coworkers   **297**

J. Ueda, W. Gallagher, A. Moualeu, M. Shinohara, K. Feigh

CONTRIBUTORS

A. Alamdari
The State University of New York at Buffalo, Buffalo, NY, United States

S. Bai
Aalborg University, Aalborg, Denmark

E. Briscoe
Georgia Institute of Technology, Atlanta, GA, United States

N. Bu
National Institute of Technology, Kumamoto College, Kumamoto, Japan

S. Christensen
Aalborg University, Aalborg, Denmark

A. Dani
University of Connecticut, Storrs, CT, United States

A. Deshpande
University of Texas, Austin, TX, United States

K. Feigh
Georgia Institute of Technology, Atlanta, GA, United States

P. Fraisse
INRIA, University of Montpellier, Montpellier, France

O. Fukuda
Saga University, Saga, Japan

W. Gallagher
NASA's Goddard Space Flight Center, Greenbelt, MD, United States

A. González
INRIA, University of Montpellier, Montpellier, France

M. Hayashibe
INRIA, University of Montpellier, Montpellier, France

N. Kirsch
University of Pittsburgh, Pittsburgh, PA, United States

V.N. Krovi
The State University of New York at Buffalo, Buffalo, NY, United States

G. Lisi
ATR Computational Neuroscience Laboratories, Kyoto, Japan

J. Morimoto
ATR Computational Neuroscience Laboratories, Kyoto, Japan

A. Moualeu
Georgia Institute of Technology, Atlanta, GA, United States

T. Niehues
University of Texas, Austin, TX, United States

Y.-L. Park
Robotics Institute, Carnegie Mellon University, Pittsburgh, PA, United States; Seoul National University, Seoul, Republic of Korea

P. Rao
University of Texas, Austin, TX, United States

H.C. Ravichandar
University of Connecticut, Storrs, CT, United States

N. Sharma
University of Pittsburgh, Pittsburgh, PA, United States

M. Shinohara
Georgia Institute of Technology, Atlanta, GA, United States

T. Tsuji
Hiroshima University, Hiroshima, Japan

J. Ueda
Georgia Institute of Technology, Atlanta, GA, United States

A. Wagner
Georgia Institute of Technology, Atlanta, GA, United States

MOTIVATION

Human factors must be well studied and understood for future assistive and healthcare technologies where the design, control, and analysis of such systems rely heavily on the behaviors of the human in the loop. Emerging areas include assistive and healthcare robotics due to the increasing societal demand to develop technologies to improve quality of life. For example, the National Robotics Initiative, announced by the US government in 2011, is to encourage and support research in human-in-the-loop robotic systems or "co-robots."

In addition to understanding the key components in general robotics, such as mechanics and programming, this emerging research requires analysis of human kinesiology, physiology, and psychology from the engineering point of view, termed *human modeling* in this book. To grasp the complex and interdisciplinary aspects of the area and apply the knowledge to practical applications, system-level studies must be performed.

Research findings relevant to co-robotics tend to be published in journals and conference proceedings across a wide variety of topics, which makes it difficult for new researchers to effectively find study materials. This book is intended to provide researchers in academia and industry with fundamental content on which to base their development in the field. The most recent outcomes in the mechanics and control of human function toward macroscale (human size) applications are presented here, organized and written by senior-level experts in their fields. While the modeling of the ocular and auditory systems is also important, these are out of the contributors' expertise and are not included.

The collected works offer a system-level investigation into human mechanisms, including topics in modeling of anatomical, musculoskeletal, neural and cognitive systems, as well as motor skills, adaptation, and integration. The intended audience is graduate-level students as well as professional engineers who are new to this field of research and need to get up to speed with the mechanical design and control aspects of the included topics. The authors/editors assume that the readers have fundamental, undergraduate-level knowledge in engineering, such as mechanical, biomedical, or electrical engineering. Each chapter emphasizes and summarizes its subject's background, research challenges, key outcomes, application, and future trends, while minimizing the use of advanced mathematics. This approach is also intended to introduce readers with physiology and biomedicine backgrounds to the engineering aspects involved. While the book is organized around a consistent theme, each chapter is written as an independent, comprehensive article.

The editors and chapter contributors have organized workshops and invited sessions at the IEEE International Conference on Robotics and Automation, IEEE Advanced Intelligent Mechatronics Conference, and ASME Dynamics and Control Conference. The effort was supported by the Mechatronics, Robotics and Bio-Systems Technical Committees in the ASME Dynamic Systems and Control Division. We would like to thank all the organizers, presenters, and staff members who helped the effort.

ABOUT THIS BOOK

This book covers two themes of human modeling:
1. Modeling of human musculoskeletal system/computational analysis of human movements and their applications.
2. Modeling of human cognitive/neuromuscular skills and their applications.

There are six themes that focus on the human musculoskeletal system and computational analysis of human movements in this book. In Chapter 1, Deshpande et al. present an approach for modeling the passive stiffness at the human metacarpophalangeal joint, and the individual contributions from the elasticity of muscle-tendon units and the capsule ligament complex; they conducted experiments with 10 human subjects and collected joint angle and finger tip force data. In Chapter 2, Alamdari and Krovi review computational modeling efforts focused on improving understanding of physical interactions of human lower extremities with their physical environment, and discuss the application settings for human-robot interactions. In Chapter 3, Bu et al. describe a hybrid motion and task modeling framework for human-robot interfaces based on electromyogram (EMG) signals, and introduce a case study of the EMG-controlled human-robot interface using task modeling. In Chapter 4, Hayashibe et al. overview home-based rehabilitation along with other state-of-the-art works, and address a personalized measure of balance that considers subject-specific variations. In Chapter 5, Sharma and Kirsch discuss the use of a forward dynamic optimization with a three-link dynamic walking model to calculate joint angle trajectories that require minimum electrical stimulation and/or motor torque inputs. In Chapter 6, Park provides an overview of soft wearable technologies with a primary emphasis on body motion, and discusses robotic systems that utilize the soft sensors and actuators integrated in various wearable systems as examples of assistive and rehabilitation technologies.

There are also five themes that focus on modeling of human cognitive and muscular skills. In Chapter 7, Lisi and Morimoto present an overview of the available noninvasive techniques followed by a detailed description of brain machine interfaces (BMIs) based on sensorimotor rhythms, and describe the main challenges in building an SMR-based decoder and its main components. In Chapter 8, Ravichandar and Dani review algorithms for human-intention estimation and introduce an algorithm to infer the intent of a human operator's arm movements based on the observations from a Microsoft Kinect sensor. In Chapter 9, Bai and Christensen overview challenges in exoskeleton design and introduce a design approach based on biomechanical simulations of human-robot interaction, which aims to effectively design an exoskeleton structure and motion controller to implement motion assistance as needed. In Chapter 10, Wagner and Briscoe examine an anticipated upcoming generation of robots that will socially interact with patients, modeling their moods, personality, likes, and dislikes, and use this information to guide the robot's assistance-related decisions. Finally, In Chapter 11, Ueda et al. describe research that aims to develop theories, methods, and tools to understand the mechanisms of neuromotor adaptation in human-robot physical interaction.

PART ONE

Modeling of Human Musculoskeletal System/Computational Analysis of Human Movements and Their Applications

CHAPTER ONE

Implementation of Human-Like Joint Stiffness in Robotics Hands for Improved Manipulation

A. Deshpande, T. Niehues, P. Rao
University of Texas, Austin, TX, United States

1. INTRODUCTION

In the past two decades, a number of robotic hands have been designed with compliance with the goal of improving interaction stability, robustness, and manipulation abilities. In the human hand, the intrinsic, passive joint stiffness of the fingers critically affects the hand functions and joint stability [1–3]. Because of its prominent role in many hand functions, a number of studies have focused on investigating the joint stiffness of the index finger metacarpophalangeal (MCP) joint [2, 4–7]. The muscle-tendon units (MTUs) contribute to the passive MCP joint stiffness by generating resistive forces when stretched [2, 3]. In addition to the MTUs, the capsule-ligament complex (CLC) at the joint provides resistance, especially to prevent joint instability. We carry out a quantitative analysis to determine the relative contributions of MTUs and CLC to the passive joint stiffness, in various finger configurations. It has been established through a number of previous studies that the resistive moment generated at the MCP joint due to the joint stiffness has a double exponential dependency on the joint angle [4, 5]. A number of previous studies assume, either explicitly or implicitly, that this joint stiffness is strictly due to the passive forces from the stretching of muscles and tendons, and that the effect of CLC can be ignored [7, 8].

Inspired by the dominant role of joint compliance in the human hands, we explore the role of mechanical springs added in parallel to the joint actuators (parallel compliance) in robotic hands. While the series compliance, introduced through series elastic actuators (SEA), has been studied extensively [9] and applied in many robotic systems [10, 11], the effects of introduction of parallel compliance on the performance of the system

3

have not been analyzed. We analyze the effects of introducing parallel compliance in improving stability, robustness, and trajectory smoothness during object grasping and manipulation in the presence of time delay.

We first develop a fundamental understanding of the effects of parallel compliance with mathematical modeling of a single joint robotic system implementing an impedance controller with time delay. This leads to identification of trade-offs in adding parallel compliance and generation of guidelines for the suitability and advantage of adding parallel compliance. The findings are experimentally validated with a tendon-driven robotic finger joint. Next we analyze the effects of parallel compliance in a system with two fingers performing object grasping and manipulation. Our focus is on object interactions with the robots, so we first develop force control strategies, specifically impedance control, for these systems. The experiments with systems with and without parallel compliance demonstrate improvements due to addition of parallel compliance in trajectory tracking accuracy and robustness to impacts.

2. MODELING OF JOINT STIFFNESS
2.1 Method

In this section, we first introduce a custom-made mechanism that allowed us to collect passive force and kinematic data from human subjects. Next, we describe information on experiment participants and the experimental design in detail. Finally, we demonstrate a subject-specific musculoskeletal modeling and data analysis.

2.1.1 Mechanism Design

We present a design of a mechanism to measure the passive moments of the MCP joint of the index finger during static and dynamic tests (Fig. 1). Here are the key features of the mechanism:

1. **Precision of the force sensing**: Attaching force/torque sensors directly on the measured fingers produces significant error caused by the local deformations of the soft tissues. Even with the braces, the sliding between the braces and the fingers increases the difficulty of measuring a reliable force under dynamic conditions. We designed a flushed contact mechanism with force sensors and extended linkages, which allows us to estimate the overall passive torque of the measured joint and resolve the issue raised by the skin deformation and sliding.

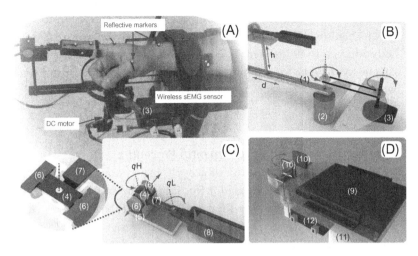

Fig. 1 (A) Rest position of the subject's hand and the full setup of the mechanism. First, we attached the markers and EMG sensors on the subject's forearm and aligned the MCP joint with the shaft of the DC motor. Then, we fixed the forearm on the testing panel with the velcro straps and the arm rest, and adjusted the palm holders to fix the hand at the zero position. Finally, we adjusted the driving arm to fit the index finger into the splint. After we secured the subject's hand on the mechanism, we manually moved the driving arm to test the setup. (B) Design of the driving arm and its subparts. The load cell holder attached on the moment arm (1) can be adjusted with the height (*h*) and the distance (*d*) to fit the different hand sizes. The DC motor (2) and the encoder (3) are connected by a chain and sprocket drive. (C) Design of a load cell holder and splint linkage. A piece called the "hammer" (4) achieves a flush contact between the splint mechanism and the load cell throughout the range of motion of the finger. The arrows indicate the sliding direction of the hammer. The hammer and the cylinder have a size tolerance 0.1 mm so that the hammer can slide along the cylinder. The hinge joint in the hammer allows a relative rotation between the hammer and linkage expanding out from the holder. The subplot shows a section view of the hammer design. (D) Design of the testing panel and adjustable stand. The palm holders (10) can be moved and fixed to the desired direction and rotate for 360 degrees through the two slots on the testing panel. The palm holder can fix the palm in place for all subjects.

2. **Accuracy of the joint kinematics**: The center of rotation of the human articular joint does not have a fixed point. Therefore, it is problematic to measure the joint angle and torque by attaching sensors directly on the joint. Infrared motion capture is a noninvasive method for estimating joint rotations accurately. The mechanism is carefully designed to integrate with the customized reflective markers and infrared motion capture system.

3. **Customizability of the device**: Subject specific design is considered in the mechanism. We designed a test panel and driving arm that can be easily adjusted to fit different sizes of the human hands. The mechanism can collect the data for different finger joints with various wrist postures by changing the configuration of moving clamps and different sizes of finger braces.

2.1.2 Human Subjects and Experimental Procedure

A total of 10 right-handed healthy subjects (6 males, 4 females) ranging in age 23 ($\pm$3.7) years were recruited for this study. The anthropometric data of the index fingers was measured for each subject (Table 1).

The subject's hand was placed into a custom-designed device for the experiments (Fig. 1). The device fixed the subject's index finger and allowed other fingers to be relaxed so that the index finger could rotate freely in the horizontal plane. Each subject performed the maximal isometric index finger flexion and extension for sEMG normalization and scaling purposes. The device drove the subject's index finger with 10-degree increments from the neutral position, defined by the encoder, to the direction of full extension and reversed the direction of finger rotation to full flexion for two cycles. The device held each finger position for 30 s during which the forces reached a steady state due to the muscle relaxation [4]. The limit of the range of motion (RoM) was decided for each subject when the subject started to feel uncomfortable close to the extremity of the rotation. To monitor the muscle relaxation for the subjects, we attached four wireless electromyographic sensors on the subject's flexor digitorum superficialis (FDS), extensor digitorum communis (EDC), biceps and triceps to monitor the muscle relaxation (Trigno, DelSys, Inc.). For EDC, we placed an electrode at the mid-forearm on a line drawn from the lateral epicondyle to the ulna styloid process; for FDS, an electrode was placed around the center point on the line joining the medial epicondyle to the ulna styloid process [12]. We placed the

Table 1 Measurements of the Anthropometric Parameters (in Millimeters) Averaged Across 10 Subjects (Mean (Standard Deviation))

	Length	Breadth	Thickness	Joint Thickness
Proximal	45.0 (3.5)	14.8 (1.1)	14.5 (1.1)	21.6 (4.9)
Intermediate	25.3 (1.7)	13.7 (1.1)	11.6 (1.2)	13.9 (1.3)
Distal	23.6 (1.8)	11.5 (1.2)	8.7 (0.7)	10.2 (0.8)

It is assumed that the finger segment has a uniform rectangle shape with rectangular cross-section. Each subject signed an informed consent form in agreement with the university's human subject policy.

electrode on the bulk of the biceps in mid-arm and four finger-breadths distal to the posterior axillary fold of the triceps (long head) [13]. A motion capture system (Vicon, Inc.) with six infrared cameras (500 Hz) and 18 reflective markers (diameter: 4.17 mm) was used to collect the three-dimensional kinematic data of the MCP joint during the experiment in order to precisely determine the MCP joint angle during movements. Each subject performed two repetitions of full range of flexion-extension motion. The forces at the finger tip, EMG signal, and the finger's kinematic data were collected simultaneously during the experiments.

2.1.3 Data Analysis

The kinematic data of the markers was synchronized with the EMG signals (Nexus 1.7.1, Vicon, Inc.). Using the marker data, the location of the instant center of rotation (iCoR) and the MCP flexion-extension angle were determined through an optimization process [14, 15]. We defined the MCP angle to be zero along the line joining the wrist and MCP joint centers, positive in flexion and negative in extension. We normalized the EMG signals of each muscle by measuring the EMG signals from the maximal voluntary isometric contraction (MVC) test before proceeding with the experiment. We processed the raw EMG signals with a fourth-order bandpass Butterworth filter (20–500 Hz), performed full-wave rectification, and then passed it through a low-pass filter with a cut-off frequency of 5 Hz to derive the linear envelope EMGs. We adopted the average of the linear envelope EMG as 100% effort of muscle activations. The data from a trial was eliminated when either one of four EMG signals exceeded the thresholds with 5% of the determinations from the maximal voluntary isometric contraction test [16].

2.1.4 Total Elastic Moment (τ_{total})

The total passive moment due to joint stiffness is given by: $\tau_{total} = l_{tip} \times F_{tip}$, where l_{tip} is the distance between the location of the force sensor and the iCoR of the MCP joint. A double exponential model, given in Eq. (1), was employed to describe the relationship between the total passive elastic moment and the MCP joint angle [4, 5, 17]:

$$\tau_{total}(\theta_m) = A(e^{-B(\theta_m - E)} - 1) - C(e^{D(\theta_m - F)} - 1) \qquad (1)$$

where θ_m is the angle of the MCP joint, and A to F are the parameters of the fitting model. We estimated the seven model parameters for each subject by using a nonlinear least squares method that minimizes the sum of square differences between the measured moment and fitting model in Matlab (Mathworks, Inc.). The slack angle (θ_{ms}), defined as the relaxed position of the index finger, was determined in the fitting model at which τ_{total} was equal to zero [6].

2.1.5 Elastic Moment From MTUs (τ_m)

The net elastic moment by the seven MTUs that cross the MCP joint (Table 2), $\tau_{MTU}(\theta_m)$ varies with the MCP joint angle and is given by Eq. (2):

$$\tau_{MTU}(\theta_m) = \sum_{i=1}^{7} R_i(\theta_m) \cdot F_{pi}(\theta_m), \quad i = 1, \ldots, 7 \tag{2}$$

where $R(\theta_m)$ is the vector of the moment arms of the seven MTUs with respect to the MCP joint angle and $F_p(\theta_m)$ is the vector of passive forces generated by the seven MTUs in response to the stretch due to change in θ_m. The moment arms for the seven MTUs vary as θ_m changes and we used the model derived for the ACT Hand MCP joint to determine the values of moment arms [18]. We assumed that the moment arms are proportional to the volume of index finger [19] and calculated the subject-specific moment arms of the seven MTUs by scaling the ACT hand moment arms

Table 2 Results of Scaled Parameters for the Seven Muscles ($n = 10$)

Muscle	Abbreviation	F_{mo} (N)	l_{mo} (mm)	l_s (mm)	α(deg)
Extensor digitorum communis	EDC	18.9 ± 3.2	61.9 ± 4.4	289.9 ± 20.9	3
	EDC[a]	(18.26)	(70)	(322)	
Extensor indicis	EI	22.5 ± 3.8	52.1 ± 3.7	167.7 ± 12	6
	EI[a]	(21.7)	(58.9)	(186)	
First palmar interosseous	PI[b]	18.3	30.7	25	6.3
Flexor digitorum profundus	FDP	75.9 ± 17.9	67.2 ± 7	265.2 ± 19.1	7
	FDP[a]	(68.3)	(74.9)	(293.5)	
First lumbrical	LU[b]	2.7	68	55.4	1.2
Flexor digitorum superficialis	FDS	68.1 ± 16.1	72.7 ± 3.9	247.5 ± 18	6
	FDS[a]	(61.24)	(83.5)	(275)	
First dorsal interosseous	DI[b]	36.6	38.9	31.7	9.2

[a]Shows the generic values of four extrinsics from [20, 21].
[b]Shows the constant values of three intrinsics adopted from [22, 23].

Table 3 Scaling Factors of the Moment Arm (r_v), l_{mto} (r_V), and F_{mo} (r_f and r_e) for 10 Subjects

Factor	1	2	3	4	5	6	7	8	9	10
r_v	0.68	1.26	0.85	0.84	0.87	0.88	0.80	0.98	0.84	0.62
r_V	0.43	0.79	0.53	0.53	0.55	0.55	0.50	0.62	0.53	0.39
r_f	1.50	1	1.21	1.19	0.64	1.25	1.29	1.12	1.22	0.71
r_e	1.22	1	1.09	0.93	0.97	1.35	1.02	1.05	1.04	0.69

with a ratio of the volume of subject's index finger (v) to the ACT hand index finger (V):($r_v = \frac{v}{V}$, Table 3). The scaled moment arms were also used to calculate the length change of MTUs (l_{mt}) due to change in θ_m as: Δl_{mt_i} (θ_m) = $R_i \times \theta_m$, where $i = 1,...,7$ refers to the number of the MTU.

To determine the passive stiffness generated by these seven MTUs as the MCP joint is moved passively through its full RoM, the MTUs are assumed to be composed of two nonlinear springs, representing the muscle and tendon, connected in series as shown in Fig. 2. Four parameters define the static passive force-length relationships in the Hill-type MTU model, namely, the maximal isometric force (F_{mo}), optimal muscle fiber length (l_{mo}), tendon slack length (l_{ts}), and pennation angle (α^m) [24–26]. Values for these parameters have been determined in the previous works through cadaver studies and modeling techniques [20–22, 27–29], and in this study we adapted the F_{mo}, l_{mo}, and l_{ts} values for each subject. Because of small pennation angles for the seven muscles [20, 22], we maintained the same values of pennation angles for all of the subjects.

The voltages generated by the EMG signals from the MVC were used to scale F_{mo} values for the four extrinsic muscles. First, we identified a subject (Subject 2) whose index finger volume matches closely (79.08%) with the model presented by Holzbaur [21], and we assigned the F_{mo} values of the extrinsic muscles (FDP, FDS, EDC, and EI) from Holzbaur's model to Subject 2's model. Then we calculated the ratio of the EMG value from MVC in flexion and extension for each subject and Subject 2. The EMG ratios in flexion and extension were used to scale the F_{mo} of the two flexors and two extensors between subjects (r_f and r_e in Table 3).

We calculated the subject-specific nominal MTU lengths, l_{mto}, by linearly scaling the l_{mto} values from Holzbaur's model with the volume ratio (r_V) for each subject (Table 3) [19]. The tendon slack lengths (l_{ts}) were functionally adjusted for each subject by implementing the numerical optimization method described in a previous study [28]. The muscle fiber lengths (l_m)

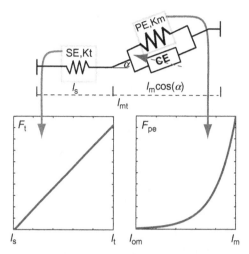

Fig. 2 A schematic showing the muscle and tendon element of the Hill-type model. Each MTU is modeled as two nonlinear springs connected in series. The force generated by the MTU for the given change in length is calculated by using the passive force-length relationships for the muscles and tendon. The passive force-length relationship in the muscle is given by $f_m(\tilde{l}_m) = \dfrac{e^{10(\tilde{l}_m - 1)}}{e^5}$, $F_m = F_{mo} f_m \cos(\alpha_m)$, where $\tilde{l}_m = \dfrac{\Delta l_m}{l_{mo}}$. The passive force-length relationship in the tendon is given by $f_t(\epsilon_t) = 0$ when $\epsilon_t \leq 0$, $f_t(\epsilon_t) = 1480.3\epsilon_t^2$ when $0 < \epsilon_t < 0.0127$, $f_t(\epsilon_t) = 37.5\epsilon_t - 0.2375$ when $\epsilon_t \geq 0.0127$, $F_t = F_{mo} \cdot f_t$, where $\epsilon_t = \dfrac{\Delta l_t}{l_{ts}}$. With two nonlinear springs in series the force generated by the two is equal: $F_m = F_t$ and the total length change is the sum of the length changes: $\Delta l_{mt} = \Delta l_m + \Delta l_t$ [30].

were randomly selected as inputs in the fully flexed, fully extended and relaxed positions. The optimal muscle length values were then calculated for each subject as: $l_{mo} = l_{mto} - l_{ts}$ at the relaxed position after the tendon slack lengths were updated. Table 2 shows the statistical results of the scaling parameters.

2.1.6 MTU Contribution

The total passive joint moment (τ_{total}) at the MCP joint is assumed to be composed of the elastic moment from the stretching of the seven MTUs (τ_{MTU}) and the passive moment from the CLC (τ_{CLC}). Values for τ_{total} and τ_{MTU} are determined by following the steps explained previously and τ_{CLC} is estimated from Eq. (3)

$$\tau_{CLC} = \tau_{total} - \tau_{MTU} \tag{3}$$

To evaluate the contributions of the MTUs to the joint stiffness, we computed the mechanical work of the passive moments at the MCP joint using Eq. (4)

$$
\begin{aligned}
W &= W_f + W_e \\
&= \left| \int_{\theta_{ms}}^{\theta_{mf}} \tau(\theta)d\theta \right| + \left| \int_{\theta_{ms}}^{\theta_{me}} \tau(\theta)d\theta \right|
\end{aligned}
\tag{4}
$$

where θ_{me} and θ_{mf} are the values of the MCP angle in full extension and full flexion, respectively, θ_{ms} is the slack angle, and W_f and W_e represent work done in flexion and extension, respectively. We calculated the work done by τ_{total} and τ_{MTU} as W_{MTU} and W_{total}, respectively, using Eq. (4) and then computed the contribution: $\eta = \dfrac{W_{MTU}}{W_{total}} \times 100$. We also calculated the MTU contribution to the total work in flexion: $\eta_f = \dfrac{W_{MTU_f}}{W_{total_f}} \times 100$ and also in extension: $\eta_e = \dfrac{W_{MTU_e}}{W_{total_e}} \times 100$. One sample t-test and the power analysis were used to test our hypothesis as a post hoc ($\alpha = 0.05$ and $power = 0.8$).

2.2 Results

2.2.1 Model of Total Passive Moment

For all the subjects, the slack angle θ_{ms} is located at $\theta_m > 0$ degrees (Table 4). The total passive moment in full extension (τ_{total_e}) is larger (438.1 ± 184.61 N-mm) than that in full flexion (τ_{total_f}, 288.41 ± 71.41 N-mm). The double exponential model defined by the parameters given in Table 4 closely fits each subject's passive moment–angle data, with the R^2 value greater than 0.9 and RMSE value smaller than 50 N-mm. The fit between the experimental data and the model for Subject 7 is shown in Fig. 3.

2.2.2 Model of Passive Moment From MTUs

The passive forces generated by the extrinsic MTUs, including the flexors (FDP and FDS) and extensors (EDC and EI), increase exponentially with the stretching length, and the resulting moment has an exponential dependency on the joint angle (Fig. 4A and C). The passive muscle moment is zero for flexors in flexion and extensors in extension. The passive forces generated by the intrinsic MTUs have nonmonotonic dependencies on the stretch length and the resulting joint moments are nonmonotonic with respect to the MCP angle change (Fig. 4B and D).

Table 4 Results From Fitting the Double Exponential Model of the MCP Joint Moment and Angle Data

	Experimental Results					Fitting Model							
Subject	θ_{me} (degree)	θ_{ms} (degree)	θ_{mf} (degree)	τ_{total_e} (N-mm)	τ_{total_f} (N-mm)	A	B	C	D	E	F	RMSE	R^2
1	− 59.22	11.79	97.21	573.00	− 292.20	0.27	0.06	13.28	0.03	72.83	− 6.74	23.86	0.93
2	− 57.30	21.24	102.48	334.70	− 278.10	0.01	0.08	4.11	0.05	73.95	16.32	15.71	0.9
3	− 68.55	9.02	108.54	372.10	− 222.00	0.19	0.05	23.82	0.02	73.55	− 1.42	15.12	0.94
4	− 79.45	26.77	98.79	381.90	− 259.90	5.54	0.03	1.71	0.07	74.51	15.34	36.74	0.74
5	− 69.43	9.20	98.44	470.70	− 209.58	3.00	0.04	28.22	0.02	70.18	− 18.27	29.25	0.9
6	− 59.24	11.96	98.79	682.50	− 433.70	3.60	0.05	2.28	0.03	47.43	− 51.86	29.18	0.92
7	− 65.36	32.97	102.65	558.10	− 329.70	1.01	0.05	3.39	0.05	70.96	13.68	17.62	0.95
8	− 55.72	8.75	93.25	477.36	− 280.31	0.53	0.05	15.83	0.03	73.51	− 11.43	36.81	0.93
9	− 67.41	23.60	103.11	609.79	− 213.63	4.54	0.03	22.32	0.02	75.49	− 5.73	32.89	0.94
10	− 64.73	19.45	101.63	358.93	− 364.99	1.44	0.05	16.72	0.03	71.23	− 4.65	41.43	0.93

The ranges of the MCP joint angles and the passive moments recorded from the experiment and the coefficients of the fitting model for the 10 subjects are given. The MCP angle is measured from the line joining the wrist and MCP joint centers. θ_{ms} is the slack angle of the MCP joint, $\theta_m \in [\theta_{me} \ \theta_{mf}]$ is the RoM, τ_{total_e} and τ_{total_f} represent the maximal passive joint moment in full extension and full flexion, respectively.

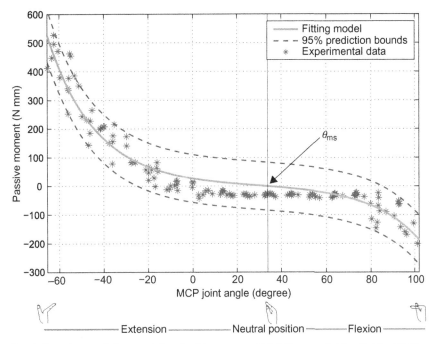

Fig. 3 Experimental data and the double exponential fitting model for Subject 7. The 95% confidence interval predicts the mean total moment of the MCP joint from the experimental data. The slack angle (θ_{ms}) determines the relative contributions of the extensors and flexors to the MTU passive moment. A high positive value of θ_{ms} means a larger range of extension than that with a low positive value of θ_{ms}, leading to larger stretch and passive moments by the flexors (FDS and FDP).

The extrinsic MTUs stretched by longer lengths and generated higher passive forces than the intrinsic MTUs. For example, in the case of Subject 7 the extrinsic MTUs generated passive forces ranging from 2 to 15 N with stretching lengths over 14 mm (Fig. 4E and F). On the other hand, intrinsic MTUs showed limited length excursions and force generation capacity. Even for the strongest intrinsic muscle DI, the maximal value of stretching length and passive force is only 2.34 mm and 0.267 N, respectively.

2.2.3 Relative Contributions of MTUs and CLC

For all the subjects, the sum of the passive moments due to the stretching of the MTUs increases exponentially as the MCP joint is flexed or extended. The increase in the passive moment due to the stretch in the MTUs in flexion and extension of the MCP joint (τ_{MTU}) is significantly lower $(p < .001)$ than the total increase in passive moment (τ_{total}), leading to a large value of

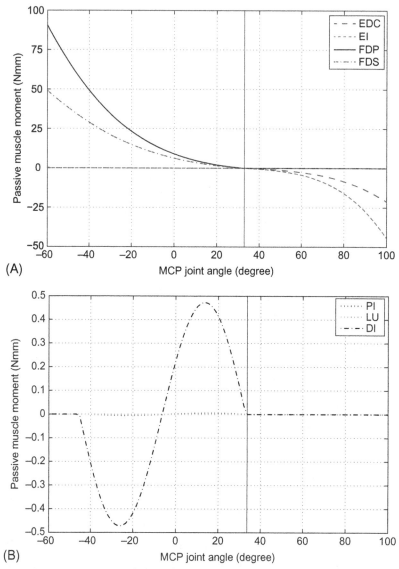

Fig. 4 Passive moments and forces of the extrinsic MTUs (A, C, and E) and the intrinsic MTUs (B, D, and F) for Subject 7. (A) Extrinsic MTU moments w.r.t. MCP joint angle. (B) Intrinsic MTU moments w.r.t. MCP joint angle.

(Continued)

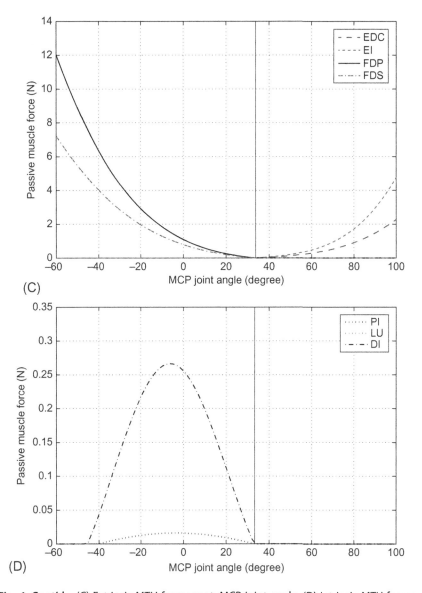

Fig. 4, Cont'd   (C) Extrinsic MTU forces w.r.t. MCP joint angle. (D) Intrinsic MTU forces w.r.t. MCP joint angle.

(Continued)

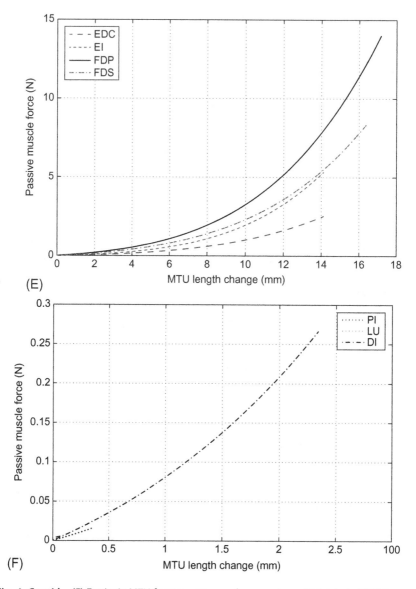

(E)

(F)

Fig. 4, Cont'd   (E) Extrinsic MTU forces w.r.t. muscle excursions. (F) Intrinsic MTU forces w.r.t. muscle excursions.

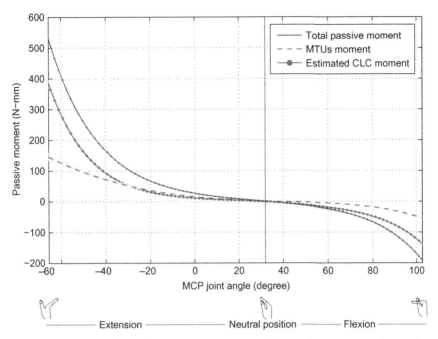

Fig. 5 Variations in the total passive moment τ_{total}, the passive moments due to the MTUs τ_{MTU} and the CLC τ_{CLC} for Subject 7. The moment due to the CLC was determined from the double exponential model and the MTU model using Eq. (3). The gray areas around the red and green curves represent the 95% confidence intervals of the MTU moment and the estimated CLC moment.

passive moment due to the CLC (τ_{CLC}) (Fig. 5). The passive moment due to the CLC also increases exponentially as the MCP joint is flexed or extended.

Fig. 6 shows that the work done by the lengthening of the MTUs is significantly lower than the total work done due to finger flexion and extension ($p < .001$). Thus the contribution of work η is less than 50% of the total work. For all the subjects, the contribution of the MTUs to the total work in flexion η_f is greater than that in extension η_e, although in both cases the contribution from the CLC dominates ($p < .001$) (Fig. 7).

2.3 Summary of Joint Stiffness Modeling

The purpose of this study was to investigate the relative contributions of the MTUs and CLC to the total passive stiffness at the index finger MCP joint in flexion and extension. Our results showed that the MTUs produced less than 50% of the contribution to the total work and moment due to stiffness at the MCP joint.

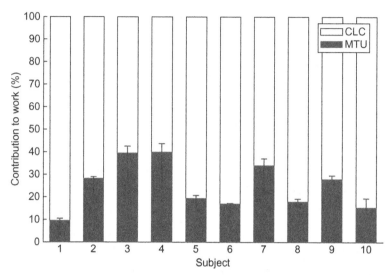

Fig. 6 Relative contribution of the MTUs and CLC to the total work done to the passive moment through the range of motion for each subject. The contribution of the MTUs to the total work, η (24.87 ± 10.61), is significantly less than 50% of the total work, $t(9) = -7.49$, $p < .001$, *power* = 0.68.

The work done due to the passive moment generation was used as a metric for the determination of the contributions to stiffness from the MTUs and CLC. The CLC contributed to more than 50% of the total work (Fig. 6), and this trend is maintained in flexion and extension (Fig. 7). The MTUs' contribution in flexion is higher than that in extension because of larger RoM in flexion at which extensors have higher torque generation abilities with longer muscle excursions (Table 4). The contribution of the MTUs to the joint stiffness varies with the MCP joint angle.

Based on the results of this study we designed a robotic joint with human-like joint stiffness. The analysis and control with such joints is presented in the following sections.

As shown in the earlier section, the joint compliance dominates the muscle-tendon compliance in the human hands. Based on this result, we explore the role of mechanical springs added in parallel to the joint actuators (parallel compliance) in robotic hands [31].

2.4 Analysis of a Robotic Joint With Parallel Compliance

We begin by modeling a generic robotic system with one actuated joint and parallel compliance. Assuming that the actuator is backdriveable and an infinitely stiff transmission, these two can be lumped together as a single

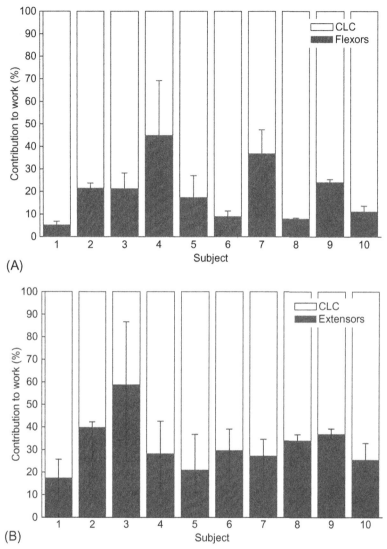

Fig. 7 Relative contribution to the work of the flexors and CLC in extension, and extensors and CLC in flexion. The contribution of the flexors to the total work in extension, η_f (31.81 $\pm$ 11.69), is significantly less than 50%, $t(9) = -4.92, p < .001, power = 0.67$. On the other hand, the contribution of the extensors to the total work in flexion, η_e (19.88 $\pm$ 12.9), is also significantly less than 50%, $t(9) = -7.38, p < .001, power = 0.68$. (A) Work contribution in extension direction η_f. (B) Work contribution in flexion direction η_e.

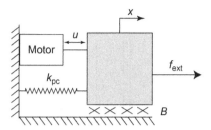

Fig. 8 Schematic of a 1-DOF prismatic robotic joint. k_{pc} is the added parallel compliance, u is the control force input to the system, x is the displacement of the joint, and f_{ext} is the external force being applied to the joint.

system (Fig. 8). The equation of motion of this combined system can be expressed as:

$$M\ddot{x} + B\dot{x} + k_{pc}x = u + f_{ext} \tag{5}$$

where M and B are the combined mass and damping of the joint and direct-drive actuator, x is the displacement of the system, k_{pc} is the stiffness of the parallel compliance element, and u is the control force input to the system.

As we are interested in robots that interact with the environment, standard position control strategies are not suitable. Impedance control, on the other hand, allows simultaneous control of both the position and forces being applied by the robot. Intuitively, impedance control enforces a fixed desired stiffness k_{des} and damping k_d that work to drive the system toward some desired reference trajectory. In this way, during environmental interactions, the robot will act as a purely passive mechanical system to ensure stability and a predictable position to force relationship. We design an impedance control law using joint position feedback with a feed-forward term to compensate for the parallel stiffness:

$$u = k_{des}(x_d - x) - k_d\dot{x} + k_{pc}x \tag{6}$$

With this control, the new equation of motion is:

$$M\ddot{x} + (B + k_d)\dot{x} + k_{des}x = k_{des}x_d + f_{ext} \tag{7}$$

Therefore, increasing k_{pc} will not affect the total joint stiffness of the system (k_{des}) or the theoretical second-order response. Instead, adding parallel compliance will act to replace *software* stiffness with *mechanical* stiffness.

2.5 Effect of Time Delay

In real systems, factors such as system complexity, noncollocation of sensors and actuators, sensor noise filtering, etc., introduce time delays in the controller. To analyze this effect, we incorporate a fixed time delay t_d in the feedback portion of the control input. In the frequency domain, the equation of motion can be written as:

$$(Ms^2 + Bs + k_{pc})X(s) = k_{des}X_d(s) - e^{-st_d}(k_{des} - k_{pc} + k_d s)X(s) \qquad (8)$$

For analysis, we linearized the system using a third-order Padé approximation of the exponential time-delay term. Depending on the order of approximation, the resultant transfer function of the system will have at a minimum an order of three. We chose the approximation such that the system order was five. As analytic solutions of such a high order system are not possible, we use graphical methods to analyze the system.

The system's parameters are chosen while keeping in mind low-inertia systems such as robotic fingers performing precision tasks such as grasping and manipulation, where high stiffness gains are often required. The system in Eq. (7) is initially modeled to contain no parallel compliance, no mechanical damping, and a high desired stiffness (k_{des}), resulting in a high natural frequency of $\omega_n = 67$ Hz, and a designed damping ratio of $\zeta = 0.707$ for a fast response and low overshoot.

Feedback delays change the characteristics of the dynamical system. Pole analysis of the resulting system with an increasing time delay t_d (Fig. 9) shows

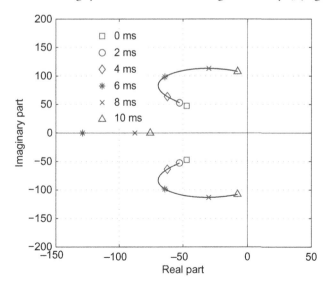

Fig. 9 Effect of increasing controller time delay. Poles of the system with no parallel compliance are shown for time delay ranging from 2 to 10 ms in steps of 2 ms.

that when the desired value of compliance is generated purely by the controller, large feedback delays reduce system stability.

2.6 Effect of Introducing Parallel Compliance

The theoretical representation of the closed-loop system (Eq. 7) shows that the same desired joint stiffness k_{des} can be achieved with and without the addition of a mechanical system in the system. The addition of mechanical stiffness effectively replaces the software stiffness to maintain the desired joint stiffness. We varied parallel stiffness from 0% to 100% of the desired stiffness. Note that at 100% of the desired stiffness ($k_{pc} = k_{des}$), the software stiffness is effectively zero.

The pole movements (Fig. 10) show that adding parallel compliance moves the underdamped poles toward more stable regions. In addition, an overdamped pole moves nearer to the origin, becoming more dominant in the system dynamics and stabilizing the system even more. The frequency response of the system (Fig. 11) shows that time delay increases the peak of the magnitude response, moving the system toward instability. Adding parallel compliance reduces that peak and increases stability. The bandwidth of the system is limited by the stiffness of the system. As this system is

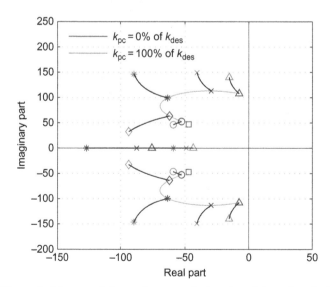

Fig. 10 Effects of adding parallel compliance on the location of the poles of a time-delay system.

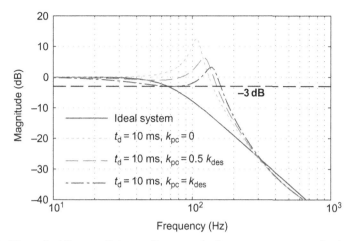

Fig. 11 Effect of adding parallel compliance on the frequency response of a time-delay system. Starting with the ideal system ($t_d = 0$), a time delay is introduced ($t_d = 10$ ms). Then, parallel compliance of different values ($k_{pc} = 0.5k_{des}$ and $k_{pc} = k_{des}$) are added to the system to compare their effects.

designed for a fixed joint stiffness k_{des}, the bandwidth is already limited. In our system with no time delays or very small delays (<2 ms), the ratio of software stiffness to mechanical stiffness does not affect the overall joint stiffness and hence the bandwidth of the system. But as time delays become larger, based on the Bode plots, the bandwidth increases at the cost of instability (large phase shifts). However, increasing the stiffness of the parallel compliance element brings the bandwidth closer to the ideal system. This characteristic becomes important in the case of impact stability. As an impact contains all frequencies, in systems with time delay the addition of parallel stiffness should improve impact response stability.

The analysis shows that relying purely on software stiffness control in a system with controller time delay can lead to system instability. Addition of parallel compliance leads to increased stability and disturbance rejection capabilities in the system even in the presence of large time delays. Making modifications to other mechanical parameters, such as the system's inertia or damping, could also be used to stabilize the system, but this has its own trade-offs. Increased inertia results in a more stable system at the cost of sluggish performance. In Fig. 12, we can see that increased parallel damping B produces a response similar to that of an increased k_{pc}, stabilizing the negative effects of time delay. However, mechanical damping may also bring with it undesired nonlinear joint friction, which can make accurate force control

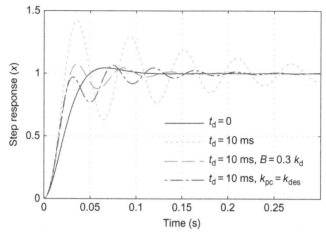

Fig. 12 Step responses comparing the effects of damping B and parallel compliance k_{pc}. Starting with the ideal system ($t_d = 0$), a time delay is introduced ($t_d = 10$ ms). Then, mechanical joint damping ($B = 0.3\ k_d$) and parallel compliance ($k_{pc} = k_{des}$) are added to the system to compare the effects of the two parameters.

more difficult. In the end, the choice is left to the mechanical designer depending on the design criteria and task requirements. In our case, we are interested in robotic hands performing precision tasks such as grasping and manipulation while keeping mass and friction as low as possible. Therefore, we will focus our study on the effects of parallel compliance.

2.7 Guidelines for Designing Parallel Compliance for a System

The right amount of parallel compliance for a system can be designed for optimal performance. A few basic guidelines for introducing compliance are:

1. **Actuator saturation:** For robotic hands, compact actuators with high torque capabilities are the usual choice. Actuator saturation occurs during tasks with high force requirements. With higher actuator loads, motor velocities will also be limited. To design the right parallel compliance for such robots, the actuator saturation limit has to be monitored. If the robotic joint has to follow a certain trajectory x_d, with a desired output force F_d and a control saturation limit of F_{sat}, then the optimal parallel compliance would be:

$$k_{pc} \leq \max \left(\frac{F_{sat} - F_d}{x_d - x_{pc0}} \right) \tag{9}$$

where $x_d - x_{pc0}$ is the desired maximum RoM with respect to the resting position of the parallel compliance. This also means that, based on certain tasks (eg, grasping), the resting position of the parallel compliance can be designed to reduce motor effort and energy requirements. In some special cases, actuator saturation can also be exploited by using well-designed parallel compliance to limit the RoM of the joint instead of hard mechanical stops.

2. **Controller gains:** To ensure the system maintains stability in the case of compliant element failure, the system is analyzed if k_{pc} suddenly becomes zero. This results in the following stability criteria:

$$k_{des} \geq k_{pc} \tag{10}$$

Enforcing this relationship gives the system a measure of safety, but is not strictly necessary for stability as long as the parallel compliance element remains functional.

3. MULTIFINGER MANIPULATION WITH PASSIVE JOINT STIFFNESS

An important goal for robotic hands is the ability to robustly grasp and manipulate objects. To fully understand the effects of parallel compliance on robotic hand performance, we present the task of two-finger grasping. Because of the natural instability of two-fingered pinching, active control must be used to robustly stabilize the grasp while also performing some desired object motions or environment interactions.

3.1 System Model

We start with a model of the entire system, shown in Fig. 13. The dynamic equations of motion for each tendon-driven finger can be represented as

$$\mathbf{M}(\mathbf{q})\ddot{\mathbf{q}} + \mathbf{C}(\mathbf{q},\dot{\mathbf{q}}) + \mathbf{K}_{pc}\mathbf{q} = \mathbf{R}\mathbf{f}_\ell + \mathbf{J}(\mathbf{q})^T\mathbf{f}_{ext} \tag{11}$$

$$\mathbf{M}_m\ddot{\boldsymbol{\ell}} + \mathbf{f}_{fric} + \mathbf{f}_\ell = \mathbf{f}_{\ell,cmd} \tag{12}$$

When grasping an object, the external force vector $\mathbf{f}_{ext,i}$ for each finger $i(i = 1,2)$ is equal to the contact force being applied to the object, $\mathbf{f}_{c,i}$, which can be expressed in the global (x,y) frame as

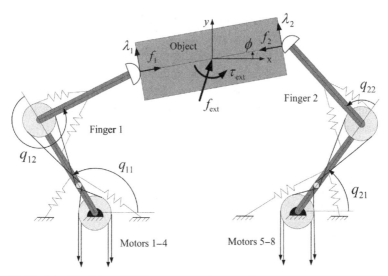

Fig. 13 Model of dual two DOF fingers grasping an object.

$$\mathbf{f}_{c,i} = \begin{bmatrix} \cos(\phi) & -\sin(\phi) \\ \sin(\phi) & \cos(\phi) \end{bmatrix} \begin{bmatrix} (-1)^{i+1} f_i \\ \lambda_i \end{bmatrix} \tag{13}$$

where f_i and λ_i are the normal and tangential forces, oriented as in Fig. 13. In the controller, these contact forces are modeled as point contacts between the fingertips and the objects.

The object dynamics are modeled as

$$M_{obj} \begin{bmatrix} \ddot{x} \\ \ddot{y} \end{bmatrix} = \mathbf{f}_{c1} + \mathbf{f}_{c2} + \begin{bmatrix} f_{ext,x} \\ f_{ext,y} \end{bmatrix} \tag{14}$$

$$I_{obj} \ddot{\phi} = \frac{1}{2} w_{obj}(\lambda_2 - \lambda_1) + \tau_{ext} \tag{15}$$

where M_{obj}, I_{obj}, and w_{obj} are the object's mass, inertia about the z-axis, and width, respectively. For this system, it is assumed that the object's COM is located midway between the contact points. The terms $f_{ext,x}$, $f_{ext,y}$, and τ_{ext} are the external forces and moment being exerted on the object.

3.2 Grasp Stability Analysis

In reality, the contact points are not point contacts but rolling contacts; the fingertips are circular with some radius r_{tip}, and the object is rectangular with total width w_{obj}. To analyze the grasp stability properties of this two-fingered grasp, we use the methods proposed in [32]. We assume the forces at the contact points are equal and opposite gripping forces, the contact points are diametrically opposed, and the COM is located on a line joining the two contact points.

From Montana's work, the requirements for borderline stability are as follows:

$$k_{1a}^{-1} + k_{2a}^{-1} \geq w_{\text{obj}}, \quad k_{1b}^{-1} + k_{2b}^{-1} \geq w_{\text{obj}} \tag{16}$$

where the k variables are the principal relative curvatures between the fingertips and the grasped object. For this scenario, these curvatures are $k_{1a} = k_{1b} = k_{2a} = k_{2b} = 1/r_{\text{tip}}$. Hence, the grasp is borderline stable if $r_{\text{tip}} \geq \frac{1}{2} w_{\text{obj}}$.

In the experimental setup to follow, this inequality does not hold ($r_{\text{tip}} \approx 1.3$ cm and $w_{\text{obj}} \approx 4.4$ cm). Therefore, the active stabilization presented in the object-grasping controller will need to be robust enough to overcome the borderline instability of the grasp.

3.3 Manipulation Controller Design

For robust performance in the presence of unknown disturbances in the environment, we implement object impedance control [33]. Letting $\mathbf{z} = [x, y, \phi]^T$ represent the object's position and orientation, we define the desired contact forces to be exerted on the object by the fingertips to be an impedance-like behavior:

$$\mathbf{f}_{\text{imp}} = \mathbf{W}^+ (\mathbf{K}_d (\mathbf{z}_{\text{des}} - \mathbf{z}) + \mathbf{B}_d (\dot{\mathbf{z}}_{\text{des}} - \dot{\mathbf{z}})) \tag{17}$$

Here, $\mathbf{W}^+$ is the pseudoinverse of the grasp matrix, which is found from the relation

$$\begin{bmatrix} \mathbf{f}_{\text{obj}} \\ \tau_{\text{obj}} \end{bmatrix} = \mathbf{W} \begin{bmatrix} \mathbf{f}_{c1} \\ \mathbf{f}_{c2} \end{bmatrix} \tag{18}$$

where $\mathbf{f}_{\text{obj}}, \tau_{\text{obj}}$ contain the forces and moments applied to the object by the fingers, and $\mathbf{W} \in \mathbb{R}^{3 \times 4}$. Matrices $\mathbf{K}_d = diag(K_x, K_y, K_\phi)$ and $\mathbf{B}_d = diag(B_x, B_y, B_\phi)$ contain the object stiffness and damping gains, respectively. Then, the four desired joint torques are:

$$\tau_{q, \text{des}} = \mathbf{J}^T \left(\mathbf{f}_{\text{imp}} + \begin{bmatrix} \mathbf{f}_{\text{grip}, 1} \\ \mathbf{f}_{\text{grip}, 2} \end{bmatrix} \right) + \begin{bmatrix} \mathbf{K}_{\text{pc}, 1} \mathbf{q_1} \\ \mathbf{K}_{\text{pc}, 2} \mathbf{q_2} \end{bmatrix} \tag{19}$$

where $\mathbf{J} = diag(\mathbf{J}_1(q_1)\,\mathbf{J}_2(q_2))$ contains the Jacobian matrices for both fingers, and $\mathbf{f}_{\mathrm{grip},i}$ ($i = 1,2$) are vectors of the gripping forces produced by each finger which will be used to produce a desired force f_{des} in the null-space of $\mathbf{W}$.

Force feedback is used to maintain accurate grip forces in the presence of model errors or other disturbances. Assuming the applied grip force is being measured using a force sensor as f_s, we can define f_{int} as a PI force feedback term,

$$f_{\mathrm{int}} = f_{\mathrm{des}} + K_{f,p}\left(f_{\mathrm{des}} - f_s\right) + K_{f,i} \int \left(f_{\mathrm{des}} - f_s\right) \tag{20}$$

Using Eq. (13), with $\lambda_i = 0$ and $f_i = f_{\mathrm{int}}$ for each finger, we find $\mathbf{f}_{\mathrm{grip},i}$ ($i{=}1,2$) to be applied by the controller. These grip forces will act in the directions of f_1 and f_2 in Fig. 13.

Object position and orientation information is being sensed using the motion-capture system, which has an inherent time lag of over 15 ms. Attempting to differentiate the motion capture data to obtain velocity values for the damping terms results in significant stability issues. Instead, we use the motor velocities $\dot{\ell}_i$ found from filtered motor encoder data, and then transform them to object velocities.

$$\dot{z} = \mathbf{W}^T \mathbf{J} \begin{bmatrix} \mathbf{R}^{T+}\dot{\ell}_1 \\ \mathbf{R}^{T+}\dot{\ell}_2 \end{bmatrix} \tag{21}$$

Note that this transformation is only valid with noncompliant tendons, as tendon stretching is not accounted for.

The motion capture time lag also causes issues with the already naturally unstable object orientation stabilization. In practice, by combining an integral term with the proportional $K_\phi(\phi_{\mathrm{des}} - \phi)$ term, the grasp can be maintained more robustly in the presence of time delay. Finally, the calculated desired joint torques $\tau_{q,\mathrm{des}}$ found in Eq. (19) are sent to a motor-level control law that determines the appropriate tendon forces to achieve $\tau_{q,\mathrm{des}}$ and simultaneously avoids tendon slacking [31].

3.4 Experimental Results

The following experiments are designed to determine the effects of parallel joint compliance, with trajectory tracking and robustness to impact as the performance metric. The control law in Section 3.3 is designed to produce the same effective impedance behavior with or without parallel compliance. Therefore, we are able to isolate the effects of parallel joint compliance in the system with all other conditions being equal. The amount of parallel compliance added to the system is equivalent to that in the single finger experiments, $\mathbf{K}_{pc,i} = diag(220, 110) \ \dfrac{\text{N} \cdot \text{mm}}{\text{rad}}$ for fingers $i = 1,2$.

3.4.1 Object Grasping: Trajectory Tracking

First, object manipulation tracking experiments are performed comparing trajectory tracking with and without the presence of parallel compliance elements (Fig. 14). The desired object trajectory in this case is a circular path, coupled with changes in the object's orientation. The commanded position and object orientation trajectories are at different frequencies to produce a different motion every cycle.

The tracking errors are similar in each case, but differences are seen in the smoothness of the motions in the case with parallel compliance. By using parallel compliance, stability is improved and disturbances can be handled more robustly by the combined efforts of the control system and mechanical compliance. These results are repeatable and consistent for any commanded object motions.

3.4.2 Object Grasping: Impact Testing

To analyze the robustness to impact, approximately identical impacts were delivered to the system with and without parallel compliance while attempting to maintain constant grasp force and object position/orientation. Results with parallel compliance (Fig. 15, left column) show significant improvements in settling time, especially with regard to the inherently unstable object orientation. Then, when a larger impact force is delivered (Fig. 15, right column), the case without parallel compliance shows instability and loss of contact, while the fingers with parallel compliance elements show resistance to the impact and maintain a stable grasp.

Impacts simulate an impulse function being applied to the system, which will contain all frequencies and tends to excite the less stable natural frequencies in the system. This makes impact testing a good method to analyze the system's stability across the entire frequency spectrum and identify unstable resonant frequencies. In the noncompliant cases, sufficiently high impacts

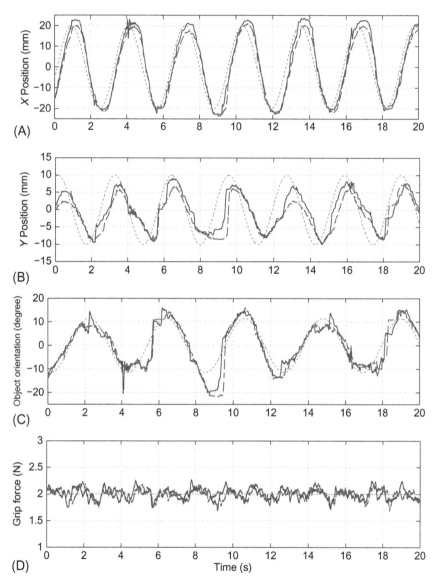

Fig. 14 Object grasping trajectory tracking comparison without parallel compliance (*solid line*) and with parallel compliance (*dashed line*), given identical desired trajectories (*dotted line*).

excite an unstable natural frequency, causing grasp instability and failure. In the Bode plots derived in the previous section in Fig. 11, it was shown that introduction of parallel compliance elements lowers the resonant frequencies caused by time delays. The experimental impact results here validate this observation with a more involved system.

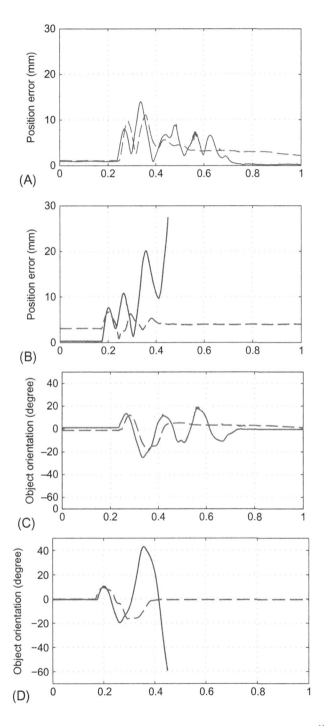

Fig. 15

(Continued)

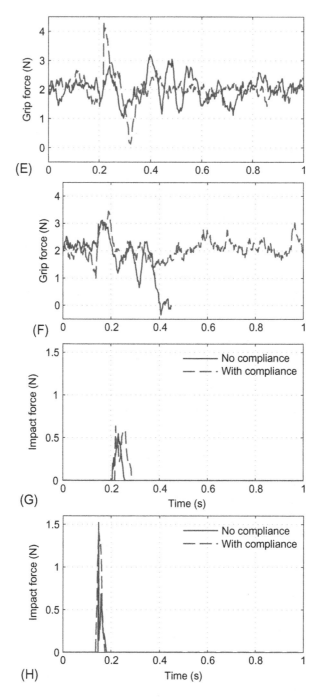

Fig. 15, Cont'd Object grasping disturbance responses when subjected to two levels of impact forces for cases with and without parallel compliance. The left column shows results for relatively lower impact forces (see bottom plots for impact force measurements), while the right column shows results for larger impact forces. The cases without compliance showed lower grasp stability, and in tests with larger impacts (right column) the grasp failed due to loss of contact at $t \approx 0.4$ s.

4. DISCUSSION

In summary, the elasticity of the MTUs only contributes a fraction of the total stiffness at the MCP joint in the hand and the majority of the joint stiffness is a result of the passive properties of the CLC. Based on the results from this paper, the studies that assume that the MTUs are the sole contributors to the joint stiffness may have to be updated. The results of this study may lead to updated designs of joint implants, and are also important for the identification and treatment of hand injuries and joint defects. The identification of the critical role of CLC in defining joint stiffness may lead to improvements in the computational models of the hand and also to guidelines for the design of artificial hands. Further study is needed to understand the roles of the passive stiffness from MTUs and CLC, and active stiffness due to muscle co-contractions in hand movement control.

Through mathematical analysis and experimentation we have demonstrated that the integration of parallel compliance leads to improved stability and disturbance rejection in robotic hands during grasping and manipulation in the presence of feedback time delays. Time delays are unavoidable in robotic systems and can lead to instability, especially for systems with low inertia and a high desired stiffness, such as robotic fingers performing precision manipulation tasks. Modeling and analysis of a generic 1-DOF robotic joint bring out the effects of introduction of parallel compliance, especially in mitigating the destabilizing effects of time delay. Validation was performed on an experimental tendon-driven joint with an impedance controller, and cases with parallel compliance resulted in more stable and robust responses. We have brought out the trade-offs between improvements in stability and increased actuator load, and have generated guidelines for choosing optimal parallel compliance.

In order to extend this analysis to more complex systems, we mathematically modeled a 2-DOF tendon-driven robotic finger with parallel compliance, and then a set of two fingers grasping an object. We also developed a testbed with two tendon-driven robotic fingers for conducting grasping and manipulation experiments, and designed an impedance controller for this system. In the manipulation experiments, we chose an object larger than the minimum bound for borderline grasp stability, so that a challenging control problem is addressed and the robustness of the system is clearly demonstrated. The experiments provide confirmation that the advantageous stabilizing effects seen in a 1-DOF joint will also apply to complex systems performing more difficult tasks. Results showed that introducing parallel compliance improves the stability, produces smoother trajectory tracking, and improves robustness to disturbances for robotic hands during a two-fingered grasping task. These results demonstrate, for the first time, the effects of parallel compliance on the system performance and are an important step toward reaching the goal of human-like hand dexterity.

REFERENCES

[1] N. Shrivastava, M. F. Koff, A. E. Abbot, V. C. Mow, M. P. Rosenwasser, R. J. Strauch, Simulated extension osteotomy of the thumb metacarpal reduces carpometacarpal joint laxity in lateral pinch1, J. Hand Surg. 28 (5) (2003) 733–738.

[2] A. Minami, K. An, W. P. Cooney, R. L. Linscheid, E. Y. S. Chao, Ligament stability of the metacarpophalangeal joint: a biomechanical study, J. Hand Surg. 10 (2) (1985) 255–260.

[3] K. Tamai, J. Ryu, K. N. An, R. L. Linscheid, W. P. Cooney, E. Y. S. Chao, Three-dimensional geometric analysis of the metacarpophalangeal joint, J. Hand Surg. 13 (4) (1988) 521–529.

[4] A. Esteki, J. M. Mansour, An experimentally based nonlinear viscoelastic model of joint passive moment, J. Biomech. 29 (4) (1996) 443–450.

[5] J. S. Knutson, K. L. Kilgore, J. M. Mansour, P. E. Crago, Intrinsic and extrinsic contributions to the passive moment at the metacarpophalangeal joint, J. Biomech. 33 (2000) 1675–1681.

[6] D. G. Kamper, T. George Hornby, W. Z. Rymer, Extrinsic flexor muscles generate concurrent flexion of all three finger joints, J. Biomech. 35 (12) (2002) 1581–1589.

[7] J. Qin, D. Lee, Z. Li, H. Chen, J. T. Dennerlein, Estimating in vivo passive forces of the index finger muscles: exploring model parameters, J. Biomech. 43 (7) (2010) 1358–1363.

[8] J. T. Dennerlein, Finger flexor tendon forces are a complex function of finger joint motions and fingertip forces, J. Hand Ther. 18 (2) (2005) 120–127.

[9] G. A. Pratt, M. M. Williamson, in: Series elastic actuators, vol. 1, IEEE/ASME International Conference of Intelligent Robots and Systems (IROS), 1995, pp. 399–406.

[10] M. A. Diftler, J. S. Mehling, M. E. Abdallah, N. A. Radford, L. B. Bridgwater, A. M. Sanders, R. S. Askew, D. M. Linn, J. D. Yamokoski, F. A. Permenter, et al., Robonaut 2—The first humanoid robot in space, in: International Conference on Robotics and Automation, IEEE, 2011, pp. 2178–2183.

[11] M. Grebenstein, A. Albu-Schäffer, T. Bahls, M. Chalon, O. Eiberger, W. Friedl, R. Gruber, S. Haddadin, U. Hagn, R. Haslinger, et al., The DLR hand arm system, in: IEEE International Conference on Robotics and Automation, IEEE, 2011, pp. 3175–3182.

[12] F. Danion, S. Li, V. M. Zatsiorsky, M. L. Latash, Relations between surface EMG of extrinsic flexors and individual finger forces support the notion of muscle compartments, Eur. J. Appl. Physiol. 88 (1) (2002) 185–188.

[13] A. Perotto, E. F. Delagi, Anatomical Guide for the Electromyographer: The Limbs and Trunk, Charles C. Thomas Pub, Ltd, Springfield, IL, 2005.

[14] K. Halvorsen, M. Lesser, A. Lundberg, A new method for estimating the axis of rotation and the center of rotation, J. Biomech 32 (11) (1999) 1221–1227.

[15] P.-H. Kuo, J. Hayes, A. Deshpande, Design of a motor-driven mechanism to conduct experiments to determine the passive joint properties of the human index finger,

in: Proceedings of the ASME International Design Engineering Technical Conference & Computers and Information in Engineering Conference, 2011.

[16] A. Nordez, P. McNair, P. Casari, C. Cornu, Acute changes in hamstrings musculo-articular dissipative properties induced by cyclic and static stretching, Int. J. Sports Med. 29 (5) (2008) 414–418.

[17] A. Silder, B. Whittington, B. Heiderscheit, D. G. Thelen, Identification of passive elastic joint moment-angle relationships in the lower extremity, J. Biomech. 40 (12) (2007) 2628–2635.

[18] A. D. Deshpande, R. Balasubramanian, J. Ko, Y. Matsuoka, Acquiring variable moment arms for index finger using a robotic testbed, IEEE Trans. Biomed. Eng. 57 (8) (2010) 2034–2044.

[19] A. M. Kociolek, P. J. Keir, Modelling tendon excursions and moment arms of the finger flexors: anatomic fidelity versus function, J. Biomech. 44 (10) (2011) 1967–1973.

[20] R. L. Lieber, B. M. Fazeli, M. J. Botte, Architecture of selected wrist flexor and extensor muscles, J. Hand Surg. 15 (2) (1990) 244–250.

[21] K. R. S. Holzbaur, W. M. Murray, S. L. Delp, A model of the upper extremity for simulating musculoskeletal surgery and analyzing neuromuscular control, Ann. Biomed. Eng. 33 (6) (2005) 829–840.

[22] M. D. Jacobson, R. Raab, B. M. Fazeli, R. A. Abrams, M. J. Botte, R. L. Lieber, Architectural design of the human intrinsic hand muscles, J. Hand Surg. 17 (5) (1992) 804–809.

[23] J. Z. Wu, K. N. An, R. G. Cutlip, K. Krajnak, D. Welcome, R. G. Dong, Analysis of musculoskeletal loading in an index finger during tapping, J. Biomech. 41 (3) (2008) 668–676.

[24] R. V. Gonzalez, T. S. Buchanan, S. L. Delp, How muscle architecture and moment arms affect wrist flexion-extension moments, J. Biomech. 30 (7) (1997) 705–712.

[25] S.L. Delp, Surgery Simulation: A Computer Graphics System to Analyze and Design Musculoskeletal Reconstructions of the Lower Limb, PhD Thesis, Stanford University, Stanford, CA (1990).

[26] F. E. Zajac, Muscle and tendon: properties, models, scaling, and application to biomechanics and motor control, Crit. Rev. Biomed. Eng. 17 (4) (1989) 359.

[27] T.M. Greiner, Hand anthropometry of US army personnel, tech. rep. (1991).

[28] K. Manal, T. S. Buchanan, Subject-specific estimates of tendon slack length: a numerical method, J. Appl. Biomech. 20 (2) (2004) 195–203.

[29] S. L. Delp, F. C. Anderson, A. S. Arnold, P. Loan, A. Habib, C. T. John, E. Guendelman, D. G. Thelen, OpenSim: open-source software to create and analyze dynamic Simulations of Movement, IEEE Trans. Biomed. Eng. 54 (11) (2007) 1940–1950.

[30] P. H. Kuo, A. Deshpande, Contribution of passive properties of muscle-tendon units to the metacarpophalangeal joint torque of the index finger, in: Proceedings of the International Conference on Biomedical Robotics and Biomechatronics, 2010.

[31] T. D. Niehues, P. Rao, A. D. Deshpande, Compliance in parallel to actuators for improving stability of robotic hands during grasping and manipulation, Int. J. Robot, Res, 2014.

[32] D. J. Montana, Contact stability for two-fingered grasps, IEEE Trans. Robot. Autom. 8 (4) (1992) 421–430.

[33] S. A. Schneider, R. H. Cannon Jr., Object impedance control for cooperative manipulation: theory and experimental results, IEEE Trans. Robot. Autom. 8 (3) (1992) 383–394.

CHAPTER TWO

A Review of Computational Musculoskeletal Analysis of Human Lower Extremities

A. Alamdari, V.N. Krovi
The State University of New York at Buffalo, Buffalo, NY, United States

1. INTRODUCTION

In recent years, numerous computational tools have been developed for kinematic and dynamic analyses of neuromusculoskeletal (NMS) systems, building on an articulated-multibody systems framework. Such NMS analysis tools allow monitoring of internal human variables such as muscle fiber lengths, joint forces, reactions of muscles/tendons/joints, metabolic power consumption, and mechanical work. Examples include both commercial tools such as LifeMod, SIMM, and AnyBody (Anybody Technology), as well as the more recent open-source tools such as OpenSim to track human movements, compute muscle forces, and analyze normal and pathological gaits [1–3]. This software includes all of the necessary computational components for deriving equations of motion for dynamical systems, performing numerical integration, and solving constrained nonlinear optimization problems.

Body anthropometry, soft-tissue properties, bone geometry, muscle paths, and muscle-tendon architecture can be substantially dissimilar among people, and consequently muscle and joint function can be different among them [120,121]. Relatively few studies have used a subject-specific model (personalized model) to simulate gait disorders resulting from conditions such as cerebral palsy, stroke, and knee osteoarthritis [4]. Most studies, however, have used generic musculoskeletal models based on estimates derived from average adult anatomy [5–7]. The value of employing subject-specific musculoskeletal models for evaluation of muscle and joint functions depends on the aim of the model. If orthopedic surgery is the main purpose, then using the subject-specific model will be unavoidable. For example, making small changes in the muscle's moment arm might change the moment

37

generated by the muscle [8]. On the other hand, if the main purpose is to study the muscle coordination of locomotion in a healthy person, then scaled-generic models may suffice. To validate these musculoskeletal models, computed-muscle forces are usually validated against electromyogram (EMG) signals of muscle activities, or internal contact forces measured from instrumented knee joint [9] or hip joint [10] replacements.

Our particular focus will be on computational modeling efforts focused on improving understanding of physical interactions of the human lower extremities with their physical environment (eg, walking, standing, bicycling). Walking is the most fundamental human motion with great complexities in terms of the central nervous system (CNS) for controlling, intermittent contact, dynamic stability, and nonlinearity of the dynamic system. There have been extensive experimental studies in the biomechanics literature that give detailed descriptions of dynamic human walking [11,12]. Furthermore, adding dynamics constraints to kinematically computed motion results in a realistic and smooth human motion in interactions with the environment.

Such efforts have been devoted to the development and simulation of different models for human walking, ranging from real-time control of humanoid robots with high degrees of freedom in the robotics field, to high-fidelity NMS models in biomechanics, and pathological gait studies. Simulation of human motion with high degrees of freedom is a challenging problem from both analytical and computational perspectives. Significant literature focuses on developing natural and human-like walking with increasing fidelity of mechanical models and efficient numerical algorithms. This review paper will trace these developments in computational modeling of human walking, beginning with: (i) skeletal models in which all effects of muscles are modeled merely as torques applied to joints (commonly seen for real-time control of biped robots); transitioning to (ii) musculoskeletal models where muscle groups are included in the multibody dynamic systems (applications in biomechanics); and finally (iii) the NMS model in which human movements are produced by the muscular and skeletal systems, coupled with the neural excitations (more recent biomechanical applications).

On an allied front, there is significant interest in studying the interactions of the human lower extremities with other articulated mechanical systems, ranging from exercise equipment such as treadmills, bench-presses, and bicycles to foot-pedal controlled complex machinery such as cranes. In addition to the innate variability of geometry, performance, and function across individuals in a population (due to sex/age/race), significant variability is also introduced by the diversity and adjustable features within such articulated mechanical equipment [13]. System-level interactions between users and equipment make it difficult for designers to select the "best" set of options/parameters to realize desired performance outcomes a priori [14,15]. Hence, this creates a need for interactive simulation of personalized musculoskeletal models coupled with articulated multibodies under a variety of "what-if" scenarios to realize the optimal system-level motor performance. To this end, we discuss various efforts at designing of virtual environments, leveraging tools from musculoskeletal analysis, optimization, and simulation-based design to permit designers to rapidly evaluate and systematically customize and match human-machine interactions.

The rest of the chapter is organized as follows. Sections 2 and 3 focus on human gait cycle and the biomechanics of human walking, while Section 4 is a review of experimental kinematic approaches for studying human walking including vision-based and nonvision-based techniques. Additionally, the development of quantitative multibody dynamic systems consisting of skeletal, musculoskeletal, and NMS models based on optimization or control-based methods is reviewed. Computational musculoskeletal interactions with articulated systems are presented in Section 5 and, finally, Section 6 presents the concluding remarks.

2. HUMAN WALKING GAIT CYCLE

Human walking can be described as a cyclic pattern of body movements which advances an individual's position. Assuming that all walking cycles are about the same, studying the walking process can be simplified by investigating one walking cycle.

In general, each of these walking cycles is composed of two phases: the single-support phase and the double-support phase. During the single-support phase, one leg is on the ground and the other leg is experiencing a swinging motion. The double-support phase starts once the swinging leg meets the ground and ends when the support leg leaves the ground [12]. Walking is generally distinguished from running in that only one foot at a time leaves contact with the ground.

Assuming we start with the right leg, in the first walking step from the vertical position of the human, the right leg is moved forward and placed on the ground. The first steady walking step involves lifting the left leg with single leg support of the right leg until the left leg is placed on the ground again. The second steady walking step is similar to the first steady walking step, but this step has single leg support of the left leg until the right leg is

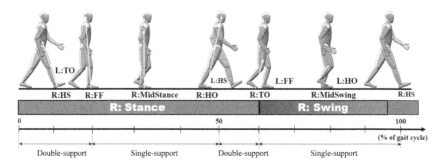

Fig. 1 Human walking gait cycle, and stance and swing phases of the right leg.

lifted and placed on the ground again. The successive repetitions of steady walking steps result in a continued locomotion in the sagittal plane.

The gait cycle is the time interval between two successive occurrences of one of the repetitive events of locomotion. The human gait cycle is split into two separate regions representing the period of time when the foot is in contact with the ground, the stance phase (shown with R: Stance, Fig. 1), and the period of time when the limb is not in contact with the ground, the swing phase (shown with R: Swing, Fig. 1). During the stance phase, the foot contacts the ground, the mass of the body is supported, and then the body is propelled forward during the later stages of stance. The stance phase itself involves five events as illustrated in Fig. 1: (i) heel-strike (HS), (ii) foot-flat (FF), (iii) midstance (MS), (iv) heel-off (HO), and (v) toe-off (TO) [11].

i. *Heel-strike*: The beginning instant of the gait cycle is represented as initial contact of one foot with the ground, usually termed HS or foot-strike.

ii. *Foot-flat*: The instant that the rest of the foot comes down to contact the ground and usually is where full body weight is being supported by the leg.

iii. *Midstance* is defined when the center of mass is directly above the ankle joint center. This is also used as the instant when the hip joint center is above the ankle joint.

iv. *Heel-off* occurs when the heel begins to lift off the ground in preparation for the forward propulsion of the body.

v. *Toe-off* happens as the last event of contact during the stance phase.

Events of a gait cycle remarkably occur in similar sequences and are independent of time. That is why the cycle is commonly described in terms of percentage, rather than the time elapsed. Initial HS is designated as 0% and the subsequent HS of the same foot as 100% (0–100%). During a normal gait cycle, hip, knee, and ankle joints experience a range of motion [16]. Fig. 2A illustrates these ranges of movement during different speeds. Hip

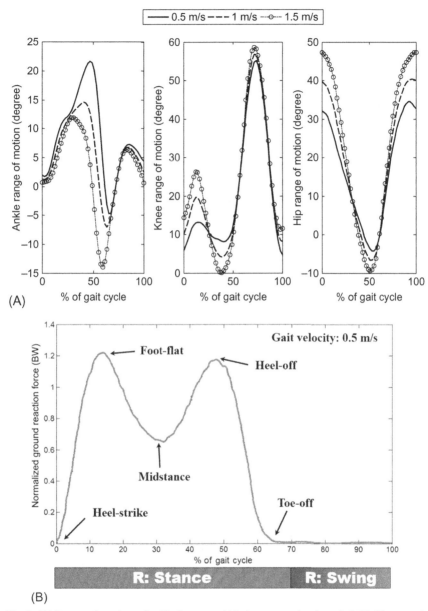

Fig. 2 (A) Range of motion of ankle, knee, and hip in a normal gait cycle [17], (B) ground reaction forces in a normal gait cycle.

movement can be categorized into two basic motions: first, hip extension, which happens during the stance phase and has the primary role of stabilization of the trunk, and, second, hip flexion, which happens during the swing phase. During the stance phase, the knee is the basic determinant of limb stability, and in the swing phase, knee flexibility is the primary factor in the limb's freedom to advance. At MS, total body weight is transferred into the flexed knee. The range of motion of the ankle is not large, but it is critical for progression and shock absorption during stance.

3. BIOMECHANICS OF NORMAL HUMAN WALKING

Normal human walking can be explained with a strategy similar to the double pendulum. During forward motion, the leg that leaves the ground swings forward from the hip. This sweep is the first pendulum. Then the leg strikes the ground with the heel and rolls through to the toe in a motion described as an inverted pendulum. The motion of the two legs is coordinated so that one foot or the other is always in contact with the ground.

While walking, the center of mass of the body raises to its highest point during the MS event when one leg passes the vertical, and then drops to its lowest point as the legs spread apart. Essentially, the kinetic and potential energy are constantly being exchanged.

To find the required moment/force at each joint during walking, an inverse dynamic analysis must be implemented. Kinetic data and external forces should be recorded during experimentation for this purpose. Ground reaction force (GRF) is the main force acting on the body during human movements. Since the body mass is being moved in all three directions, a 3D force vector consisting of a vertical component and two shear components will act upon the contact area. These shear forces are usually resolved into anterior-posterior and medial-lateral directions [12]. The two shear forces are small compared to the vertical GRF. Herein, only the vertical GRF is investigated, and the two shear components are not discussed.

Fig. 2B shows a typical vertical GRF profile of a single walking step. The vertical GRF at the moment of contact with the ground (HS) will be zero and will rise sharply up to almost body weight in a fraction of a second. At the instant of FF, the body mass is moving downwards and landing on the leg. In order to decelerate this downward motion and at the same time support the body weight, it will be necessary to apply a vertical force larger than body weight on the foot. This instant for the subject shows 120% body weight being applied to the foot.

At MS the movement of the center of mass of the body is actually upward. This movement creates an upward acceleration that allows a force of less than a body weight to support the body. This subject shows 63% body weight at MS. At HO, the body mass is accelerated forward and upward, ready for the stance phase of the other leg. This means that more than body weight will be required to support the body. Finally, TO is the instant where contact with the ground is lost and the force returns to zero. The M-shaped graph, or double peak graph, is typical for normal gait and shows the fluctuation of force relative to body weight [12].

During stance phase, the forces applied to the foot are backwards as the body lands and then forwards in late stance as the body lifts up and moves more rapidly in the forward direction (shown with *yellow* (*light gray* in print version) *arrow* in Fig. 1) [18]. We can assume the line of action of the resultant force is passing through the position of whole-body center of mass.

4. QUANTITATIVE HUMAN WALKING MODELS

Gait analysis has been used for more than a century to provide quantitative information on the kinematics and kinetics of human walking. Early gait analysis efforts usually focused on experimental observations to study kinematics and kinetics. Studies involving human subjects may be associated with discomfort or even injuries. Therefore, it is not always possible to use human subjects for experimentation. One feasible way to proceed is to combine the power of computational modeling with available noninvasive instruments to determine information that is not readily attainable from an experiment. Thus, these challenges illuminate the importance of articulated-multibody system modeling. However, it is only in the past couple of decades that a more complete understanding of muscles and joint function via high-fidelity computational musculoskeletal models and analysis has emerged [18–20]. Herein, a review of human walking modeling and simulation is presented.

One important challenge for clinicians is to understand the relationships between the observed motion patterns and muscle behavior [21]. Unfortunately, no fully satisfactory data analysis tools are currently available to perform patient data interpretation. Modeling tools have been previously developed but are not entirely satisfactory because of a lack of integration of the real underlying functional joint behavior [22]. In order to improve our understanding about the relationships between muscle behavior and motion data, modeling tools must guarantee that the joint kinematics in the model are correctly validated to ensure meaningful interpretation. Once a proper generic model is available, properly adjusted, and validated, it could then be used in further simulation to obtain clinically relevant muscle information [22]. Sholukha et al. [23] presented a model-based method that allows fusing accurate joint kinematic information with motion analysis data collected using either marker-based stereophotogrammetry (ie, bone displacement collected from reflective markers fixed on the subject's skin) or markerless single-camera hardware. They described a model-based approach for human motion data reconstruction by a scalable method for combining joint physiological kinematics with limb segment poses, and finally presented physiologically acceptable human kinematics.

4.1 Kinematics of Human Walking

To understand kinematics of human walking, it is necessary to track human motion. The first set of tools developed for analyzing human movement in computer simulation was based on forward and inverse kinematics [24]. These tools use empirical and biomechanical knowledge of human motion in order to compute realistic motions. In forward kinematics the state vector of an articulated human movement over time and interpolation techniques are used to generate in-between positions to generate smooth motion. Inverse kinematic algorithms can be used to solve some constraints such as foot penetration to the ground. Most of the kinematic approaches used for generating synthetic human locomotion rely on biomechanical knowledge, and combine forward and inverse kinematics for computing motions [25]. The kinematic techniques presented here rely on a certain understanding of the basic walking motion mechanisms. One of the main advantages of these models is the high-level parameter (eg, velocity, acceleration, steplength) they provide leading to the generation of families of different gaits. Another advantage is the low cost of such computations.

The goal of this approach is to work directly on experimental data from captured motions. Using optical technologies, it is possible to store the positions and orientations of markers located on the human body. A further computation provides the link between the synthetic skeleton and the real skeleton, in order to adapt data to the new morphology.

In general, there are different types of human movement tracking systems including nonvision-based tracking systems and vision-based tracking systems (marker and markerless).

4.1.1 Nonvision-Based Tracking System

Nonvision-based systems employ sensor technology attached to the human body to collect human movement information. These sensors are commonly categorized as mechanical, inertial (eg, accelerometer and gyroscopes) [26], acousto-inertial, radio or microwave [27], and magnetic [28] sensing. These systems do not suffer from the "line-of-sight" problem.

Skeletal movement can also be measured directly through stereo-radiograph, bone pins [29], and X-ray fluoroscopic techniques [30]. While these methods provide direct measurement of skeletal movement, they are invasive or expose the human subject to radiation. Real-time magnetic resonance imaging (MRI) provides noninvasive and harmless in vivo measurement of bones, ligaments, and muscle positions [31]. However, all these methods impede natural patterns of movement and care must be taken when attempting to extrapolate these types of measurements to natural patterns of locomotion.

4.1.2 Vision-Based Tracking System

These techniques use optical sensors, for example, cameras, to track human movements, which are captured by placing identifiers on the human body. Herein, we review two current systems for tracking of human movement: (i) marker-based and (ii) markerless-based vision systems.

Marker-Based Tracking Systems

Marker-based vision systems have attracted the attention of researchers in medical science and engineering. In this system, the movement of the markers is used to determine the relative movement between two adjacent segments with the goal of precisely defining the movement of the joint. These systems are able to minimize the uncertainty of a subject's movements, due to the unique appearance of markers. This basic theory is embedded in current, state-of-the-art optical motion trackers. Qualisys, Vicon, Codamotion, and Polaris systems are examples of motion capture systems. Qualisys (Qualisys Motion Capture Systems, Gothenburg, Sweden) consists of several cameras, each emitting a beam of infrared light. Small reflective markers are placed on an object to be tracked. Infrared light is flashed and then picked up by the cameras. The system then computes a 3D position of the reflective target, by combining 2D data from several cameras. A Vicon system is also used to calculate joint centers and segment orientations by optimizing skeletal parameters from the trials. For example, Davis et al. reported a study of using a Vicon system for gait analysis [32].

Codamotion (Charnwood Dynamics Ltd) is another active visual tracking system which is precalibrated for 3D measurement, without the need to recalibrate. The Polaris system is also particularly useful when background lighting varies and is unpredictable.

Markerless Vision Systems

Accurate measurement of 3D human body kinematics can be achieved using the markerless motion capture (MMC) tracking system for a subject-specific model. Eliminating the need for markers would also considerably reduce patient preparation time and enable simple, time-efficient, and potentially more meaningful assessments of human movement in research and clinical practice. The feasibility of precisely measuring 3D human body kinematics for the lower [33] and upper limbs using a MMC system is demonstrated in literature [34]. For example, a recently described point cluster technique employs an overabundance of markers placed on each segment to minimize the effects of skin movement artifact [35].

4.2 Dynamics of Human Walking

Generally, there are two ways to enforce the equations of motion for gait simulation: forward dynamics and inverse dynamics. Forward dynamics calculates the motion from given forces and joint torques and muscle excitations by integrating equations of motion with specified initial conditions. In contrast, inverse dynamics computes associated joint torques that lead to a prescribed motion for the system. In this framework, the main question is to find the joint torques and muscle forces that will result in the desired motion [122].

Broadly speaking, in the literature, human locomotion modeling is divided into three categories as illustrated in Fig. 3: (i) skeletal models in which all effects of muscles are modeled as simply as torques applied to joints, (ii) musculoskeletal models where muscle groups are included in the system dynamics, and (iii) NMS models in which human movements produced by the muscular and skeletal systems are controlled by the CNS. Features of various models are explained and their advantages and disadvantages are discussed in detail.

4.2.1 Skeletal Modeling

The skeletal model has been used quite extensively in the robotics field, particularly in humanoid robots [36–38]. In this model, the muscle group at a joint is lumped and represented by a joint torque. Therefore, skeletal models are commonly used in biped walking simulation due to their simplicity and

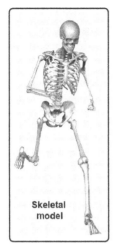  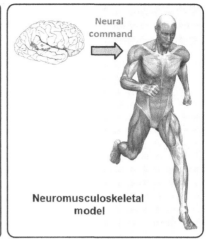

Skeletal model   Musculoskeletal model   Neuromusculoskeletal model

Fig. 3 Classification of human modeling.

computational efficiency. The number of DOFs for these simplified mechanical models is quite different. The simplified mechanical model can be categorized as a planar or spatial model based on the geometry. For a planar model, the gait motion is assumed to be in the sagittal plane because of the complexity of studying the lateral motion [39–45]. On the other hand, for a spatial model, both sagittal and lateral walking motions are considered [46–50].

Several attempts have been made in the literature to develop realistic and smooth human walking using a rigid-link multibody system such as the inverted pendulum model [36,51], passive dynamic walking [52], zero-moment-point (ZMP) method [37,53], optimization- [49], and control-based [54] approaches.

Skeletal Modeling Using Inverted Pendulum Model

As we mentioned earlier, potential energy and kinetic energy are being traded periodically during human walking. Therefore, the simple inverted pendulum model can be used to simulate biped locomotion. This approach uses a simple pendulum model with lumped body mass at the center of gravity (COG). This model always has a closed-form analytical solution for the trajectories of the COG of the model. However, it is difficult to generate natural and smooth biped walking through this method. To resolve this problem, an enhanced inverted pendulum model named the angular-momentum-inducing inverted pendulum model was presented by Kudoh and Komura [50] to generate continuous gait motion. The method can easily handle angular momentum around the COG. Using this method it is possible to plan motion paths for biped robots without discontinuity in the acceleration, even during switching from single-support phase to double-support phase, and vice versa, in both sagittal and frontal planes.

To solve the continuity of motion, the gravity-compensated inverted pendulum model was proposed to generate a natural human gait pattern [55]. This model accommodates the effect of the free leg dynamics based upon its predetermined trajectory. One mass was assigned to the free leg, and another mass for the rest of the body. The mass for the free leg was assumed to be concentrated at the foot. The trajectory of the COG was analytically obtained by solving the linear equations of the two-mass inverted pendulum. This model was developed further by considering the dynamic effect of the swinging leg as a two-mass inverted pendulum model and multiple-mass inverted pendulum model. The mechanical model consisted of 12 DOFs without a trunk, and the designed walking motion was constrained in the sagittal plane [43].

Skeletal Modeling Using Passive Dynamic Walking

The idea of passive dynamics walking is that a skeletal model can only be driven by gravity to walk down a slope automatically with no actuation and control. They can walk downhill with human-like gaits. This concept was first proposed by McGeer, and studied a compass-like structure to perform a passive sagittal walking down a slope by gravity-induced motion without any actuation and control [56,57]. The leg swings naturally as a pendulum, and conservation of angular momentum governs the contact of the swing foot with the ground. Therefore, the model is simple and energy-efficient. Passive dynamics walking has progressed from 2D sagittal models to 3D spatial models. In addition, some form of actuation and control is added to the model to extend the passive dynamics walking on level ground [52]. Small active power sources are introduced to substitute for gravity to extend the passive dynamics walking to level ground. Kuo [58] extended the planar passive dynamics walking to a 3D biped motion allowing for tilting side to side. Collins et al. [59] built the first 3D passive dynamics walking machine with knees. The model has a four-link sagittal model with a knee joint, curved feet, a compliant heel, and mechanically constrained arms to achieve a harmonious and stable gait. Furthermore, the 3D passive dynamics walking model was recently used to study arm swing in human walking [60].

Skeletal Modeling Using ZMP Method

Another method to develop realistic and natural human walking using a rigid-link multibody dynamic system is ZMP trajectory generation. This method is a fast and efficient anthropomorphic gait simulation method. ZMP is a significant dynamic equilibrium criterion and plays a major role in stability analysis of dynamic human walking [53]. Zero-moment point is a concept related to dynamics and control of humanoid robots. It specifies the point with respect to which the dynamic reaction force at the contact of the foot with the ground does not produce any moment in the horizontal direction on the sagittal plane, that is, the point where the total of horizontal inertia and gravity forces equals zero. The concept assumes the contact area is planar and has sufficiently high friction to keep the feet from sliding. The history of ZMP and clarification of some basic concepts was first reviewed in Ref. [61]. Kajita et al. [38] combined the inverted pendulum model with the ZMP-based method to plan walking motion for a biped robot. The multibody dynamics of the robot was represented by a 3D inverted pendulum model from which the ZMP was calculated efficiently. Hirai et al. [62] presented the development of a Honda humanoid robot that had 26 DOFs: 12 DOFs in 2 legs and 14 DOFs in 2 arms. The general procedure for the ZMP method was to first plan a desired ZMP trajectory and then derive the hip or torso motion required to achieve that ZMP trajectory. In this process, the whole-body walking motion was decoupled into three parts: (i) GRF control, (ii) ZMP control, and (iii) foot landing position control.

In ZMP, the basic idea is to enforce the mechanism tracing the prescribed ZMP locations. The key point of this approach is that the dynamics equations are used only to formulate the balance ZMP constraint rather than generation of the entire motion trajectory directly.

Skeletal Modeling Using Optimization-Based Approaches

In biped walking simulation using optimization approaches, it is ideal to optimize the entire walking motion including all physical details with accurate human models. However, such a detailed and accurate model needs powerful computational resources. Therefore, the mathematical gait models used in the literature are generally simplified. Chow and Jacobson [63] first used an optimal programming for gait motion simulation; then, a more complicated 3D skeletal model was successfully developed for gait simulation with optimization-based approaches [47,49].

The complex walking motion includes seven phases in a complete gait cycle [12]. For optimization-based simulation, the gait cycle is simplified so that it only covers a partial gait motion. In the literature, a simplified gait cycle for skeletal modeling based on an optimization method has been generally separated into four groups: (1) single-support swing motion [63]; (2) single-support with instantaneous double-support [40,48]; (3) single-support and double-support—a step [41,49]; and (4) complete gait cycle [64].

In general, there are two approaches for gait optimization of a skeletal model: forward dynamics optimization and inverse dynamics optimization [123]. For a forward dynamics optimization problem, forces and torques are the design variables. The optimal gait is calculated by minimizing a human performance measure or muscle activities subject to physical constraints. During optimization iterations, motion is obtained by integrating the equations of motion with initial conditions. The main problem for forward dynamics optimization is the high computational cost of integration of equations of motion [39,65]. The benefit of this method is that the forces are optimized directly and the motion is generated from equations of motion during optimization.

For inverse dynamics optimization of a skeletal model, the design variables are the joint angle profiles. The optimization problem is solved for optimal gait motion. During an optimization iteration, the forces are directly calculated from equations of motion so that their numerical integration is avoided. The inverse dynamics optimization is computationally efficient because the equations of motion are not integrated in the solution process [40,47,66,67]. Therefore, inverse dynamics is better to be used in the optimization process instead of forward dynamics to avoid integration of the equations of motion [68].

In optimization problems, one needs to decide which physical quantities should be treated as unknowns (design variables), which objective functions should be used to drive the human motion, and which constraints should be imposed for a specific task. Therefore, different formulations result in various optimization problems, and their solution procedures and numerical performance are quite different. The following performance measures are commonly used in the literature for skeletal models to simulate walking motion: mechanical energy, dynamic effort, jerk, stability, and maximum absolute value of joint torque. The constraints associated with human walking simulation are categorized as: (i) physical constraints such as joint angle limits, joint torque limits and (ii) characteristic constraints such as continuity and unilateral contact constraints. The hip, knee, and ankle are constrained to follow measured data [69].

A complete gait cycle in the sagittal plane on level ground by using the inverse dynamics optimization method was simulated in Ref. [44]. Walking was formulated as an optimal motor task subject to multiple constraints with minimization of mechanical energy expenditure over a complete gait cycle being the performance criterion. The mechanical model has a seven-segment linkage, and Fourier series approximated joint profiles.

In Ref. [67], an eight-segment 3D model was used to simulate normal walking for the gait cycle. The hip, knee, and ankle extension/flexion motions were measured from experiments and treated as constraints. This input movement was reconstructed to a kinematically and dynamically consistent 3D movement. In another study, a seven-link planar biped robot for single-support in the sagittal plane was solved for optimal gait motion [39]. The optimization problem of gait simulation was treated as a continuous problem and solved by optimal control methods. The method was based on the implementation of the Pontryagin maximum principle used as a mathematical optimization tool. It applied to mechanical systems with kinematic tree-like topology such as serial robots, walking machines, and articulated biosystems. In this process, the optimal control equations needed to be derived. The algorithm was not efficient when a large-DOF mechanical model was used. This study was extended to a nine DOFs model that moved in the sagittal plane; then an optimization method was used for cyclic, symmetric gait motion of a skeletal model to minimize the actuating torques [45]. Bessonnet et al. [46] extended their 2D model to a 3D skeletal model with 13 DOFs. The optimal control problem was formulated as a nonlinear programming problem where the dynamic effort was considered as a human performance measure to be minimized. In the parameter optimization problem which was used, ordinary differential equations were discretized into algebraic equations, and the time-dependent constraints were simply imposed at time grid points.

Xiang et al. [49] presented a new methodology to simulate one-step spatial digital human walking motion using a 55-DOFs skeletal model. The proposed methodology was based on an optimization formulation that minimizes the dynamic effort of people during walking. The formulation considered symmetric and periodic normal walking and GRFs. Recursive Lagrangian dynamics and analytical gradients for all the constraints and the objective function were incorporated in the optimization formulation. The predicted walking motion was verified with six walking determinants obtained from motion capture experiments. The formulation also showed high-fidelity in predicting joint torques and GRFs.

Skeletal Modeling Using Control Algorithm

Biped locomotion modeling with tracking control is another approach presented in the biped walking literature. Tracking control chooses a proper input force/torque to drive the biped to follow a desired preplanned motion. The key issue is to obtain the desired walking trajectories before utilizing the tracking control. This can be achieved by (i) generation of the desired walking trajectory by a motion capture method which constructs a database from human motion experimentation [70] and (ii) synthesizing desired walking trajectories using an inverted pendulum model or ZMP-based methods. Various control algorithms have been implemented in the literature such as feedback control [55] and intelligent control techniques [54] (neural network [71], fuzzy logic [72], genetic algorithms [73]) and their hybrid forms (neuro-fuzzy networks [74], neuro-genetic and fuzzy-genetic algorithms) in the area of humanoid robotic systems.

Optimal control drives the model from the initial state to the final state while minimizing a cost function. The standard optimal control problem is to find the control history $\tau(t)$ that minimizes the performance measure in the time interval. The optimal control of biped walking is equivalent to the continuous forward optimization problem in which the continuous input joint torques are treated as unknowns in the formulation [45,63].

4.2.2 Musculoskeletal Modeling

The musculoskeletal model has been used quite extensively in the biomechanics field. Many human walking features cannot be represented by the rigid-link mechanical model (skeletal model). Biomechanics gait analysis using a musculoskeletal model can give more details about the physiology of human walking. In contrast to the skeletal model, the musculoskeletal model aims to predict the motion and forces at the muscle level. This is crucial for pathological studies, and it deepens our understanding of muscle excitation during the walking motion [5,18,75].

Muscle contraction dynamics govern the transformation of muscle activation, to muscle force. Once the muscle begins to apply force, the tendon (in series with the muscle) transfers force from the muscle to the bone. This force is named the musculo-tendon force. The joint moment is the sum of the musculo-tendon forces multiplied by their corresponding moment arms. The force in each musculo-tendonous unit contributes in the total moment about the joint. The musculoskeletal geometry determines the moment arms of the muscles (muscle force is dependent on muscle length, ie, the classic muscle "length-tension curve"). It is important to note that the moment arms of muscles are not constant values, but change as a function of joint angles [20,76].

Musculoskeletal modeling has been applied to a broad range of problems in movement science, including: (a) understanding how geometry and muscle-tendon properties independently affect a muscle's ability to develop moment about a joint [8,77]; (b) evaluating a muscle's capacity to accelerate the body joints in various tasks such as walking, jumping, and cycling [6]; and (c) analyzing how orthopedic surgical procedures, such as muscle-tendon transfers, alter the lengths and moment arms of muscles [78]. For example, Riewald et al. [78] used musculoskeletal modeling, neuromuscular control, and forward dynamic simulation to investigate the role of rectus femoris tendon transfer surgery on balance recovery after support-surface perturbations for children with cerebral palsy. However, the most common use of modeling has been in the determination of muscle and joint loading [18,20]. Accurate knowledge of muscle forces could improve the diagnosis and treatment of patients with movement disabilities.

The ability to predict patient-specific joint contact and muscle forces accurately could improve the treatment of walking disorders. Muscle synergy analysis, which decomposes a large number of muscle EMG signals into a small number of synergy control signals, could reduce the dimensionality and thus redundancy of the muscle and contact force prediction process. In Ref. [79], authors investigated whether use of subject-specific synergy controls can improve optimization prediction of knee contact forces during walking. In an optimization problem, the sum of squares of muscle excitations was minimized to investigate how synergy controls affect knee contact force predictions.

Alternatively, inverse dynamics approaches begin by measuring the position of markers and the external forces acting on the body. In gait analysis, for example, the position of markers attached to the participants' limbs can be recorded using a camera-based video system and the external forces recorded using a force platform. The tracking targets on adjacent limb segments are used to calculate relative position and orientation of the segments, and from these, the joint angles are calculated. These data are differentiated to obtain velocities and accelerations. The accelerations and the information about other forces exerted on the body (eg, the recordings from a force plate) can be input to the equations of motion to compute the corresponding joint reaction forces and moments. If the musculoskeletal geometry is included, muscle forces can then be estimated from the joint moments. However, inverse dynamics has important limitations: (i) it is difficult to measure and estimate inertia and mass of each segment, (ii) differentiating displacement data is ill conditioned and sensitive to noises, (iii) co-contraction of muscles is very common and the resultant joint reaction forces and moments are net values, (iv) there are multiple muscles spanning each joint, and the transformation from joint moment to muscle forces yields many possible solutions and cannot be readily determined, and (v) there is no verified model which gives inverse transformation from muscle forces to muscle excitation.

One sensible way to proceed is to combine the power of computational modeling with available measurements to determine information that is not readily obtainable from an experiment. Muscle and joint function can be determined when the following information is available: (i) accurate measurements of the forces applied to the body by the ground, (ii) accurate measurements of body segmental motion, and (iii) accurate knowledge of muscle and joint contact loading. In gait analysis experiments, force platforms are used to measure GRFs, while video-based motion capture techniques are applied to monitor the 3D positions and orientations of the body segments. X-ray fluoroscopy [30] and MRI [80] are also used to record dynamic joint motion in vivo.

The synthesis of human movement involves accurate reconstruction of movement sequences, modeling of musculoskeletal kinematics, dynamics and actuation, and characterization of reliable performance criteria. Task-based methods used in robotics can provide novel musculoskeletal modeling methods and accurate performance predictions. Khatib et al. [81] presented a new method for the real-time reconstruction of human motion trajectories using direct marker tracking with new human performance measure, and a task-driven muscular effort minimization criterion. Dynamic motion reconstruction through the control of a simulated human model was able to follow the captured marker trajectories in real time.

The musculoskeletal model is mechanically redundant as several muscles span each joint and many combinations of muscle forces can produce a net joint moment. For example, more than 15 muscles control 3 degrees of freedom at the hip [18]. Also, biarticular muscles cross two joints and so contribute to the net moments exerted about both joints simultaneously. It is therefore not possible to discern the actions of individual muscles from calculations of net joint moments alone.

This musculoskeletal redundancy allows for an infinite number of combinations of muscle activation patterns for performing a task. Typically, the redundancy resolution is resolved by assuming some optimization criterion, for example, minimizing muscle tension/stress [82], to determine a unique solution for muscle activation pattern among an infinite number of solutions that satisfy biomechanical constraints such as joint torques, joint contact forces [79], and joint impedance [83]. The most typical biomechanical constraints on muscle activation patterns are based on experimentally measured kinematics (eg, joint angles) and kinetics (eg, GRFs), and finally defining single or multiple optimization criteria. Such inverse approaches identify optimal solutions that may capture major features of experimentally measured muscle activation patterns [75].

Musculoskeletal Modeling Using Optimization Methods

Progress in using musculoskeletal models in the recent decade has been accelerated because of the enormous increases in computing power and the availability of more efficient and robust algorithms for modeling and numerical simulation. Computation approaches such as optimization methods have been widely used to simulate and analyze human motions. With the development of optimization techniques these methods have become more attractive. The methods can handle large-scale models and can optimize any human-related performance measure simultaneously. More design variables can be included in the optimization formulation so more natural human walking simulation can be achieved. For human walking simulations, the methods can produce optimal motions and joint force profiles subjected to all the necessary constraints [69].

As mentioned earlier, in forward dynamics optimization of a musculoskeletal model, muscle forces are the design variables for the optimization problem. During optimization iterations, motion is obtained by integrating the equations of motion with initial conditions. The benefit of this method is that the forces are optimized directly and the motion is generated from equations of motion during optimization [5,84].

Musculoskeletal modeling of human motion using optimization is usually a large-scale nonlinear programming problem. In recent years, there has been much progress in research on simulation of human walking, especially in utilizing optimization techniques for large-scale musculoskeletal systems [5,84,85]. Human locomotion is so efficient and the CNS tries to minimize the metabolic cost, that is, the energy expended per unit distance traveled. In Ref. [86], mechanics and energetics predictions in forward dynamics simulations of human walking using different Hill-type muscle energy models are compared. The following performance measures are commonly used in the literature for musculoskeletal models to simulate walking motion: dynamic effort, mechanical energy, muscle activation, fatigue, and metabolic energy. In reality, human motion may be governed by multiple performance measures, and multiobjective optimization methods can be used for gait simulation.

In clinical studies, muscle forces are the main concerns instead of the net joint torques. In optimization problems, muscle forces can be included in the formulation in two ways: (i) a static optimization in which one can partition the joint torques into each muscle group using the equilibrium equation at each joint [75,84] and (ii) a dynamic optimization formulation in which the muscle forces can be treated as design variables for dynamic motion prediction, then the equilibrium equations between the muscles forces and the net joint torques are simply imposed as equality constraints in the optimization formulation [5,18,84], though it requires more computational effort. However, the performance measures evaluated over time, such as total muscular effort [81] or metabolic energy consumption [84], can be considered in the second formulation but not in the static optimization.

Computer modeling and simulation of human movement using the forward dynamics optimization method with a musculoskeletal model were reviewed in Refs. [6,18]. Muscle modeling and computational issues were presented in detail. Also, the forward dynamic optimization approach was illustrated by simulating human jumping, walking, and pedaling motions.

A 3D musculoskeletal model with 23 degree-of-freedom mechanical linkage and 54 muscles was developed by Anderson and Pandy [84] for normal symmetric walking on level ground using the forward dynamics optimization method. Muscle forces were treated as design variables, and metabolic energy expenditure per unit distance was considered as a human performance measure to be minimized. Muscle metabolic energy was calculated by summing five terms: the resting heat, activation heat, maintenance heat, shortening heat, and the mechanical work done by all the muscles in the model. Then, the model was used to analyze human walking, and many insights on muscle functions for normal and pathological gait were obtained in Ref. [87].

As discussed earlier, musculo-tendon forces and joint reaction forces are typically estimated using a two-step method: first computing the musculo-tendon forces by a static optimization procedure and then deducing the joint reaction forces from the force equilibrium. However, this method does not allow studying the interactions between musculo-tendon forces and joint reaction forces in establishing this equilibrium. So, the joint reaction forces are usually overestimated. Moissenet et al. [88] introduces a new 3D lower-limb musculoskeletal model based on a one-step static optimization procedure allowing simultaneous musculo-tendon, joint contact, ligament, and bone forces estimation during gait.

Neptune et al. [5] proposed a method which used muscle-actuated forward dynamics optimization to drive the model to follow the measured joint angle profiles and GRFs. The simulation analysis showed that a simple neural control strategy involving five muscle activation modules was sufficient to perform the basic subtasks of walking (ie, body support, forward propulsion, and leg swing). The musculoskeletal model in the sagittal plane consisted of 7 rigid segments driven by 13 muscle groups. A complete gait cycle from right HS to the subsequent right HS was simulated. The difference of kinematics and GRF between simulation and experiment was minimized, and the muscle actuations were treated as design variables in the optimization formulation.

The effects of different performance criteria on predicted gait patterns using a 2D musculoskeletal model was studied in Ref. [89]. In this study the mechanical model consisted of seven rigid segments with nine DOFs to simulate single step walking and generate cyclic motion. Eight muscle groups were included in each lower extremity. In the optimization problem state variables, controls, and muscle activations were all treated as design variables. The objective of this paper was to shed some light on the effects of the cost function choice on the predicted kinematics and muscle recruitment patterns of gait. A series of predictive simulations of gait are performed utilizing a family of cost functions representative of a large range of performance criteria traditionally adopted in the literature. It was found that a fatigue-like cost function predicted a more realistic normal gait.

Optimization-based approaches determine the optimal solution among the infinite number of solutions for muscle forces by defining biomechanical constraints. However, deviation from the optimal patterns for muscle activation may also satisfy these constraints. Characterizing all these viable deviations in muscle activity from an optimal solution would facilitate interpretation of experimental variations in muscle activation patterns. In addition, it would facilitate interpretation of how those biomechanical constraints affect not just the optimal solutions, but the set of all possible solutions. A few attempts to define feasible muscle activation ranges for a given movement have been made previously [90–92].

Musculoskeletal Modeling Using Control-Based Approaches

In musculoskeletal modeling based on the control-based approaches, there are two main issues: (a) experimental data collection and (b) appropriate controller use to track measured trajectories. In general, there are many factors that should be considered to accomplish proper experiments on human subjects: (i) suitable subjects should be recruited, (ii) experimental protocol should be set up, (iii) instruments should be calibrated, noise should be filtered, and errors should be quantified, (iv) experience and skills are required to perform good experiments on human subjects, (v) appropriate controllers need to be selected and designed to provide feedback on joint torques, (vi) the nonlinear dynamics system should be simplified, and finally (vii) the controllability and stability of the controller should be considered and tested. Once the controller is tuned and validated, the control-based method can be used to obtain joint torques and muscle forces based on the measured experimental data [75].

Control methods can also be built into the optimization problem to compute muscle forces such as the computed-muscle-control method developed by Thelen and Anderson [75]. The objective of this study was to develop an efficient methodology for generating muscle-actuated simulations of human walking that closely reproduce experimental measures of kinematics and GRFs. In this method, forward dynamics and feedback control were employed to obtain the joint actuation torques, which drive the kinematic trajectories of the model toward a set of desired trajectories obtained from experiments. Then, the muscle forces were computed by using a static optimization algorithm. This method is more efficient and stable than the dynamic muscle optimization algorithm using a forward dynamics optimization approach.

Optimal control also drives the model from the initial state to the final state while minimizing a cost function. Optimal control of a musculoskeletal model is equivalent to the continuous forward optimization problem in which the continuous input joint torques are treated as unknowns in the formulation [7].

4.2.3 NMS Modeling

NMS modeling deals with the modeling of human movements generated by the muscular and skeletal systems while it is controlled by the CNS. This model is important for different purposes including: (i) studying how the nervous system controls limb movements in both unimpaired people and those with pathologies such as spasticity caused by stroke or cerebral palsy [93,94], (ii) studying functional electrical stimulation of paralyzed muscles, and (iii) designing prototypes of myoelectrically controlled limbs.

The human CNS system is quite complicated as it determines human behavior. As we reviewed earlier, both optimization algorithms and the control-based approaches can be used to accurately approximate the CNS so as to predict and analyze human movement. In optimization methods, the CNS is considered as one of the several human performance measures, and information from experiments assist as constraints. Thus, it can mimic the entire CNS behavior system. The weakness of the method is that having completely smooth, natural, and repeated motion is difficult to obtain because the objective function only represents a partial CNS unless more constraints are used to drive the motion [69]. Therefore, the optimization algorithms are good at answering how people behave if some conditions are changed by solving an optimization problem with different inputs.

In contrast, the control-based methods first duplicate complete human motion from experiments, and then a controller is used to represent the CNS to drive the model through the measured trajectories. The beauty of the method is that the CNS is better approximated to re-create the recorded human motion. Therefore, natural and subject-specific motions can be accurately tracked and simulated. It is suitable to study pathological gait and associated muscle forces to reveal the physical insights behind the motion. Thus, the control-based method is good at answering why people behave the way they do by tracking the corresponding experimental data [93,95].

Buchanan et al. [96] provided an overview of forward dynamic NMS modeling in which the estimation of muscle forces and joint moments, and movements from measurement of neural command, are discussed in detail. In the first step of their four-step process (as illustrated in Fig. 4), muscle activation dynamics governs the transformation from the neural command to a measure of muscle activation—a time-varying parameter between 0 and 1. The neural command can be taken from EMGs in which the magnitudes of the EMG signals will be changed when the neural command calls for increased or decreased muscular effort. In the second step, muscle contraction dynamics characterize how muscle activations are transformed into muscle forces. The third step requires a model of the musculoskeletal geometry to transform muscle forces to joint moments. Finally, the equations of motion allow joint moments to be transformed into limb acceleration and joint movement.

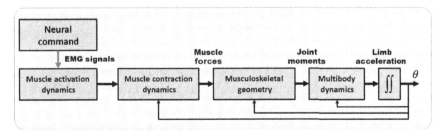

Fig. 4 NMS model for studying human movement.

However, the forward dynamics approach has some limitations, including: (i) estimation of muscle activation from EMG signals is difficult; (ii) the transformation from muscle activation to muscle force is not completely understood (although one way for direct determination of muscle forces from EMGs is using optimization), as there is no known correlation between the level of a measured EMG signal and the amount of force that the muscle might be producing during a dynamic contraction; and (iii) determination of the muscle-tendon moment arms and lines of action are still challenging problems.

A neuromusculoskeletal tracking (NMT) method was developed by Seth and Pandy [95] to estimate muscle forces from observed motion data. This NMT method consisted of two phases: (i) skeletal motion tracking control and (ii) optimal neuromuscular tracking control. In the first stage, the skeletal motion tracker calculates the joint torques needed to actuate a skeletal model and track observed segment angles and ground forces in a forward simulation of the motor task. In the second stage, an optimal neuromuscular tracker was used to determine the optimal muscle excitations and muscle forces by tracking the joint torques obtained in the first stage. The proposed NMT approach was compared with conventional approaches such as inverse dynamics analysis and forward dynamics optimization. It was concluded that NMT gave more accurate and more efficient simulation than the inverse and forward dynamics method. It requires three orders-of-magnitude less CPU time than parameter optimization. The speed and accuracy of this method make it a proper tool for estimating muscle forces using experimentally obtained kinematics and ground force data.

5. COMPUTATIONAL MUSCULOSKELETAL ANALYSIS INTERACTION WITH ARTICULATED SYSTEMS

In the biomechanics field, many computational tools have been developed for kinematic and dynamic analysis of musculoskeletal systems, building on an articulated-multibody systems framework [124]. Constrained musculoskeletal models can be constructed modularly by placing constraints on anatomical components. Such musculoskeletal analysis tools allow monitoring of internal human variables such as muscle lengths, forces, reactions of muscles/tendons/joints, metabolic power consumption, and mechanical work.

Examples include both commercial tools such as LifeMod, SIMM, and AnyBody, as well as open-source software such as OpenSim to follow human movements, compute muscle-tendon forces, and analyze the normal and pathological gaits [1–3]. Among these software packages, the AnyBody modeling system offers a convenient tool for modeling and analyzing various musculoskeletal systems [97]. The AnyBody musculoskeletal model is established as a constrained articulated-multibody system with rigid skeletal bones connected with multiple muscles which actuate the system. The governing equations of motion can be obtained as the constrained dynamic equations of this articulated-multibody system. The indeterminacy in muscle-tendon force distribution is resolved by employing optimization approaches. In AnyBody, redundancy resolution takes the form of minimization of the maximal muscle activity subject to equality constraints and nonnegative muscle force constraints. For example, in Ref. [3], a biomedical model of a thumb is developed by real CT scans of four bony sections, and actuated by nine realistic muscle-tendons in AnyBody, and then the indeterminacy problem is solved by optimization-based approaches.

A graphical user interface (GUI) is also developed to facilitate user interaction with AnyBody settings (eg, using radio buttons and sliders) allowing performance of parametric studies by manually varying the appropriate design variables [15]. The optimization process, data manipulation, and interfacing to AnyBody are handled using MATLAB, and the results of the optimization are displayed back within the GUI (see Figs. 5 and 6).

Contrary to AnyBody, OpenSim uses a forward dynamics approach in solving the musculoskeletal question. OpenSim is open-source software, which includes all computational components for deriving equations of motion for articulated-multibody systems, performing numerical integration, and solving constrained nonlinear optimization problems. This software also offers accessibility to control algorithms such as computed-muscle control, actuators, and analyses (eg, muscle-induced accelerations). OpenSim integrates all these components into a modeling and simulation platform. Users can extend OpenSim by writing their own plug-ins for analysis or control, or to represent NMS elements [99].

Simbody is an application programming interface which serves as the dynamics engine behind OpenSim. It can incorporate robust, high-performance, minimal coordinate multibody dynamics into a broad range of domain-specific end-user applications. Simbody provides a diverse set of tools to handle the modeling and computational aspects of multibody dynamics, to ensure correct and efficient deployment. Simbody also includes contact modeling, numerical integration and differentiation, constraint stabilization and redundancy handling, etc. [100].

The rapid model-based design, control systems, and powerful numerical method strengths of MATLAB/Simulink can be combined with the simulation

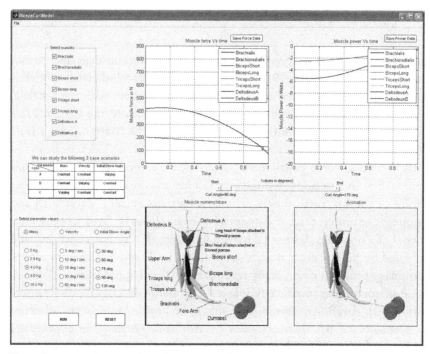

Fig. 5 Parametric bicep curl study on a simplified upper-arm/shoulder musculoskeletal model with the graphical user interface [15].

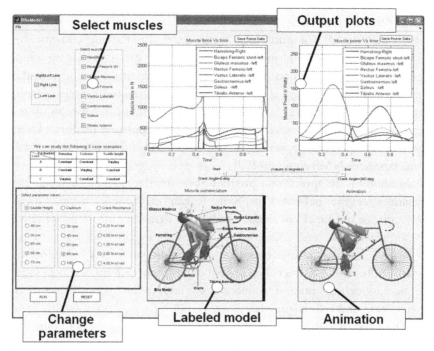

Fig. 6 Performing a bicycling motion study on the lower extremity/hip musculoskeletal model [98].

and human movement dynamics strengths of OpenSim by developing a new interface between the two software tools. In Ref. [101], OpenSim is integrated with Simulink using the MATLAB S-function mechanism, and the interface is demonstrated using both open-loop and closed-loop control systems.

Exoskeletons are a new class of articulated-multibody systems used for motion assistance for the elderly or disabled individuals and their performance is realized while in intimate contact with individual musculoskeletal systems. They are also intended to improve rehabilitation for people with disabilities caused by strokes [14], muscle disease, spinal cord injuries [102], etc. The state of the art for lower-limb exoskeletons presented by Dollar and Herr [14] showed that having knowledge of the biomechanics of walking is important to build an exoskeleton that can interact with the user with minimal chances of harm. Alamdari et al. [103] comprehensively surveyed the clinic- and home-based rehabilitation devices for upper and lower limbs therapy to recognize the situations which need human-robot interaction to be considered. For training patients under rehabilitation with an exoskeleton, physical human-robot interaction is a major concern for a safe and comfortable usage. For example, in Refs. [104,105] as shown in Fig. 7A, a cable-driven end-effector-based exoskeleton named PACER is directly coordinated with a human arm, and in Refs. [102,106] as shown in Fig. 7B, a cable-driven exoskeleton named ROPES is in intimate contact with human lower limbs. For modeling, analyzing and deep understanding of human-robot interaction, the musculoskeletal model of the human body

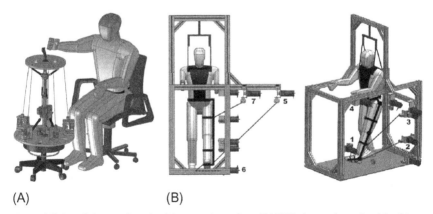

(A)   (B)

Fig. 7 (A) Parallel articulated-cable exercise robot (PACER): home-based cable-driven mechanism for upper extremities rehabilitative exercises [104,105], (B) *robotic physical exercise and system* (ROPES): a cable- driven robotic rehabilitation system for lower extremity [102,106] which is driven by seven motors installed on the frame labeled 1–7.

as well as multibody dynamics of the exoskeleton need to be integrated and modeled [15,125] Therefore, the model is able to estimate the muscle activities in cooperative motions and enables the design analysis and optimization of robotic exoskeletons.

Safety is one of the top priorities when designing any kind of exoskeleton [107], as they interact closely with humans. In the past, one safe way to design and test an exoskeleton was to build two robots prior to their usage; if there is an unwanted dangerous torque on the master robot, then no human would get harmed [108]. The other criteria when designing exoskeletons are the range of motion and magnitude of effort [109]. These two criteria define the difficulties faced when matching to human biomechanics. It is difficult to detect the axis of the human joints, to mimic all degrees of freedom, and to avoid the relative motion between the exoskeleton and the human due to nonoptimal fixation during exercise.

Nowadays, exoskeleton testing can be accomplished virtually using multibody modeling [110]. Multibody dynamic modeling can be seen as a powerful tool to design exoskeletons by simulating both the musculoskeletal system and exoskeleton dynamics, enabling the prediction, in a noninvasive way, of the efforts performed by the exoskeletons and forces applied to the human body [111]. Virtually designing the exoskeleton directly on a human musculoskeletal model helps to constrain the exoskeleton kinematics to the human kinematics. Ferrati et al. [112] analyzed an existing exoskeleton by reproducing it virtually and constraining it to a human musculoskeletal model. This means that the rehabilitation can be virtually quantified, and therefore several designs can be tested according to the injury. To virtually prototype exoskeletons on human musculoskeletal models opens the possibilities of custom exoskeletons for rehabilitation. Agarwal et al. [113] argued that virtually designing an exoskeleton model directly on a musculoskeletal model allows the introduction of biomechanical, morphological, and controller measures to quantify the performance of the device.

It is important to know how much force the exoskeleton generates before manufacturing the real exoskeleton. One sensible way to predict the force is using a virtual prototyping environment during the design stage. In Ref. [114], a combined human–exoskeleton model is generated, and dynamic analysis is performed under different constraints using AnyBody software to predict the effect of connecting the user to the exoskeleton. This dynamic analysis makes it possible to calculate the human joint torque and the interaction force for the musculoskeletal model and the exoskeleton. To accomplish this, the musculoskeletal model generated by AnyBody software is scaled with human subject data collected using motion capture data, and the designed exoskeleton is converted into an STL model in SolidWorks and merged with the musculoskeletal model in AnyBody software.

In Ref. [115], an integrated musculoskeletal-exoskeleton system is proposed for the optimal design of exoskeletons. The human-robot interaction system is implemented in AnyBody, and kinematic and dynamic simulation of the system is conducted in this software for a cooperative motion of the exoskeleton and human arm for lifting a payload. The design parameters of the cable-driven exoskeleton are formulated as an optimization problem. The activities of three selected muscles of the upper limbs are displayed, which are considered as the major extensor and flexors of arm motion. The muscle activity is evaluated by taking the mean activity of major muscles. These individual muscle forces and elbow flexion moments can serve as performance measures. Such performance measures allow the designer to directly evaluate the effectiveness of the exoskeleton design. These measures are considered for analysis because of the following reasons: (i) individual muscle forces show which muscles play a significant role in performing the given experiment and hence how modifications in design can relieve them; and (ii) elbow flexion moment signifies the load carried by the human elbow joint, and thus gives an idea of the external load acting on the joint [15].

The effects of backrest inclination of a seated person in an adjustable car seat, and also the effects of vertical vibration frequency of the suspension system on the muscle activity in the dynamic environment are assessed using the musculoskeletal model in AnyBody software [116]. This study shows that the vibration frequency significantly affects the muscle activity of the lumbar area, and likewise, the inclination degree of the backrest significantly affects the muscle activities of the right leg and the abdomen. The combination of vibration and forward inclination of the backrest can be used to maximize the muscle activity of the leg, similar to the abdomen and lumbar muscles. Similarly, Rasmussen et al. [117], Grujicic et al. [118] used a detailed musculoskeletal computer model to examine the influence of car seat design/adjustments and the consequences of variations in the seat pan angle, friction coefficient of the seat surface on spinal joint forces, soft tissue contact normal and shear stresses and muscular activity. Ma et al. [119] analyzed the muscle activities and joint forces of the lower limb with knee normal and knee lock postures under vertical whole-body vibration, based on inverse dynamics and using the AnyBody Modeling System.

In Ref. [112], a virtual exoskeleton for lower-limb rehabilitation was constrained to a human musculoskeletal model in OpenSim software to study its behavior and to evaluate the design parameters. By using OpenSim built-in tools and a human musculoskeletal model, four different models were implemented in order to study the behavior of the system in different operating conditions, and in particular the interactions that are generated between the human musculoskeletal structure and the exoskeleton.

6. CONCLUSION

In this paper, we surveyed computational musculoskeletal modeling, analysis, and simulation of human lower extremities. We illustrated that with increasing the complexity and level of detail of human models, a need for high-performance numerical time-varying schema has emerged. Hence, in this chapter, we first reviewed the experimental kinematic approaches for studying human walking, including vision-based and nonvision-based techniques. Then, we divided biomechanical dynamic models into three categories: (i) skeletal models in which all effects of muscles are modeled as simply as torques applied to joints, (ii) musculoskeletal models where muscle groups are included in the system dynamics, and (iii) NMS models in which human movements produced by the muscular and skeletal systems are controlled by the CNS.

The skeletal model has been used extensively in the robotic field to develop realistic and smooth human walking using rigid multibody systems via inverted pendulum modeling, passive walking dynamics, ZMP, optimization, and control-based approaches. On the other hand, the musculoskeletal and NMS models have been used quite extensively in biomechanical fields. We reviewed current methods for numerical simulation of musculoskeletal modeling including optimization and control-based approaches. We realized that inverted pendulum, ZMP, and passive dynamics walking models are based on the idealized models with few DOFs, in contrast, optimization- and control-based methods are important approaches for simulating natural human walking with higher DOFs and physical details. The control-based methods are suitable for applications in neurological studies of human movement, while the optimization-based methods are advantageous for conducting "what-if" scenarios.

Finally, we surveyed efforts on improving the understanding of the physical interaction of human limbs with their physical environment. We realized that for optimal design of exoskeletons, an integrated musculoskeletal-exoskeleton system is necessary to understand how much force the exoskeleton model generates before manufacturing the real exoskeleton. We also realized that, in order to simulate in the real operating conditions, a virtual exoskeleton could be properly constrained to a human musculoskeletal model within a biomechanical simulation platform. Then, analysis of simulation results might suggest potential kinematic and dynamic modifications of the real system in order to test the new design.

REFERENCES

[1] J.A. Reinbolt, A. Seth, S.L. Delp, Simulation of human movement: applications using OpenSim, Procedia IUTAM 2 (2011) 186–198.

[2] A. Seth, et al., OpenSim: a musculoskeletal modeling and simulation framework for in silico investigations and exchange, Procedia IUTAM 2 (2011) 212–232.

[3] J.Z. Wu, et al., Modeling of the muscle/tendon excursions and moment arms in the thumb using the commercial software anybody, J. Biomech. 42 (3) (2009) 383–388.

[4] J. Higginson, et al., Muscle contributions to support during gait in an individual with post-stroke hemiparesis, J. Biomech. 39 (10) (2006) 1769–1777.

[5] R.R. Neptune, D.J. Clark, S.A. Kautz, Modular control of human walking: a simulation study, J. Biomech. 42 (9) (2009) 1282–1287.

[6] M.G. Pandy, Computer modeling and simulation of human movement, Annu. Rev. Biomed. Eng. 3 (1) (2001) 245–273.

[7] M.G. Pandy, F.C. Anderson, D. Hull, A parameter optimization approach for the optimal control of large-scale musculoskeletal systems, J. Biomech. Eng. 114 (4) (1992) 450–460.

[8] M. Mansouri, et al., Rectus femoris transfer surgery affects balance recovery in children with cerebral palsy: a computer simulation study, Gait Posture 43 (2016) 24–30.

[9] H.J. Kim, et al., Evaluation of predicted knee-joint muscle forces during gait using an instrumented knee implant, J. Orthop. Res. 27 (10) (2009) 1326–1331.

[10] L. Modenese, A. Phillips, A. Bull, An open source lower limb model: hip joint validation, J. Biomech. 44 (12) (2011) 2185–2193.

[11] E. Ayyappa, Normal human locomotion, Part 1: basic concepts and terminology, J. Prosthet. Orthot. 9 (1) (1997) 10–17.

[12] D.A. Winter, Biomechanics and Motor Control of Human Movement, fourth ed., John Wiley and Sons, New Jersey, 2009.

[13] L.-F. Lee, et al., Case studies of musculoskeletal-simulation-based rehabilitation program evaluation, IEEE Trans. Robot. 25 (3) (2009) 634–638.

[14] A.M. Dollar, H. Herr, Lower extremity exoskeletons and active orthoses: challenges and state-of-the-art, IEEE Trans. Robot. 24 (1) (2008) 144–158.

[15] P. Agarwal, et al., Simulation-based design of exoskeletons using musculoskeletal analysis, in: ASME 2010 International Design Engineering Technical Conferences and Computers and Information in Engineering Conference, American Society of Mechanical Engineers, New York, 2010.

[16] A. Roaas, G.B. Andersson, Normal range of motion of the hip, knee and ankle joints in male subjects, 30–40 years of age, Acta Orthop. 53 (2) (1982) 205–208.

[17] B. Stansfield, et al., Regression analysis of gait parameters with speed in normal children walking at self-selected speeds, Gait Posture 23 (3) (2006) 288–294.

[18] M.G. Pandy, T.P. Andriacchi, Muscle and joint function in human locomotion, Annu. Rev. Biomed. Eng. 12 (2010) 401–433.

[19] F.E. Zajac, R.R. Neptune, S.A. Kautz, Biomechanics and muscle coordination of human walking: Part I: introduction to concepts, power transfer, dynamics and simulations, Gait Posture 16 (3) (2002) 215–232.

[20] F.E. Zajac, R.R. Neptune, S.A. Kautz, Biomechanics and muscle coordination of human walking: part II: lessons from dynamical simulations and clinical implications, Gait Posture 17 (1) (2003) 1–17.

[21] A. Dallmeijer, et al., Association between isometric muscle strength and gait joint kinetics in adolescents and young adults with cerebral palsy, Gait Posture 33 (3) (2011) 326–332.

[22] S.V.S. Jan, Introducing anatomical and physiological accuracy in computerized anthropometry for increasing the clinical usefulness of modeling systems, Crit. Rev. Phys. Rehabil. Med. 17 (4) (2005) 249–274.

[23] V. Sholukha, et al., Model-based approach for human kinematics reconstruction from markerless and marker-based motion analysis systems, J. Biomech. 46 (14) (2013) 2363–2371.

[24] F. Multon, et al., Computer animation of human walking: a survey, J. Vis. Comput. Animat. 10 (1) (1999) 39–54.

[25] R. Boulic, N.M. Thalmann, D. Thalmann, A global human walking model with real-time kinematic personification, Vis. Comput. 6 (6) (1990) 344–358.

[26] H. Zhou, H. Hu, N. Harris, Wearable inertial sensors for arm motion tracking in home-based rehabilitation, in: IAS, 2006.

[27] C. Patten, F.B. Horak, D.E. Krebs, Head and body center of gravity control strategies: adaptations following vestibular rehabilitation, Acta Otolaryngol. 123 (1) (2003) 32–40.

[28] T. Molet, R. Boulic, D. Thalmann, A Real Time Anatomical Converter for Human Motion Capture, Springer, New York, 1996.

[29] T. Beth, et al., Characteristics in human motion—from acquisition to analysis, in: IEEE International Conference on Humanoid Robots, 2003.

[30] S. Tashman, W. Anderst, In-vivo measurement of dynamic joint motion using high speed biplane radiography and CT: application to canine ACL deficiency, J. Biomech. Eng. 125 (2) (2003) 238–245.

[31] G.G. Handsfield, et al., Relationships of 35 lower limb muscles to height and body mass quantified using MRI, J. Biomech. 47 (3) (2014) 631–638.

[32] R.B. Davis, et al., A gait analysis data collection and reduction technique, Hum. Mov. Sci. 10 (5) (1991) 575–587.

[33] T. Persson, A marker-free method for tracking human lower limb segments based on model matching, Int. J. Biomed. Comput. 41 (2) (1996) 87–97.

[34] S. Corazza, L. Mündermann, T.P. Andriacchi, The evolution of methods for the capture of human movement leading to markerless motion capture for biomechanical applications, J. Neuroeng. Rehabil. 3 (1) (2006) 1–11.

[35] L. Mündermann, S. Corazza, T.P. Andriacchi, Accurately measuring human movement using articulated ICP with soft-joint constraints and a repository of articulated models, in: IEEE Conference on Computer Vision and Pattern Recognition, 2007, CVPR'07, IEEE, Minneapolis, MN, 2007.

[36] S. Kajita, O. Matsumoto, M. Saigo, Real-time 3D walking pattern generation for a biped robot with telescopic legs, in: Proceedings 2001 ICRA. IEEE International Conference on Robotics and Automation, 2001, IEEE, 2001.

[37] K. Harada, et al., An analytical method for real-time gait planning for humanoid robots, Int. J. Humanoid Rob. 3 (01) (2006) 1–19.

[38] S. Kajita, et al., Biped walking pattern generation by using preview control of zero-moment point, in: Proceedings. ICRA'03, IEEE International Conference on Robotics and Automation, 2003, IEEE, 2003.

[39] G. Bessonnet, P. Sardain, S. Chessé, Optimal motion synthesis–dynamic modelling and numerical solving aspects, Multibody Syst. Dyn. 8 (3) (2002) 257–278.

[40] C. Chevallereau, Y. Aoustin, Optimal reference trajectories for walking and running of a biped robot, Robotica 19 (05) (2001) 557–569.

[41] X. Mu, Q. Wu, Synthesis of a complete sagittal gait cycle for a five-link biped robot, Robotica 21 (05) (2003) 581–587.

[42] M. Rostami, G. Bessonnet, Sagittal gait of a biped robot during the single support phase. Part 2: optimal motion, Robotica 19 (03) (2001) 241–253.

[43] A. Albert, W. Gerth, Analytic path planning algorithms for bipedal robots without a trunk, J. Intell. Robot. Syst. 36 (2) (2003) 109–127.

[44] L. Ren, R.K. Jones, D. Howard, Predictive modelling of human walking over a complete gait cycle, J. Biomech. 40 (7) (2007) 1567–1574.

[45] T. Saidouni, G. Bessonnet, Generating globally optimised sagittal gait cycles of a biped robot, Robotica 21 (02) (2003) 199–210.

[46] G. Bessonnet, et al., Parametric-based dynamic synthesis of 3D-gait, Robotica 28 (04) (2010) 563–581.

[47] H.J. Kim, et al., Dynamic motion planning of 3D human locomotion using gradient-based optimization, J. Biomech. Eng. 130 (3) (2008) 031002.

[48] D. Tlalolini, Y. Aoustin, C. Chevallereau, Design of a walking cyclic gait with single support phases and impacts for the locomotor system of a thirteen-link 3d biped using the parametric optimization, Multibody Syst. Dyn. 23 (1) (2010) 33–56.

[49] Y. Xiang, et al., Optimization-based dynamic human walking prediction: one step formulation, Int. J. Numer. Methods Eng. 79 (6) (2009) 667–695.

[50] S. Kudoh, T. Komura, C^2 continuous gait-pattern generation for biped robots, in: Proceedings 2003 IEEE/RSJ, International Conference on Intelligent Robots and Systems, 2003 (IROS 2003), IEEE, 2003.

[51] S. Kajita, T. Yamaura, A. Kobayashi, Dynamic walking control of a biped robot along a potential energy conserving orbit, IEEE Trans. Robot. Autom. 8 (4) (1992) 431–438.

[52] S. Collins, et al., Efficient bipedal robots based on passive-dynamic walkers, Science 307 (5712) (2005) 1082–1085.

[53] P. Sardain, G. Bessonnet, Forces acting on a biped robot. Center of pressure-zero moment point, IEEE Trans. Syst. Man Cybern. Syst. Hum. 34 (5) (2004) 630–637.

[54] D. Katić, M. Vukobratović, Survey of intelligent control techniques for humanoid robots, J. Intell. Robot. Syst. 37 (2) (2003) 117–141.

[55] J.H. Park, K.D. Kim, Biped robot walking using gravity-compensated inverted pendulum mode and computed torque control, in: Proceedings, IEEE International Conference on Robotics and Automation, 1998, IEEE, 1998.

[56] T. McGeer, Passive dynamic walking, Int. J. Robot. Res. 9 (2) (1990) 62–82.

[57] T. McGeer, Passive walking with knees, in: Proceedings, 1990 IEEE International Conference on Robotics and Automation, IEEE, 1990.

[58] A.D. Kuo, Stabilization of lateral motion in passive dynamic walking, Int. J. Robot. Res. 18 (9) (1999) 917–930.

[59] S.H. Collins, M. Wisse, A. Ruina, A three-dimensional passive-dynamic walking robot with two legs and knees, Int. J. Robot. Res. 20 (7) (2001) 607–615.

[60] S.H. Collins, P.G. Adamczyk, A.D. Kuo, Dynamic arm swinging in human walking, Proc. R. Soc. B (2009). p. rspb20090664.

[61] M. Vukobratović, B. Borovac, Zero-moment point—thirty five years of its life, Int. J. Humanoid Rob. 1 (01) (2004) 157–173.

[62] K. Hirai, et al., The development of Honda humanoid robot, in: Proceedings, 1998 IEEE International Conference on Robotics and Automation, IEEE, 1998.

[63] C. Chow, D. Jacobson, Studies of human locomotion via optimal programming, Math. Biosci. 10 (3) (1971) 239–306.

[64] Y. Xiang, Optimization-Based Dynamic Human Walking Prediction, ProQuest, Ann Arbor, 2008.

[65] G. Bessonnet, S. Chesse, P. Sardain, Optimal gait synthesis of a seven-link planar biped, Int. J. Robot. Res. 23 (10–11) (2004) 1059–1073.

[66] G. Bessonnet, P. Seguin, P. Sardain, A parametric optimization approach to walking pattern synthesis, Int. J. Robot. Res. 24 (7) (2005) 523–536.

[67] B. Koopman, H.J. Grootenboer, H.J. de Jongh, An inverse dynamics model for the analysis, reconstruction and prediction of bipedal walking, J. Biomech. 28 (11) (1995) 1369–1376.

[68] B.-I. Koh, et al., Limitations of parallel global optimization for large-scale human movement problems, Med. Eng. Phys. 31 (5) (2009) 515–521.

[69] Y. Xiang, J. Arora, K. Abdel-Malek, Physics-based modeling and simulation of human walking: a review of optimization-based and other approaches, Struct. Multidiscip. Optim. 42 (1) (2010) 1–23.

[70] J. Pettre, J.P. Laumond, A motion capture-based control-space approach for walking mannequins, Comput. Anim. Virtual Worlds 17 (2) (2006) 109–126.

[71] J. Hu, J. Pratt, G. Pratt, Stable adaptive control of a bipedal walking robot with CMAC neural networks, Proceedings, 1999 IEEE International Conference on Robotics and Automation, IEEE, 1999.

[72] K. Low, Fuzzy position/force control of a robot leg with a flexible gear system, in: Proceedings, ICRA'02. IEEE International Conference on Robotics and Automation, IEEE, 2002.

[73] M.Y. Cheng, C.S. Lin, Genetic algorithm for control design of biped locomotion, J. Robot. Syst. 14 (5) (1997) 365–373.

[74] J.-G. Juang, Fuzzy neural network approaches for robotic gait synthesis, IEEE Trans. Syst. Man Cybern. B Cybern. 30 (4) (2000) 594–601.

[75] D.G. Thelen, F.C. Anderson, Using computed muscle control to generate forward dynamic simulations of human walking from experimental data, J. Biomech. 39 (6) (2006) 1107–1115.

[76] T.S. Buchanan, et al., Estimation of muscle forces and joint moments using a forward-inverse dynamics model, Med. Sci. Sports Exerc. 37 (11) (2005) 1911.

[77] M.G. Hoy, F.E. Zajac, M.E. Gordon, A musculoskeletal model of the human lower extremity: the effect of muscle, tendon, and moment arm on the moment-angle relationship of musculotendon actuators at the hip, knee, and ankle, J. Biomech. 23 (2) (1990) 157–169.

[78] S.A. Riewald, S.L. Delp, The action of the rectus femoris muscle following distal tendon transfer: Does it generate knee flexion moment? Dev. Med. Child Neurol. 39 (2) (1997) 99–105.

[79] J.P. Walter, et al., Muscle synergies may improve optimization prediction of knee contact forces during walking, J. Biomech. Eng. 136 (2) (2014) 021031.

[80] F. Sheehan, F. Zajac, J. Drace, In vivo tracking of the human patella using cine phase contrast magnetic resonance imaging, J. Biomech. Eng. 121 (6) (1999) 650–656.

[81] O. Khatib, et al., Robotics-based synthesis of human motion, J. Physiol. Paris 103 (3–5) (2009) 211–219.

[82] D.G. Thelen, F.C. Anderson, S.L. Delp, Generating dynamic simulations of movement using computed muscle control, J. Biomech. 36 (3) (2003) 321–328.

[83] D.W. Franklin, D.M. Wolpert, Computational mechanisms of sensorimotor control, Neuron 72 (3) (2011) 425–442.

[84] F.C. Anderson, M.G. Pandy, Dynamic optimization of human walking, J. Biomech. Eng. 123 (5) (2001) 381–390.

[85] S.L. Delp, J.P. Loan, A computational framework for simulating and analyzing human and animal movement, Comput. Sci. Eng. 2 (5) (2000) 46–55.

[86] R.H. Miller, A comparison of muscle energy models for simulating human walking in three dimensions, J. Biomech. 47 (6) (2014) 1373–1381.

[87] F.C. Anderson, et al., Contributions of muscle forces and toe-off kinematics to peak knee flexion during the swing phase of normal gait: an induced position analysis, J. Biomech. 37 (5) (2004) 731–737.

[88] F. Moissenet, L. Chèze, R. Dumas, A 3D lower limb musculoskeletal model for simultaneous estimation of musculo-tendon, joint contact, ligament and bone forces during gait, J. Biomech. 47 (1) (2014) 50–58.

[89] M. Ackermann, A.J. van den Bogert, Optimality principles for model-based prediction of human gait, J. Biomech. 43 (6) (2010) 1055–1060.

[90] S. Martelli, et al., Computational tools for calculating alternative muscle force patterns during motion: a comparison of possible solutions, J. Biomech. 46 (12) (2013) 2097–2100.

[91] M.H. Sohn, J.L. McKay, L.H. Ting, Defining feasible bounds on muscle activation in a redundant biomechanical task: practical implications of redundancy, J. Biomech. 46 (7) (2013) 1363–1368.

[92] C.S. Simpson, M.H. Sohn, J.L. Allen, L.H. Ting, Feasible muscle activation ranges based on inverse dynamics analyses of human walking, J. Biomech. 48 (12) (2015) 2990–2997.

[93] B.J. Fregly, M.L. Boninger, D.J. Reinkensmeyer, Personalized neuromusculoskeletal modeling to improve treatment of mobility impairments: a perspective from European research sites, J. Neuroeng. Rehabil. 9 (2012) 18.

[94] M. Sartori, D. Farina, D.G. Lloyd, Hybrid neuromusculoskeletal modeling to best track joint moments using a balance between muscle excitations derived from electromyograms and optimization, J. Biomech. 47 (15) (2014) 3613–3621.

[95] A. Seth, M.G. Pandy, A neuromusculoskeletal tracking method for estimating individual muscle forces in human movement, J. Biomech. 40 (2) (2007) 356–366.

[96] T.S. Buchanan, et al., Neuromusculoskeletal modeling: estimation of muscle forces and joint moments and movements from measurements of neural command, J. Appl. Biomech. 20 (4) (2004) 367.

[97] M. Damsgaard, et al., Analysis of musculoskeletal systems in the AnyBody modeling system, Simul. Model. Pract. Theory 14 (8) (2006) 1100–1111.

[98] M.S. Narayanan, et al., Virtual musculoskeletal scenario-testing case-studies, in: Virtual Rehabilitation, IEEE, 2008.

[99] S.L. Delp, et al., OpenSim: open-source software to create and analyze dynamic simulations of movement, IEEE Trans. Biomed. Eng. 54 (11) (2007) 1940–1950.

[100] M.A. Sherman, A. Seth, S.L. Delp, Simbody: multibody dynamics for biomedical research, Procedia IUTAM 2 (2011) 241–261.

[101] M. Mansouri, J.A. Reinbolt, A platform for dynamic simulation and control of movement based on OpenSim and MATLAB, J. Biomech. 45 (8) (2012) 1517–1521.

[102] A. Alamdari, V.N. Krovi, Design and analysis of a cable-driven articulated rehabilitation system for gait training. J. Mech. Robot. 8 (5) (2016). 051018-051018-12. http://dx.doi.org/10.1115/1.4032274

[103] A. Alamdari, et al., A review of home-based robotic rehabilitation, in: S. Agrawal (Ed.), Encyclopedia of Medical Robotics, World Scientific, Singapore, 2016.

[104] A. Alamdari, V.N. Krovi, Parallel articulated-cable exercise robot (PACER): novel home-based cable-driven parallel platform robot for upper limb neuro-rehabilitation, in: ASME 2015 International Design Engineering Technical Conferences and Computers and Information in Engineering Conference, Boston, MA, 2015.

[105] A. Alamdari, V.N. Krovi, Modeling and control of a novel home-based cable-driven parallel platform robot: pacer, in: Proceedings of the 2015 IEEE/RSJ International

Conference on Intelligent Robots and Systems (IROS 2015), Hamburg, Germany, 2015.

[106] A. Alamdari, V.N. Krovi, Robotic physical exercise and system (ROPES): a cable-driven robotic rehabilitation system for lower-extremity motor therapy, in: ASME 2015 International Design Engineering Technical Conferences and Computers and Information in Engineering Conference, Boston, MA, 2015.

[107] P. Beyl, et al., Pleated pneumatic artificial muscle-based actuator system as a torque source for compliant lower limb exoskeletons, IEEE/ASME Trans. Mechatron. 19 (3) (2014) 1046–1056.

[108] E. Rocon, J.L. Pons, Exoskeletons in Rehabilitation Robotics: Tremor Suppression, in: Springer Tracts in Advanced Robotics, vol. 69, Springer Science & Business Media, Berlin, Heidelberg, 2011.

[109] D. Galinski, J. Sapin, B. Dehez, Optimal design of an alignment-free two-DOF rehabilitation robot for the shoulder complex, in: 2013 IEEE International Conference on Rehabilitation Robotics (ICORR), IEEE, 2013.

[110] S. Hernandez, et al., Refinement of exoskeleton design using multibody modeling: an overview, in: CCToMM Mechanisms, Machines, and Mechatronics Symposium, 2015.

[111] M. Laitenberger, et al., Refinement of the upper limb joint kinematics and dynamics using a subject-specific closed-loop forearm model, Multibody Syst. Dyn. 33 (4) (2014) 413–438.

[112] F. Ferrati, R. Bortoletto, E. Pagello, Virtual modelling of a real exoskeleton constrained to a human musculoskeletal model, in: Biomimetic and Biohybrid Systems, Springer, Berlin, Heidelberg, 2013, pp. 96–107.

[113] P. Agarwal, et al., A novel framework for virtual prototyping of rehabilitation exoskeletons, in: 2013 IEEE International Conference on Rehabilitation Robotics (ICORR), IEEE, 2013.

[114] K. Cho, et al., Analysis and evaluation of a combined human-exoskeleton model under two different constraints condition, in: Proceedings of the International Summit on Human Simulation, 2012.

[115] L. Zhou, et al., Design and optimization of a spring-loaded cable-driven robotic exoskeleton, in: Proceedings of the 25th Nordic Seminar on Computational Mechanics, 2012.

[116] W. Li, et al., Biomechanical response of the musculoskeletal system to whole body vibration using a seated driver model, Int. J. Ind. Ergon. 45 (2015) 91–97.

[117] J. Rasmussen, S. Tørholm, M. de Zee, Computational analysis of the influence of seat pan inclination and friction on muscle activity and spinal joint forces, Int. J. Ind. Ergon. 39 (1) (2009) 52–57.

[118] M. Grujicic, et al., Musculoskeletal computational analysis of the influence of car-seat design/adjustments on long-distance driving fatigue, Int. J. Ind. Ergon. 40 (3) (2010) 345–355.

[119] C. Ma, et al., Analysis of lower-limb muscle activities during whole body vibration with different standing postures, in: 2010 3rd International Conference on Biomedical Engineering and Informatics (BMEI), IEEE, 2010.

[120] M.G. Sangachin, L.A. Cavuoto, Obesity-related changes in prolonged repetitive lifting performance, Appl. Ergon. 56 (2016) 19–26.

[121] M.G. Sangachin, W.W. Gustafson, L.A. Cavuoto, Effect of active workstation use on workload, task performance, and postural and physiological responses, IIE Trans. Occup. Ergon. Hum, Factors, 2016 (just-accepted).

[122] J.A. Ambrosio, Multibody Dynamics Approaches to Biomechanical Applications to Human Motion Tasks, Springer, Vienna, 2013, pp. 259–289.

[123] J.A. Ambrósio, A. Kecskeméthy, Multibody dynamics of biomechanical models for human motion via optimization, in: Multibody Dynamics, Springer, Netherlands, 2007, pp. 245–272.

[124] M. Tändl, T. Stark, D. Raab, N.E. Erol, A. Kecskeméthy, F. Löer, An integrated simulation environment for human gait analysis and evaluation, Mater. Werkst. 40 (1–2) (2009) 43–53.

[125] A. Alamdari, Cable-driven articulated rehabilitation system for gait training, Doctoral dissertation, State University of New York at Buffalo, 2016.

CHAPTER THREE

EMG-Controlled Human-Robot Interfaces: A Hybrid Motion and Task Modeling Approach

N. Bu*, T. Tsuji†, O. Fukuda‡
*National Institute of Technology, Kumamoto College, Kumamoto, Japan
†Hiroshima University, Hiroshima, Japan
‡Saga University, Saga, Japan

1. INTRODUCTION

Surface electromyogram (EMG) signals are electric recordings easily measured on the skin surface that contain information on muscle activities. Pioneering research studies on EMG-controlled prostheses were conducted during the 1950s and 1960s [1–4]. Over the past decades, significant progress has been made in EMG signal processing, feature extraction, motion classification, and many other related fields. Some comprehensive reviews can be found in [5–8].

EMG-controlled human-robot interfaces (HRIs) have been extensively studied to devise myoelectric control systems for multifunctional powered upper limb prostheses [6,8], voice prostheses [9,10], human–assisting robots [11], and other technologies. Advances in device hardware have led to significant developments of powered lower limb prostheses [12] and multi-articulated prosthetic arms and/or hands. Progress in myoelectric control is still necessary in order to perform simultaneous movements of multiple joints toward improvement of multidegree operability and realization of a high level of dexterity [13–16].

An extensive range of myoelectric control schemes can be found in the literature, and experimental studies have been conducted with intact subjects and/or amputees to evaluate performance and usability of these schemes [17–26]. Despite the large number of research studies, pattern recognition of EMG features is the main methodology adopted to derive control information from EMG signals. A set of distinguishing features is extracted from EMG signals and then applied to classifiers to identify intended motions from a variety of predefined classes of motions. Control information can be consequently generated from the classified motions. Performing motion classification on successive or overlapped segmentations of EMG signals makes online myoelectric control available [27].

It is generally accepted that performance of such EMG-controlled HRIs depends greatly on motion classification accuracy. It is important to note that EMG is an exceedingly complicated and nonstationary signal, whereby the feature patterns vary significantly depending on tasks and conditions of users. EMG signals are significantly affected by numerous issues, such as electrode shifts, varying arm postures, variations in electrode-skin impedances, and muscle fatigue. In addition, studies on motion classification across different days of use indicate that accuracies decrease with increasing time [28,29]. With regard to practical applications, additional efforts are required to improve classification accuracy, performance stability, and elements that are essential for robustness and reliability of myoelectric control schemes.

In order to overcome these problems, the research community continues to explore novel representative feature sets, more powerful classifiers, and adaptive learning and retraining algorithms. On the other hand, post-processing methods have been developed to refine or correct the classifier's decision by removing misclassifications.

A popular postprocessing method, known as the majority vote (MV) decision rule, was proposed by Englehart et al. [19,27]. The MV counts the most recent outputs of a classifier and chooses the value that occurs most frequently as the corresponding estimator for motion classification. This method smooths the classifier's outputs so that misclassifications, such as transient jumps in the estimated motions, may be eliminated. However, a delay due to the window lengths would be superposed on the response time of a myoelectric control scheme.

In an alternative approach, Tsuji et al. first introduced an entropy-based decision rule to reduce misclassification [30]. The entropy of a classifier's outputs provides a measurement of the amount of uncertainty of the classification results. When entropy exceeds a predefined threshold, the motion decision rule suspends the judgment [11,31]. Similarly, a confidence metric was developed with a linear discriminant analysis (LDA) classifier in order to reject the classifier's decisions if values are lower than class-specific thresholds [32]. Rejection rules are also found in myoelectric control schemes based on binary classification frameworks [33,34]. In addition, a self-correcting method proposed by Amsüss et al. [35] eliminates misclassification according to correctness evaluation of an LDA classifier's decision. The probability of correctness of a decision is estimated using a neural network (NN) based on the mean global muscle activity of the forearm and the maximum likelihoods (MLs) that are calculated by the LDA classifier. In these rejection methods, if a classifier's output does not surpass the threshold check of entropy, confidence, or correctness, either a no-motion outcome or the last accepted classification result is usually used as the decided motion.

Simon et al. [36] developed a novel postprocessing strategy, a decision-based velocity ramp, in order to minimize the effect of misclassification. Every motion class has its own speed coefficient. After a change in the classifier's decision, this method attenuates the motion speed. If the classifier's decision is in favor of a class, the speed coefficient of the class increases; otherwise it decreases. Decreasing the initial speed of all motions can depress impact of misclassifications, which always occur intermittently.

It was also found that task information can enhance classification performance. For example, in "reach-to-grasp" movements, object-related EMG patterns are specific and consistent [37]. Liarokapis et al. adopted this task specificity in their motion-decoding models to discriminate motions under different grasp strategies [38,39].

In our previous studies [40–42], task models were incorporated in the EMG pattern classification schemes in order to improve reliability and stability of EMG-controlled HRIs. In these studies, motion prediction is based on the idea that each user exhibits a particular behavior under certain conditions, and a user task model representing these relations can be manually constructed or derived from a database. Prediction of a user's intention, preferences, and forthcoming actions can be used to reduce the number of candidates, to assign high priority to some candidates, or to rerank the candidates concerned in pattern classification. Such information can significantly ease motion classification and improve the classification accuracy. Contextual information of ongoing tasks was modeled with a Petri net (PN) and a Bayesian network (BN). It was demonstrated that incorporation of task information enhanced the total classification performance and improved stability of the EMG-controlled HRIs.

In this chapter, we introduce a general interface framework of hybrid motion and task modeling for motion classification in order to improve reliability and stability of EMG-controlled HRIs. The EMG modeling part deciphers user motions/intentions from EMG signals. Any EMG pattern classification scheme can be applied in the proposed framework, if it produces probabilities of motions, or if its outputs can be transferred into probabilistic metrics. In parallel to the EMG modeling, a task model predicts occurrence probability for the forthcoming motion, based on contextual information of a task. An essential problem is the combination of the outputs of the EMG modeling and task modeling parts. The present interface framework uses a product rule to combine the probabilities for motion candidates in order to reach a motion decision.

The main body of this chapter is organized as follows: Section 2 introduces basic concepts of EMG motion classification. The task modeling techniques are explained in Section 3. A case study and experimental results are presented in Section 4. Finally, Section 5 presents a discussion and a summary for this chapter.

2. EMG MOTION CLASSIFICATION

2.1 EMG Acquisition and Analysis

The skeletal muscles account for approximately 40% of the weight of the human body. They are attached to the skeleton, functioning as actuators. Voluntary control of length and tension of muscles can result in various human body movements. Skeletal muscles contract when stimulated by the nervous system. Electromyography (EMG) measures the electrical activities in response to neural stimulation of the muscle.

2.1.1 Measuring Methods

There are two types of EMG, namely intramuscular EMG and surface EMG, according to the electrodes used for signal acquisition. Intramuscular EMG is measured with needle or fine-wire electrodes inserted into muscles, allows the detection of electric potentials very close to the source, and provides a high quality measurement of motor unit action potentials. Intramuscular EMG is mainly used in clinical diagnostic tests, but recently some research groups have attempted to use it for motion classification [43,44]. Alternatively, surface EMG electrodes are placed on the skin surface above muscles of interest, providing information on muscle activities in a large volume that is dominated mostly by superficial muscles. Advantages of surface EMG techniques, such as its noninvasiveness, recording simplicity, and abundance of muscular activity information, have led to its extensive use in academic research studies, clinical applications, and commercial prosthetic control systems. Recent surface techniques, based on two-dimensional electrode grids, allow detailed observation of superficial motor units so that the motion classification accuracy can be certainly improved [45,46].

Surface EMG electrodes can be roughly divided into two categories. The first includes active electrodes using built-in amplifiers at the electrode site to improve the impedance characters. This configuration decreases noise, such as movement artifacts, and thus increases the signal-to-noise ratio. The other category includes passive electrodes, which should be connected to an external amplifier. Surface electrodes can also be grouped as wet (gelled) and dry electrodes. Wet surface electrodes incorporate an electrolyte gel layer to reduce skin-electrode impedance. Wet electrodes can either be disposable or reusable, and proper skin preparation is required to reduce the contact impedance as much as possible. However, these electrodes may cause skin irritation and allergic reactions. On the other hand, dry electrodes have the advantage of not requiring a gel layer or other conducting media between skin and the electrode surface. These electrodes are easy to use and have good reusability. Dry electrodes are typically used with an active configuration to ensure functioning.

The electrode placement should be determined in accordance with adequate understanding of the anatomy of the human body and knowledge of the targeted muscle(s). In a bipolar configuration, two electrodes are always placed on the belly of the muscles along the direction of the muscle fibers. A reference electrode is used in addition to the bipolar electrode pair, acting as a ground for this signal. It should be placed on an electrically neutral area, such as joints, bony prominence, or the seventh cervical (C7) spinous process. A large variation in both the size of the conductive area of electrodes and the interelectrode distance can be found in the literature. It is recommended that the electrode size should lie in the range of 5–10 mm, and the bipolar electrodes are set with a 20-mm center-to-center distance [47].

A schematic diagram of an EMG signal-conditioning system is shown in Fig. 1. EMG signals attenuate significantly after propagation through the skin and subcutaneous fat tissue. Before amplification, the surface EMG signal generates typical voltage levels that range between tens of microvolts and several millivolts. For signal conditioning, the EMG signal is generally amplified by a factor of at least 100–10,000, and then input to a storage or computer device for subsequent feature analysis. The differential amplification technique has been extensively used to minimize noise and artifacts from EMG signals. This process eliminates common mode voltages, such as power line interferences, or contributions from far-away sources whose field reaches both of the recording electrodes with the same signal intensity and at the same time. On the other hand, it maintains the signal features that are

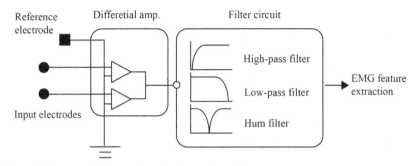

Fig. 1 Schematic diagram of an EMG signal-conditioning system.

different in the two electrodes as the signal of interest, and outputs after amplification. The decibel (dB) is the most common unit used to express the amplifier gain, which is the ratio of the value of the EMG signal after amplification (*A*) to the signal value before amplification (*B*). It is calculated using the formula

$$20 \times \log_{10} \frac{A}{B}$$

Thus, if a signal is amplified by a factor of 1000, the amplifier gain is 60 dB.

When the electrode placement and the amplification configuration are completed, it is important to check the validity and quality of the EMG signals before the actual data acquisition. Signal checks generally include impedance checks, visual inspection of the raw EMG baseline quality, and verification of the correspondence between movements and amplitude changes of the EMG signals. Subsequent tuning of any parts of the aforementioned preparation is imperative in order to achieve proper detection of EMG signals.

The primary characteristics of typical bioelectric signals are given in Fig. 2. Motor unit activity lies in the frequency range of 5–2000 Hz. Due to the propagation attenuation, it is commonly accepted that surface EMG frequency ranges from 5 to 500 Hz [48]. The EMG signal is highly influenced by noise, even with proper electrode placement and differential amplification. A filtering process is usually applied to EMG signals to remove noise contaminants outside of the referred frequency band. In addition, power line interference (50 or 60 Hz) is the most common artifact. Typically, a notch filter is used to remove this interference. Noise can also be generated from low-frequency movement artifacts. A high-pass filter with

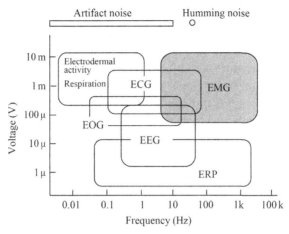

Fig. 2 Primary characteristics of typical bioelectric signals. *ECG,* electrocardiography; *EOG,* electrooculography; *EEG,* electroencephalography; *ERP,* event-related potential.

a 10–20 Hz cut-off frequency can be applied to remove movement artifacts. The cut-off frequency depends on the activity being analyzed, eg, 10 Hz for walking and 20 Hz for rapid movements.

2.1.2 EMG Analyzing Methods
Features, instead of raw EMG signals, are extracted to increase the information density, and consequently improve classification efficiency. Numerous feature extraction methods have been proposed to facilitate motion classification. Most of them fall into one of two categories, namely either time domain (TD) or frequency domain methods.

(1) Time domain methods

    Amplitude changes of EMG signals reflect muscle activation level, duration, and force information. For example, it has been known that the amplitude of EMG signals is roughly proportional to muscle strength during voluntary isometric contractions [49]. In the case of dynamic movements, the amplitude changes must be processed, considering changes in relative positions and orientations of surface electrodes with respect to the muscle volume. Some typical processing techniques for time-domain analysis are introduced in the following paragraphs.

    Rectification: Most features demand a rectification of the raw EMG signals. It can be achieved by forming the absolute values of the raw samples (full-wave) or by deletion of all negative aspects of the signal (half-wave).

Integration: Integrated EMG (iEMG) is defined as the area under the curve of the rectified EMG signal, that is, the mathematical integral of the absolute value of the raw EMG signal; iEMG can be a simple integration for the entire activity, or can be integrated and reset for a fixed time interval or after a preset amplitude value is reached.

Moving average: A moving average of full-wave rectified EMG signals is commonly used to smooth out short-term fluctuations and highlight longer-term trends. A simple moving average (unweighted mean) may derive the well-known mean absolute value EMG feature.

Linear envelope: The linear envelope is a representation of the rectified and low-pass filtered signal. Low-pass filtering can more or less smooth the rectified EMG signal with a cut-off frequency of several hertz. The result looks like the "envelope" of the original signal. The phase shift of the linear envelope is dependent on the order of the filter and the cut-off frequency.

Root mean square (RMS): The RMS value of the raw EMG provides a measure of the electrical power of the signal. The RMS is a recommended method of smoothing and estimating the overall EMG signal intensity.

Normalization: EMG amplitude is strongly influenced by acquisition conditions. It varies largely between subjects, electrode sites, and other factors. One way to overcome this problem is the use of normalization with a reference value. The main effect of normalization is that the influence of the given acquisition condition can be eliminated, and measurement values are rescaled from voltage to percent of the reference. A maximum voluntary contraction (MVC) method is commonly used to correct differences in EMG activation level between channels.

(2) Frequency domain methods

The frequency content of the EMG signal has a physiological significance, and frequency analysis has been used in studies of muscle fatigue. Frequency domain features are also utilized in motion classification providing an insight into spectral components. Typical processing methods of this category are fast Fourier transform, short-time Fourier transform (STFT), and wavelet transform (WT).

Fast Fourier transform: If the EMG signals are assumed to be locally stationary or quasi-stationary, the power spectral density is defined as the Fourier transform of the autocorrelation function of the signal. The mean and median frequencies of the power spectrum are widely used as spectral change indicators.

Short-time Fourier transform: Based on data segmentation and windowing techniques, joint time-frequency analysis can be conducted. STFT of a signal is a function of time and frequency, depending on a moving window function. The moving window slides across the recording, and at each position the power spectrum is computed for the signal within the window. STFT can extract features that effectively represent transient EMG patterns resulting from dynamic contractions.

Wavelet transform: WT is another popular time-frequency analysis method for nonstationary signals. It enables localized transient analysis of EMG signals. WT represents signals with an analysis window, whose size is chosen to be short at high frequencies and long at low frequencies. This method yields a time-frequency representation with a good time resolution at high frequencies, and a good frequency resolution at low frequencies.

One may choose some of the existing temporal and/or frequency domain processing techniques, and construct a feature extraction method according to the intended application. Generally speaking, TD features are the most popular in EMG motion classification because of their comparative computational simplicity. Many studies concerning real-time and online control prefer TD features [20,50]. On the other hand, features of the frequency domain exhibit a good performance, especially for the analysis of dynamic movements [27,51,52]. Some comparative studies have been reported in the literature [28,53–55]. Phinyomark et al. [28] investigated 50 EMG features to classify 10 upper limb motions, including 33 TD features and 17 frequency features. A comparison between TD features and time-frequency representations, STFT and WT, can be found in [56]. These results can be used as a reference for a choice of feature patterns.

When feature patterns occupy a high-dimensional space, a procedure of dimensionality reduction is crucial to avoid overloading the classifier. This part retains information that is important for motion classification while discarding irrelevant information. A lot of feature reduction/selection methods have been developed in the context of EMG motion classification, using principal component analysis (PCA) [27,57], independent component analysis (ICA) [58], nonnegative matrix factorization (NMF) [59], or nonlinear projection [60].

2.2 Motion Classification

Almost all possible classification methods that have been developed within the fields of statistics and computational intelligence have been investigated and tested in EMG motion classification. In this part, a classifier identifies intended motions from EMG feature patterns. Usually, a learning stage is involved to train or estimate parameters of the classifier with training sample data. Besides the classification accuracy, one can choose a classifier for the intended application considering various issues, such as computational complexity, training performance, and classifier adaptation with online training.

2.2.1 Statistical Classifiers

Graupe et al. classified four motions using autoregressive (AR) parameters extracted from EMG signals. The classifier achieved a simple comparison in a nearest neighbor manner [61]. Kang et al. applied a Bayesian classifier on three statistical distance measures of feature patterns for motion classification [62]. Probability was calculated based on the distance measures, and the motion with the maximum probability was selected.

LDA is one of the standard classifiers in EMG classification. LDA is a generalization of Fisher's linear discriminant [63]. The LDA classifier is simple to implement and easy to train. In addition, LDA shows competitive classification performance. The combination of LDA and the TD features, which were proposed in [17], has become the most commonly used motion classification scheme [21,28,36].

Statistical models, such as hidden Markov models (HMMs) and Gaussian mixture models (GMMs), have also demonstrated their potential in motion classification [64,65]. For example, GMM is able to form smooth approximations for probability density functions through the weighted sum of multiple Gaussian functions. A GMM-based classification scheme has been investigated by Huang et al. [65], using the Hudgins' TD feature set, RMS, and AR features, to classify six limb motions.

2.2.2 Support Vector Machine

Support vector machine (SVM) is a pattern classification algorithm with nonlinear formulation [66]. SVM maps input data, such as EMG feature patterns, into a high-dimensional feature space, where it constructs an optimal discriminant hyperplane using a nonlinear kernel function. Although standard SVM is defined in binary form, multiclass problems can be settled with a one-versus-all approach, or by combining several binary SVMs, such as in a tree structure, for example. Recently, SVM has become another popular classifier for motion classification since it can handle classes with complex nonlinear decision boundaries, and since its optimization is guaranteed with a unique global solution. However, the choice of kernel function has a significant effect on its performance, and the best choice is application dependent. An SVM-based classification scheme was developed in a four-channel-five-motion configuration [67]. Comparison with an LDA classifier and a multiple layer perceptron (MLP) NN demonstrates that SVM has an acceptable accuracy, and that it does not impose a significant computational load. Some other implementations of SVM can be found in [20,34,53,54].

2.2.3 Neural Networks-Multiple Layer Perceptron
Hudgins's group first applied an MLP to classify four upper limb motions based on time-domain features [17,68]. Inspired by these pioneering works, artificial NNs have received increasing attention and, even in recent days, many approaches have been proposed with an MLP training with a back-propagation learning rule [14,51,59,69]. The advantage of NNs is twofold: (1) NNs are able to represent both linear and nonlinear relationships, and (2) the learning ability of NNs allows them to form a flexible classifier and adjust to dynamic and changing characteristics of feature patterns. NNs are efficient and have achieved outstanding classification performance for complex applications. However, it must be noted that a large amount of data is usually required for training the classifiers.

2.2.4 Probabilistic Neural Networks
Tsuji et al. proposed a probabilistic neural network (PNN) for pattern classification, which is called log-linearized Gaussian mixture network (LLGMN) [70]. LLGMN is a three-layer feedforward NN. The structure of LLGMN is based on the GMM and a log-linear model, and this NN is able to estimate probability density functions of input patterns. Used for motion classification, LLGMN possesses an inherent advantage of achieving statistical classification using Bayes' decision rule, and shows good generalization ability. LLGMN has been successfully applied to EMG motion classification, even in cases of amputee users [11,40,42].

The authors further developed a recurrent probabilistic NN, termed a recurrent log-linearized Gaussian mixture network (R-LLGMN) [71]. R-LLGMN is based on the HMM. The recurrent connections are incorporated in the network structure to make use of temporal information of signals. R-LLGMN was investigated in an EMG classification scheme and was found to outperform the traditional MLP NNs [31]. Recently, an extension of R-LLGMN was proposed to classify high-dimensional time-series patterns. Time-series discriminant component analysis is incorporated into the network structure for dimensionality reduction. Network coefficients, which function as dimensionality reduction and time-series classification, can be obtained simultaneously according to a back-propagation through a time-based learning algorithm using the Lagrange multiplier method [72]. With the use of this PNN, it is expected that an improved motion classification performance can be achieved.

In addition to the aforementioned methods, there are other EMG motion classification approaches, including fuzzy logic [73,74], k-nearest neighbor [22], and genetic algorithms [75].

3. TASK MODELING FOR HUMAN INTERFACES

3.1 Human Task Modeling

Predicting the user's future behavior based on contextual information, eg, past observations of his/her behaviors, may have a great impact on task performance and efficiency of human-machine interfaces. For example, Darragh et al. developed a predictive typing aid to accelerate text entry [76]. This method predicts what a user is going to type from previously entered text, and the prediction is generated from an adaptively trained text model. Smirnova and Watt suggested that prediction of the candidates based on grammatical information can play an important role in accurate mathematical character classification [77]. Human motion prediction has also attracted attention in the field of human-robot collaboration, where cooperative manipulation requires that a robot anticipate human motions and adjust its movement accordingly [78,79].

In the case of EMG motion classification, referential knowledge or a prediction of the forthcoming motion provides hints or preference for some motion candidates. Consequently, this eases the classification problem. The context-based inference of motion relies on a task model that consists of a set of motion primitives, among which (partial) orders or transitions are defined to represent task sequences and motion occurrences. Two types of task modeling strategies are considered in this chapter: descriptive and statistical. The former represents a task by listing possible sequences of motions that lead from the initial state to the final state. AND-OR trees, finite state machines, and PNs [80] are classical tools used in this strategy. On the other hand, statistical prediction of a user's forthcoming motion is inherently uncertain work. It is desirable that a task model has a knowledge representation capable of dynamically capturing and modeling the uncertainty of human behaviors. Task models, based on BNs or Markov models, are able to deal with this type of information using statistical relationships between motions and the status of the user [81].

3.1.1 Descriptive Task Model Using Petri Nets

The PN is a graphic tool and a mathematical formalism for modeling of event-driven systems. It has been successfully applied to task and behavioral analysis in the robotics community, such as robot learning from demonstration [82] and task sequence planning [83,84].

A classical PN is a directed bipartite graph with two types of nodes: places and transitions. Places are represented by circles and transitions by bars. The nodes are connected via directed arcs. With the graphical notation, PN provides a flow-chart representation, thereby allowing an intuitive visualization of the system's dynamic behavior.

Let us consider a "water-drinking" task, for example. This task can be decomposed into three states, namely standby, grasping a cup, and drinking, and three motions, namely hand grasp, wrist rotation, and hand opening. One possible sequence of this task can be listed as: standby state → hand grasp → grasping-a-cup state → wrist rotation → drinking state → hand opening → standby state. The corresponding PN task model uses places, transitions, and arcs to denote states, operator motions, and task flow, respectively. Graphically, places may contain a discrete number of marks called tokens, which indicate the current state(s) or condition(s) of the task. Any distribution of tokens over the places will represent a configuration of the PN, known as a marking.

A PN task model consists of a four-tuple: $N = (P, T, F, \mu_0)$, where P is the finite set of all places $\{p_i\}$, T is the finite set of all transitions $\{t_i\}$, directed arcs are defined by $F \subseteq (P \times T) \cup (T \times P)$, and μ_0 is the initial marking. A marking μ is a vector with an element for each place p_i to indicate the number of tokens in that place.

A hierarchical structure is used to model a set of relevant subtasks with branches representing the details of each subtask. An example of the PN task model is shown in Fig. 3. In this figure, the place with a token mark denotes the standby state, and the branches connected with this place represent subtasks in the model. For example, during meals, the series of operations, such as "water-drinking" and "eating-with-chopsticks," are subtasks. The hierarchical extension makes it possible to model complex tasks.

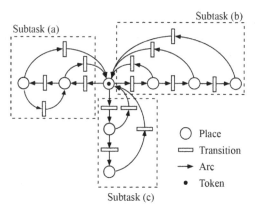

Subtask (a)

Subtask (b)

○ Place
▭ Transition
→ Arc
• Token

Subtask (c)

Fig. 3 A task model represented using a Petri net.

The task model determines a forthcoming transition (ie, a motion in the task flow) according to current task states. This prediction is then used to generate a set of bias values, or weights, which are applied to modify the outputs of an EMG motion classifier. These modification parameters are usually preset by the user, and need to be determined by trial and error. For details, please refer to [40,41].

The PN task model is an explicit description of a user's task. It is obvious that all possible sequences of motions and task states should be described *a priori* precisely. Although there are some automated synthesis methods available for structuring a PN, it is still a challenge to apply this method to complicated tasks because it is possible that one may not have a clear knowledge or an adequate repository of all branches (subtasks) in the tasks.

3.1.2 Statistical Task Model Using Bayesian Network

A BN is a graphical notation that encodes conditional dependence relationships among a set of events [85]. BN is one of the popular techniques for user modeling and predicting [86–88]. The advantages of BN lie in the following: (1) probabilistic expression in BNs is effective in dealing with uncertainty of reasoning human behavior, (2) directed arcs between nodes give an intuitive representation of causal relationships and the network structure can be devised manually or learned from data, (3) the parameters, ie, the conditional probability tables (CPTs), can be extracted from a database and can be updated whenever new samples are available.

For task modeling, a simple implementation of BN is based on dependent relationships between two consecutive motions. We can extract the conditional probabilities of motions from case data directly. Given contextual information of the previous motion, the task model predicts occurrence probabilities for candidate motions.

A BN is a directed acyclic graph, where the nodes are probability variables representing certain events. Generally, a BN can be defined as $G = (\mathbf{V}, \mathbf{A}, \mathbf{P})$, where $\mathbf{V} = \{V_1, V_2, ..., V_N\}$ is a set of nodes (variables), and where $\mathbf{A}$ stands for an assembly of directed arcs between the nodes, and $\mathbf{P}$ is a set of CPTs that are associated with each node. A directed arc from V_i to V_j, $(V_i, V_j) \in \mathbf{A}$, represents the conditional dependency between the variables, and this dependency is indicated with $P(V_j = a | V_i = b)$, which is the conditional probability for $V_j = a$ given that $V_i = b$.

We assume that a task consists of a series of motions, $m(s)$ $(s = 1, 2, ..., S)$, where $m(s)$ represents the motion of the sth step. At each step, only one motion occurs. A set of M motions is considered in the task model, $m(s) \in \{1, ..., m, ..., M\}$. During a task, motion transition takes place between two consecutive steps, say from $m(s-1)$ to $m(s)$. The transition illustrates the dependence between the motions. On the other hand, the motion at the sth step is related to one or more statuses of the motion at the $(s-1)$th step. For example, the location where $m(s-1)$ is achieved is one of these cues that may be used to predict $m(s)$.

Accordingly, a BN task model, shown in Fig. 4, is developed to model the dependent relationships among four variables, namely m_c, m_p, l_p, and h_p. Herein, m_c is the motion at the current step, m_p denotes the motion for the previous step, l_p indicates the location of motion m_p, and h_p represents the user's hand position at the previous step. $P(m_p)$ and $P(h_p)$ are the probabilities of motion and hand position at the previous step, respectively.

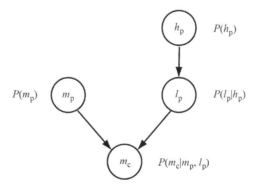

Fig. 4 The task model for motion prediction using a BN.

$P(l_p|h_p)$ represents the conditional probability of the location at the previous step given the previous hand position, and $P(m_c|m_p, l_p)$ is the conditional probability of the current motion with respect to motion and location information at the previous step. There are N locations defined in the workspace, $l_p \in \{l_1, ..., l_n, ..., l_N\}$. These locations are the positions of the items used in the task and the possible places where users are expected to achieve some particular motions.

There are three discrete nodes (m_c, m_p, and l_p) and one continuous node (h_p) in this BN. The conditional probabilities $P(m_c|m_p, l_p)$ can be estimated by counting the number of samples in a database. For example,

$$P(m_c = m' \mid m_p = m, l_p = l_n) = \frac{P(m_c = m', m_p = m, l_p = l_n)}{P(m_p = m, l_p = l_n)}$$
$$\cong \frac{N(m_c = m', m_p = m, l_p = l_n)}{N(m_p = m, l_p = l_n)}$$

where $N(m_c = m', m_p = m, l_p = l_n)$ denotes the number of samples in the database, which are $m_c = m'$, $m_p = m$, and $l_p = l_n$. $N(m_p = m, l_p = l_n)$ is the number of samples for $m_p = m$, and $l_p = l_n$. Frequencies of motion transitions as well as dependencies between motions and locations vary among individuals. A task model can be further adapted to a user based on the statistical information extracted from databases of the user's task records.

On the other hand, $P(l_p|h_p)$ is a continuous probability distribution. According to Bayes' law, $P(l_p = l_n|h_p)$ can be derived as

$$P(l_p = l_n|h_p) = \frac{P(l_p = l_n, h_p)}{P(h_p)} = \frac{P(h_p|l_p = l_n)P(l_p = l_n)}{\sum_{l_p=l_1}^{l_N} P(h_p|l_p)P(l_p)}$$

where $P(h_p|l_p)$ is the conditional probability of h_p at location l_p. The location l_p is dependent on the coordinates of h_p (see Fig. 5). Suppose that $P(h_p|l_p = l_n)$ follows a two-dimensional normal distribution with its mean at location l_n at coordinates (x_n, y_n), and with the standard deviations denoted as σ_{xn} and σ_{yn}. Thus, with the coordinates of h_p, we have

$$P(h_p|l_p = l_n) = \frac{1}{2\pi\sigma_{xn}\sigma_{yn}} \exp\left\{-\frac{1}{2}\left[\frac{(x - x_n)^2}{\sigma_{xn}^2} + \frac{(y - y_n)^2}{\sigma_{yn}^2}\right]\right\}$$

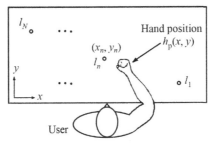

Fig. 5 Locations and the user's hand position in the workspace. Location of motion is determined according to coordinates of hand position.

Furthermore, given that $P(l_p)$ is a uniform distribution, $P(l_p = l_n | h_p)$ can be simplified as

$$P(l_p = l_n | h_p) = \frac{P(h_p | l_p = l_n)}{\sum_{l_p = l_1}^{l_N} P(h_p | l_p)}$$

When the contextual information of the previous motion step is added to the task model, belief updating is performed to estimate the probability of the user's forthcoming motion to motion prediction.

3.2 Combining EMG Modeling With Task Modeling

In the proposed hybrid framework [40–42], the motion decision is reached by combining two opinions from an EMG classifier and a task model. The outputs of the task model are predictions based on the previous motion rather than classification results based on a substantial event that resulted from the current motion. We consider the EMG pattern classifier to be a major part in the combination. It is desirable that outputs of the EMG classifier can be modified by a set of weights according to the prediction of the task model. Occurrence probability of the motion predicted by the task model may be enhanced, while the other motions are depressed.

Methods of combining classifiers have been extensively discussed in the area of pattern classification, and various applications have utilized this scheme to improve efficiency and accuracy of classification [89,90]. Suppose that the outputs of the task model and the EMG classifier are expressed as probabilistic values for motions. A product rule can then be used to generate occurrence probabilities of the candidate motions for decision.

During the sth motion, the occurrence probability of motion m, $P_m(s)$, is defined as follows:

$$P_m(s) = \frac{w_m(s) O_m(s)}{\sum_{m'=1}^{M} w_{m'}(s) O_{m'}(s)}$$

where $w_m(s)$ is the modification weight for motion m according to the belief for motion prediction $B_m(s)$ from the task model, and $O_m(s)$ is the EMG classifier's output. Given that the interval of $B_m(s)$ is $[0, 1]$, $w_m(s)$ is transformed as

$$w_m(s) = \alpha\left(B_m(s) - \frac{1}{2} \right) + \frac{1}{2}, 0 \le \alpha \le 1$$

within the interval $[(1 - \alpha)/2, (1 + \alpha)/2]$. This transformation is conducted in order to prevent misclassification due to $B_m(s) = 0$ for some motions. It should be noted that α determines the influence of motion prediction in the combination. The larger the α is, the greater the influence exists. The parameter α should be defined in advance, and it can be adjusted according to a user's preference.

4. AN EMG-CONTROLLED HUMAN-ROBOT INTERFACE USING TASK MODELING

4.1 System Description

An EMG-controlled HRI is used to verify the proposed hybrid framework. The HRI system is composed of a probabilistic NN for motion classification and a BN for motion prediction. EMG signals measured from a user's forearm are classified in order to estimate his/her (intended) motion. The motions are then used to control a prosthetic hand. The prosthetic hand can be directly attached to an amputee's residual limb. The structure of the interface system is shown in Fig. 6. This system consists of three major parts: (1) EMG motion classification, (2) motion prediction with a BN task model, and (3) motion decision.

4.1.1 EMG Motion Classification

EMG signals are processed to extract the feature patterns for classification. The EMG signals, which are measured from D pairs of electrodes, are rectified and filtered by a second-order Butterworth filter (cut-off frequency f_c). They are then digitized by an A/D converter with a sampling frequency of f_s.

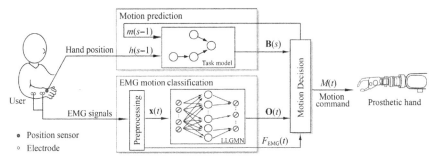

Fig. 6 Hybrid structure of an EMG-controlled human-robot interface.

The sampled data are defined as $\text{EMG}_d(t)\,(d = 1, \ldots, D)$ and are normalized to make the sum of D channels equal to 1.0

$$x_d(t) = \frac{\text{EMG}_d(t) - \text{EMG}_d^{st}}{\sum_{d'=1}^{D}\left(\text{EMG}_{d'}(t) - \text{EMG}_{d'}^{st}\right)}$$

where EMG_d^{st} is the mean value of $\text{EMG}_d(t)$, which is measured while the arm is relaxed. The feature vector $\mathbf{x}(t) = [x_1(t), x_2(t), \ldots, x_D(t)]$ is used for motion classification.

A probabilistic NN, LLGMN [70], is used as the classifier. LLGMN can be trained with a simple back-propagation training algorithm based on an ML criterion. Given an EMG feature vector $\mathbf{x}(t)\,(t = 1, \ldots, T)$, the output of LLGMN $O_m(t)\,(m = 1, \ldots, M)$ represents the posterior probabilities of motion m. The posterior probability vector $\mathbf{O}(t) = [O_1(t), O_2(t), \ldots, O_M(t)]$ is then fed into the motion decision part.

In this HRI system, we assume that the amplitude level of EMG signals changes in proportion to the muscle force. The force information $F_{\text{EMG}}(t)$ for the input vector $\mathbf{x}(t)$ is defined as

$$F_{\text{EMG}}(t) = \frac{1}{D}\sum_{d=1}^{D}\frac{\text{EMG}_d(t) - \text{EMG}_d^{st}}{\text{EMG}_d^{\max} - \text{EMG}_d^{st}}$$

where $\text{EMG}_d^{\max}$ is the mean value of $\text{EMG}_d(t)$, which is measured while maintaining the MVC of the arm. The force information is used to determine the onset and end of motions in the motion decision part.

4.1.2 Motion Prediction With a BN Task Model
The BN task model is utilized for motion prediction. The motion and the hand position of a user are added to the task model as evidence. The motion

of the previous step $m(s-1)$ is obtained from the results of the motion decision part. The hand position $h(s-1)$ is measured with a position sensor, which is attached on the user's wrist.

Based on the notation used in Section 3.1, the belief about the current motion, which is the conditional probability $P(m_c|m_p,h_p)$, can be calculated as follows:

$$P(m_c|m_p,h_p) = \frac{\sum_l P(m_c|m_p,l)P(l|h_p)}{\sum_m\sum_{l'}P(m|m_p,l')P(l'|h_p)}$$

For the first motion step, the belief for motion prediction $B_m(1)$ $(m = 1, 2, ..., M)$ is set to $1/M$. For the sth step $(s > 1)$, $B_m(s)$ is calculated for each motion based on the contextual information of the $s-1$th step,

$$B_m(s) = P(m(s)|m(s-1),h(s-1))$$

After the belief propagation, the belief vector $\mathbf{B}(s) = [B_1(s), B_2(s), ..., B_M(s)]$ is output to the motion decision part.

4.1.3 Motion Decision

In order to recognize whether a motion has really occurred or not, the force information $F_{EMG}(t)$ is compared to a predefined motion appearance threshold F_{th} and the motion is considered to have occurred if $F_{EMG}(t)$ exceeds F_{th}. Thus, the duration of the sth motion step is the period $[t_{ON}^s, t_{OFF}^s]$,

$$F_{EMG}(t) \geq F_{th}, t \in \left[t_{ON}^s, t_{OFF}^s\right]$$

where t_{ON}^s stands for the onset of the sth motion, and t_{OFF}^s is the end of the motion. The probability of motion $P_m(t)(m = 1, 2, ..., M)$ is calculated using the product rule during the sth motion, $t_{ON}^s \leq t \leq t_{OFF}^s$.

The entropy of $P_m(t)$ is then obtained in order to prevent misclassification. The entropy is defined as follows:

$$H(t) = -\sum_{m=1}^{M} P_m(t)\log P_m(t)$$

If the entropy $H(t)$ is less than the threshold H_d, the motion with the largest probability is determined as the user's intended motion $M(t)$. Otherwise, the determination is suspended. Finally, $M(t)$ is used to generate control commands for the prosthetic hand.

4.2 Motion Classification Experiments

Motion classification experiments were conducted in order to evaluate classification performance of the proposed method. Four male subjects voluntarily participated in these experiments. Subjects C and D had no experience in EMG classification experiments. The motion classification method utilized in [11] was used for comparison, which makes decisions only based on outputs of LLGMN. The entropy-based decision rule was used in this comparison method. For the sake of simplicity, the comparison method is represented as LLGMN in the rest of this chapter.

4.2.1 Experimental Conditions

In the experiments, EMG signals were measured from five pairs ($D = 5$) of electrodes (NT-511G: NIHON KOHDEN Corp.). The electrodes were attached to the user's forearm and upper arm: flexor carpi radialis (FCR), extensor carpi ulnaris, flexor carpi ulnaris, biceps brachii. Two pairs of electrodes were attached on the FCR and one pair on each of the others. The differential EMG signals were amplified by a multitelemeter (MT11: NEC Medical Systems Corp.). The cut-off frequency f_c of the Butterworth filter was 1 Hz, and EMG signals were recorded with f_s of 1 kHz (see Section 4.1 for details). Additionally, a 3D position sensor (ISOTRACK II: POLHEMUS, Inc.) was used to measure the hand positions in the workspace.

A cooking task was used in the experiments. This task consists of four operations: (1) turning on a gas ring, (2) placing an ingredient in a fry pan, (3) adding some salt to the ingredient, and (4) shaking the fry pan. Six motions were considered in this task ($M = 6$: hand open, hand grasp, wrist flexion, wrist extension, pronation, and supination). We defined six locations in the workspace ($N = 6$), as shown in Fig. 7. Three items were used in this task, ie, ingredients (l_1), a fry pan (l_2), and salt (l_4). A typical order of the motion series and location transitions in the cooking task is depicted in Table 1.

Each subject performed 15 trials of the task. The data from the first 5 trials were used to train LLGMN and the BN task model, while the other trials were used for test. The training trials were performed with the order of operations as (1)–(2)–(3)–(4). The test trials were further divided equally into two groups: Group I, where the operation order is the same as the training trials, and group II, where the operation order is (1)–(3)–(2)–(4). According to previous studies on EMG motion classification that use the same TD

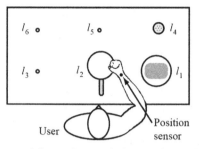

Fig. 7 Locations and items of the cooking task. l_1: Ingredients, l_2: a fry pan, and l_4: salt.

Table 1 Typical Order of the Motion Series and Location Transitions in the Cooking Task

Operation	m_p	l_p	m_c
(1)	–	–	Hand grasp
	Hand grasp	l_2	Pronation
(2)	Pronation	l_2	Hand open
	Hand open	l_1	Hand grasp
	Hand grasp	l_1	Hand open
(3)	Hand open	l_2	Hand open
	Hand open	l_4	Hand grasp
	Hand grasp	l_4	Pronation
	Pronation	l_2	Hand grasp
	Hand grasp	l_2	Hand open
(4)	Hand open	l_4	Hand grasp
	Hand grasp	l_2	Wrist flexion
	Wrist flexion	l_2	Wrist flexion
	Wrist flexion	l_2	Hand grasp
	Hand grasp	l_2	Hand open

features as described in Section 4.1 [31,69], the motion appearance threshold F_{th} was set to 0.2. In the motion decision part, the determination threshold H_d was set to 0.5 and the parameter α to 0.8.

4.2.2 Experimental Results

An example of the experimental results (subject A) is shown in Fig. 8. The operation order of this test trial is (1)–(2)–(3)–(4). This figure plots five channels of EMG signals, force information $F_{EMG}(t)$, outputs of the BN task model $B_m(s)$, the classification results of LLGMN, and the results of the proposed method. The gray areas indicate that no motion was achieved because $F_{EMG}(t)$ was less than F_{th}. Motions with a label of "0" represent decision suspension due to high entropy, and a no-motion command would be

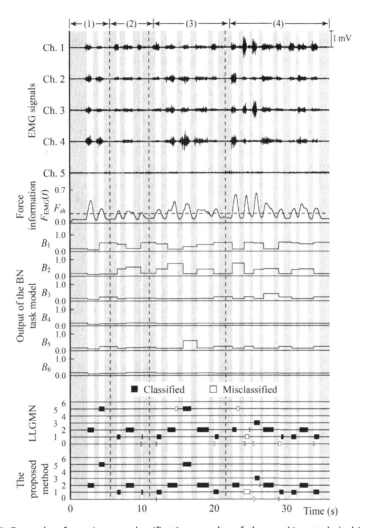

Fig. 8 Example of continuous classification results of the cooking task (subject A). The labels of classification results are as follows: 0—Decision suspension, 1—hand open, 2—hand grasp, 3—wrist flexion, 4—wrist extension, 5—pronation, and 6—supination.

generated by the interface. The correct classification rates of LLGMN and the proposed method are 85.1% and 92.9%, respectively.

From the results of LLGMN, it can be found that most of the misclassifications occur at the beginnings and endings of motion steps. These are chiefly due to the variation in EMG patterns during the transitional phases of motions. In contrast to the method that is only based on LLGMN, classification accuracy of the proposed method is substantially improved, and the results are much more stable at each motion step. Since the BN task model generates a higher belief for the motions, which are predicted to appear at the current step, most of the misclassifications made by LLGMN are corrected by combining the output of LLGMN with the belief vector $\mathbf{B}(s)$. For example, at approximately 15 and 23 s, EMG patterns of motion 2 are incorrectly classified as motion 5 by LLGMN. With the proposed method, the belief of motion 2 (B_2) is higher than other motions during these motion steps so that misclassifications are prevented.

The accuracy of the classification results for the four subjects was also investigated. Motion classification experiments were conducted using all test trials. Figs. 9 and 10 show the mean values and standard deviations of the classification rates for trials from test groups I and II, respectively. In both cases, the proposed method outperforms the comparison method. For test trials of group II, although the operation order is different from that of the training trials, improvements of classification results are confirmed for all subjects. Since the BN task model represents dependencies among variables of two consecutive motion steps, the proposed method is expected to provide high flexibility when dealing with practical motion series.

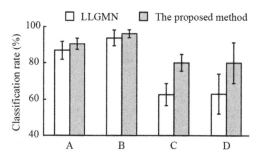

Fig. 9 Classification results of trials from test group I.

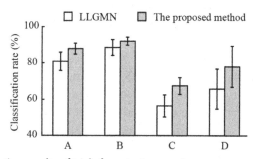

Fig. 10 Classification results of trials from test group II.

4.3 Robot Manipulation Experiments

In order to examine the performance of the proposed hybrid motion classification method, manipulation experiments with an EMG-controlled robot were conducted. The subjects were instructed to perform a meal task, which includes three operations: (1) pouring some water into a glass, (2) drinking the water, and (3) eating soup with a spoon. The schematic view of the workspace is presented in Fig. 11.

A robotic manipulator is set on the left of the workspace. The robotic manipulator consists of a prosthetic hand (Imasen Laboratory) and a robot arm (Mitsubishi Electric Corporation). The motion of the prosthetic hand is controlled with the commands $M(t)$ which are generated by the motion decision part. The prosthetic hand can achieve six different motions corresponding to the six forearm motions of users. The robot arm supports the prosthetic hand and transports it to positions in the workspace according to a user's hand position. The prosthetic hand is detachable from the robot arm, and an amputee can attach it to his/her residual limb to replace the amputated arm. This robotic manipulator was developed by Fukuda et al. [91,92], and it has been used in previous studies. For details, see Refs. [11,40]. The experimental conditions of EMG measurement and position sensing are the same as those listed in Section 4.2. The parameters for motion decision are as follows: $F_{th} = 0.22$, $H_d = 0.3$, and $\alpha = 0.8$.

Fig. 12 shows a sample session of experimental results of a manipulation trial (subject A). In the first motion step (approximately 6 s), misclassifications can be found in both methods. It should be noted, however, that no contextual information is available for the first step. Therefore, the classification results of both methods are the same. For later motion steps,

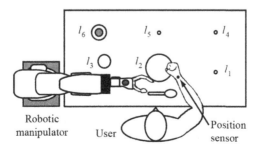

Fig. 11 Workspace of the meal task. A robotic manipulator is set on the left side. The items in the workspace are a disk and a spoon (l_2), a glass (l_3), and a bottle of water (l_6).

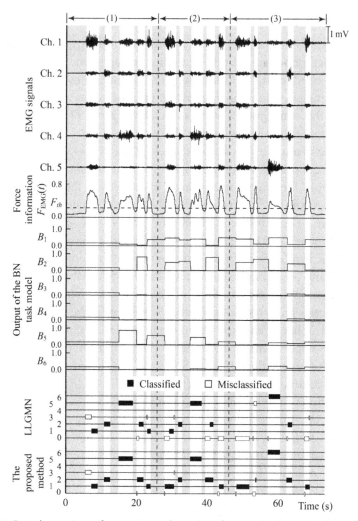

Fig. 12 Sample session of experimental results of a manipulation trial (meal task; subject A). The labels of classification results are as follows: 0—Decision suspension (no motion command is generated for the manipulator), 1—hand open, 2—hand grasp, 3—wrist flexion, 4—wrist extension, 5—pronation, and 6—supination.

the proposed method corrected most of the misclassifications of the comparison method. With the proposed method, the subject achieved the task successfully. The classification rate of the proposed method is 86.4%, with an increase of 17% from the result of LLGMN. Some scenes of the robot manipulation experiments are shown in Fig. 13.

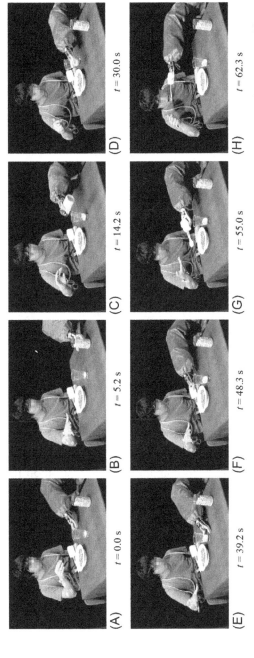

Fig. 13 Scenes of the robot manipulation experiments (subject A). (A) $t = 0.0$ s. (B) $t = 5.2$ s. (C) $t = 14.2$ s. (D) $t = 30.0$ s. (E) $t = 39.2$ s. (F) $t = 48.3$ s. (G) $t = 55.0$ s. (H) $t = 62.3$ s.

5. DISCUSSION AND SUMMARY

In this chapter, the authors have introduced a hybrid framework for EMG-controlled HRI, which is based on EMG motion classification and a task model for motion prediction. A case study was developed based on a task model using BNs, and was applied to an EMG-based HRI system as an assist to support motion classification. In the remaining part of this section, the discussion is based on the performance of the BN task model. It is assumed that similar results can be demonstrated in the case of the PN task model.

5.1 Increase of Classification Rates

The increase of classification rates was computed with experimental results reported in Section 4.2. The increase $\triangle CR$ is defined as

$$\triangle CR = CR - CR'$$

where CR is the classification rate of the proposed method and CR' is the corresponding rate of LLGMN. A summary of the increase of classification rate is listed in Table 2. As demonstrated by the results, the increase of the classification rate has been achieved by the proposed method.

Furthermore, it can be found that $\triangle CR$s of subjects C and D are larger than those of A and B. Thus, the increase ratio (IR) is evaluated as

$$IR = \frac{\triangle CR}{CR'}$$

The variation of IRs versus CR's for ten trials of each subject is plotted in Fig. 14. Generally, there is an increase in the classification rate when CR' decreases. If LLGMN makes a correct classification based on the EMG signals, the contextual information and characteristics of tasks, such as the

Table 2 Summary of the Increase of the Classification Rate (%)

Subject	Test Group I			Test Group II		
	Mean	Min.	Max.	Mean	Min.	Max.
A	3.66	0.00	7.83	7.10	0.53	15.38
B	2.28	0.00	5.63	3.47	0.39	5.39
C	17.43	9.16	28.48	11.00	2.75	14.84
D	17.10	12.80	29.51	12.26	8.73	16.88

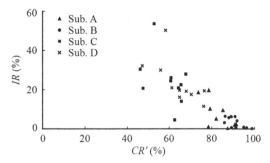

Fig. 14 *IR* versus classification rate of LLGMN (*CR'*).

transition between motions, and information of locations, do not make a significant contribution to the classification rate. When LLGMN fails, however, the motion prediction made by the BN task model helps greatly in the motion decision process. This may be an encouraging result for users, like subjects C and D, who do not have much experience in the control of EMG-based human interfaces.

5.2 Influence of the BN Task Model on Classification Results

Classification accuracies were evaluated using various values of the parameter α. This parameter was set to 0.0, 0.2, 0.4, 0.6, 0.8, and 1.0. For each subject, five trials of test data were classified. Mean values of the classification rates for each α are shown in Fig. 15. It can be found that the classification rate rises when α increases, especially for subjects C and D. In the case of $\alpha = 0.0$, the classification results are only based on LLGMN. By increasing the influence of the task model, a better classification performance is available. Remember that for the combination of the EMG classifier (LLGMN) and the motion prediction model (BN), we set the former to be the major part so that the parameter α was set as 0.8 in this study.

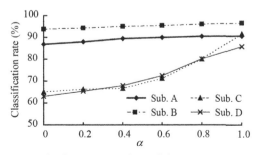

Fig. 15 Classification results for various values of the parameter α.

Since BNs extract the statistical dependency between two continuous motions, the task model outputs the conditional probabilities as the belief for motion prediction according to contextual information. The belief can then be easily combined with the output of a probabilistic NN classifier to improve stability and accuracy of the motion decision. Additionally, the probabilistic parameters in the task model can be obtained by training with a database, and online learning methods [93] are possible to keep adaptability for a user. Experiments of EMG motion classification and robot manipulation have proved the feasibility and effectiveness of the proposed method.

Nevertheless, it is still quite difficult for the proposed task model to provide a precise prediction, especially considering the uncertainty in human behavior. There are many studies that are ongoing on improving the accuracy of BN-based human modeling, but consideration of these is beyond the scope of this chapter. In this chapter, we used a weak predictor, ie, a BN with a simple structure, as a subpart in the proposed hybrid motion classification framework. The BN part does not directly provide an answer but provides a suggestion for motion decision based on the characteristics of task flows, the user's operation process, and his/her preference so that a more robust and reliable classification result can be obtained.

Finally, the structure of the BN task model was manually designed in this chapter and it was not learned from a database. In fact, the task model shown in Fig. 4 is not the only form that can be used. Motion, location, and hand position of the previous motion step constitute only a small part of the status related to a user's operation. Alternative designs of the task model using contextual information, such as the user's posture, duration of a motion, and other related features, may elicit more promising results.

REFERENCES

[1] C.K. Battye, A. Nightengale, A. Whillis, The use of myo-electric currents in the operation of prostheses, J. Bone Joint Surg. 37-B (1955) 506–510.
[2] A.H. Bottomley, Myoelectric control of powered prostheses, J. Bone Joint Surg. 47-B (1965) 411–415.
[3] D.S. Dorcas, R.N. Scott, A three-state myo-electric control, Med. Biol. Eng. 4 (1966) 367–370.
[4] P. Herberts, Myoelectric signals in control of prostheses: studies on arm amputees and normal individuals, Acta Orthop. Scand. 40 (Suppl. 124) (1969) 1–83.
[5] K. Englehart, B. Hudgins, P. Parker, Multifunction control of prostheses using the myoelectric signal, in: H.L. Teodorescu, L.C. Jain (Eds.), Intelligent Systems and Technologies in Rehabilitation Engineering, CRC Press, Boca Raton, 2001, pp. 153–208.
[6] P. Parker, K. Englehart, B. Hudgins, Myoelectric signal processing for control of powered limb prostheses, J. Electromyogr. Kinesiol. 16 (2006) 541–548.

[7] M.A. Oskoei, H. Hu, Myoelectric control systems—a survey, Biomed. Signal Process. Control. 2 (2007) 275–294.

[8] D. Farina, N. Jiang, H. Rehbaum, et al., The extraction of neural information from the surface EMG for the control of upper-limb prostheses: emerging avenues and challenges, IEEE Trans. Neural Syst. Rehabil. Eng. 22 (2014) 797–809.

[9] T. Tsuji, N. Bu, J. Arita, M. Ohga, A speech synthesizer using facial EMG signals, Int. J. Comput. Intell. Appl. 7 (2008) 1–15.

[10] Q. Zhou, N. Jiang, K. Englehart, B. Hudgins, Improved phoneme-based myoelectric speech recognition, IEEE Trans. Biomed. Eng. 56 (2009) 2016–2023.

[11] O. Fukuda, T. Tsuji, M. Kaneko, A. Ohtsuka, A human-assisting manipulator teleoperated by EMG signals and arm motions, IEEE Trans. Robot. Autom. 19 (2003) 210–222.

[12] L.J. Hargrove, A.M. Simon, A.J. Young, et al., Robotic leg control with EMG decoding in an amputee with nerve transfers, N. Engl. J. Med. 369 (2013) 1237–1242.

[13] T. Tsuji, K. Shima, Y. Murakami, Pattern classification of combined motions based on muscle synergy theory, J. Robot. Soc. Jpn. 28 (2010) 606–613.

[14] A. Ameri, E.N. Kamavuako, E. Scheme, K. Englehart, P. Parker, Real-time, simultaneous myoelectric control using visual target-based training paradigm, Biomed. Signal Process. Control 13 (2014) 8–14.

[15] N. Jiang, H. Rehbaum, I. Vujaklija, B. Graimann, D. Farina, Intuitive, online, simultaneous, and proportional myoelectric control over two degrees-of-freedom in upper limb amputees, IEEE Trans. Neural Syst. Rehabil. Eng. 22 (2014) 501–510.

[16] J. Ngeo, T. Tamei, T. Shibata, Continuous and simultaneous estimation of finger kinematics using inputs from an EMG-to-muscle activation model, J. Neuroeng. Rehabil. 11 (2014) 122.

[17] B. Hudgins, P. Parker, R.N. Scott, A new strategy for multifunction myoelectric control, IEEE Trans. Biomed. Eng. 40 (1993) 82–94.

[18] K. Englehart, B. Hudgins, P.A. Parker, M. Stevenson, Classification of the myoelectric signal using time-frequency based representations, Med. Eng. Phys. 21 (1999) 431–438.

[19] K. Englehart, B. Hudgins, A robust, real-time control scheme for multifunction myoelectric control, IEEE Trans. Biomed. Eng. 50 (2003) 848–854.

[20] P. Shenoy, K.J. Miller, B. Crawford, R.N. Rao, Online electromyographic control of a robotic prosthesis, IEEE Trans. Biomed. Eng. 55 (2008) 1128–1135.

[21] G. Li, A.E. Schultz, T.A. Kuiken, Quantifying pattern recognition-based myoelectric control of multifunctional transradial prostheses, IEEE Trans. Neural Syst. Rehabil. Eng. 18 (2010) 185–192.

[22] C. Cipriani, C. Antfolk, M. Controzzi, et al., Online myoelectric control of a dexterous hand prosthesis by transradial amputees, IEEE Trans. Neural Syst. Rehabil. Eng. 19 (2011) 260–270.

[23] T. Tommasi, F. Orabona, C. Castellini, B. Caputo, Improving control of dexterous hand prostheses using adaptive learning, IEEE Trans. Robot. 29 (2013) 207–219.

[24] E. Scheme, B. Lock, L. Hargrove, et al., Motion normalized proportional control for improved pattern recognition-based myoelectric control, IEEE Trans. Neural Syst. Rehabil. Eng. 22 (2014) 149–157.

[25] H. Kawasaki, M. Kayukawa, Learning system for myoelectric prosthetic hand control by forearm amputees, in: Proceedings of the 23rd IEEE International Symposium on Robot and Human Interactive Communication (RO-MAN), 2014, pp. 899–904.

[26] J. Ma, N.V. Thakor, F. Matsuno, Hand and wrist movement control of myoelectric prosthesis based on synergy, IEEE Trans. Human–Mach. Syst. 45 (2015) 74–83.

[27] K. Englehart, B. Hudgins, P.A. Parker, A wavelet-based continuous classification scheme for multifunction myoelectric control, IEEE Trans. Biomed. Eng. 48 (2001) 302–311.

[28] A. Phinyomark, F. Quaine, S. Charbonnier, et al., EMG feature evaluation for improving myoelectric pattern recognition robustness, Expert Syst. Appl. 40 (2013) 4832–4840.

[29] S. Amsüss, L.P. Paredes, N. Rudigkeit, et al., Long term stability of surface EMG pattern classification for prosthetic control, in: Proceedings of the 35th Annual International Conference of the IEEE Engineering in Medicine and Biology Society (EMBC), 2013, pp. 3622–3625.

[30] T. Tsuji, H. Ichinobe, K. Ito, M. Nagamachi, Discrimination of forearm motions from EMG signals by error back propagation typed neural network using entropy, Trans. Soc. Instrum. Control Eng. 29 (1993) 1213–1220.

[31] N. Bu, T. Tsuji, O. Fukuda, EMG-based motion discrimination using a novel recurrent neural network, Int. J. Intell. Syst. 21 (2003) 113–126.

[32] E.J. Scheme, B.S. Hudgins, K.B. Englehart, Confidence-based rejection for improved pattern recognition myoelectric control, IEEE Trans. Biomed. Eng. 60 (2013) 1563–1570.

[33] L. Hargrove, E. Scheme, K. Englehart, B. Hudgins, Multiple binary classifications via linear discriminant analysis for improved controllability of a powered prosthesis, IEEE Trans. Neural Syst. Rehabil. Eng. 18 (2010) 49–57.

[34] E. Scheme, K. Englehart, B. Hudgins, Selective classification for improved robustness of myoelectric control under nonideal conditions, IEEE Trans. Biomed. Eng. 58 (2011) 1698–1705.

[35] S. Amsüss, P.M. Goebel, N. Jiang, et al., Self-correcting pattern recognition system of surface EMG signals for upper limb prosthesis control, IEEE Trans. Biomed. Eng. 61 (2014) 1167–1176.

[36] A.M. Simon, L.J. Hargrove, B.A. Lock, T.A. Kuiken, A decision-based velocity ramp for minimizing the effect of misclassifications during real-time pattern recognition control, IEEE Trans. Biomed. Eng. 58 (2011) 2360–2368.

[37] N. Fligge, H. Urbanek, P. van der Smagt, Relation between object properties and EMG during reaching to grasp, J. Electromyogr. Kinesiol. 23 (2013) 402–410.

[38] M.V. Liarokapis, P.K. Artemiadis, K.J. Kyriakopoulos, E.S. Manolakos, A learning scheme for reach to grasp movements: on EMG-based interfaces using task specific motion decoding models, IEEE J. Biomed. Health Inform. 17 (2013) 915–921.

[39] M. Liarokapis, K.J. Kyriakopoulos, P. Artemiadis, A learning scheme for EMG based interfaces: on task specificity in motion decoding domain, in: P. Artemiadis (Ed.), Neuro-Robotics: From Brain Machine Interfaces to Rehabilitation Robotics, Springer, Dordrecht, 2014, pp. 3–36.

[40] O. Fukuda, T. Tsuji, K. Takahashi, M. Kaneko, Skill assistance for myoelectric control using an event-driven task model, in: Proceedings of the 2002 IEEE/RSJ International Conference on Intelligent Robots and Systems, 2002, pp. 1445–1450.

[41] T. Tsuji, K. Takahashi, O. Fukuda, M. Kaneko, Pattern classification of EMG signals using an event-driven task model, J. Robot. Soc. Jpn. 20 (2002) 771–777.

[42] N. Bu, M. Okamoto, T. Tsuji, A hybrid motion classification approach for EMG-based human-robot interfaces using Bayesian and neural networks, IEEE Trans. Robot. 25 (2009) 502–511.

[43] L. Hargrove, K. Englehart, B. Hudgins, A comparison of surface and intramuscular myoelectric signal classification, IEEE Trans. Biomed. Eng. 54 (2007) 847–853.

[44] J. Birdwell, L. Hargrove, R. Weir, T. Kuiken, Extrinsic finger and thumb muscles command a virtual hand to allow individual finger and grasp control, IEEE Trans. Biomed. Eng. 62 (2015) 218–226.

[45] H. Daley, K. Englehart, L. Hargrove, U. Kuruganti, High density electromyography data of normally limbed and transradial amputee subjects for multifunction prosthetic control, J. Electromyogr. Kinesiol. 22 (2012) 478–484.

[46] D.C. Tkach, A.J. Young, L.H. Smith, E.J. Rouse, L.J. Hargrove, Real-time and offline performance of pattern recognition myoelectric control using a generic electrode grid

with targeted muscle reinnervation patients, IEEE Trans. Neural Syst. Rehabil. Eng. 22 (2014) 727–734.
[47] R. Merletti, H.J. Hermens, Detection and conditioning of the surface EMG signal, in: R. Merletti, P. Parker (Eds.), Electromyography: Physiology, Engineering, and Noninvasive Applications, John Wiley & Sons, Inc., Hoboken, NJ, 2004, pp. 107–131.
[48] R. Begg, D.T.H. Lai, M. Palaniswami, Computational Intelligence in Biomedical Engineering, CRC Press, Boca Raton, 2007.
[49] D. Staudenmann, K. Roeleveld, D. Stegeman, J. van Dieënc, Methodological aspects of SEMG recordings for force estimation—a tutorial and review, J. Electromyogr. Kinesiol. 20 (2010) 375–387.
[50] M. Ortiz-Catalan, B. Håkansson, R. Brånemark, Real-time and simultaneous control of artificial limbs based on pattern recognition algorithms, IEEE Trans. Neural Syst. Rehabil. Eng. 22 (2014) 756–764.
[51] K. Kiatpanichagij, N. Afzulpurkar, Use of supervised discretization with PCA in wavelet packet transformation-based surface electromyogram classification, Expert Syst. Appl. 39 (2009) 7420–7431.
[52] J. Rafiee, M.A. Rafiee, F. Yavari, M.P. Schoen, Feature extraction of forearm EMG signals for prosthetics, Expert Syst. Appl. 38 (2011) 4058–4067.
[53] T. Lorrain, N. Jiang, D. Farina, Surface EMG classification during dynamic contractions for multifunction transradial prostheses, in: Proceedings of the 32nd Annual International Conference of the IEEE Engineering in Medicine and Biology Society (EMBC), 2010, pp. 2766–2769.
[54] I. Kuzborskij, A. Gijsberts, B. Caputo, On the challenge of classifying 52 hand movements from surface electromyography, in: Proceedings of the 34th Annual International Conference of the IEEE Engineering in Medicine and Biology Society (EMBC), 2012, pp. 4931–4937.
[55] A. Phinyomark, P. Phukpattaranont, C. Limsakul, Feature reduction and selection for EMG signal classification, Expert Syst. Appl. 39 (2012) 7420–7431.
[56] S. Guo, M. Pang, B. Gao, H. Hirata, H. Ishihara, Comparison of sEMG-based feature extraction and motion classification methods for upper-limb movement, Sensors 15 (2014) 9022–9038.
[57] J. Liu, Feature dimensionality reduction for myoelectric pattern recognition: a comparison study of feature selection and feature projection methods, Med. Eng. Phys. 36 (2014) 1716–1720.
[58] G.R. Naik, D.K. Kumar, Identification of hand and finger movements using multi run ICA of surface electromyogram, J. Med. Syst. 36 (2012) 841–851.
[59] G.R. Naik, H.T. Nguyen, Nonnegative matrix factorization for the identification of EMG finger movements: evaluation using matrix analysis, IEEE J. Biomed. Health Inform. 19 (2015) 478–485.
[60] J.U. Chu, I. Moon, M.S. Mun, A real-time EMG pattern recognition system based on linear-nonlinear feature projection for a multifunction myoelectric hand, IEEE Trans. Biomed. Eng. 53 (2006) 2232–2239.
[61] D. Graupe, W. Cline, Functional separation of EMG signals via ARMA identification methods for prosthesis control purposes, IEEE Trans. Syst., Man, Cybern. SMC-5 (1975) 252–259.
[62] W.J. Kang, J.R. Shiu, C.K. Cheng, J.S. Lai, H.W. Tsao, T.S. Kuo, The application of cepstral coefficients and maximum likelihood method in EMG pattern recognition, IEEE Trans. Biomed. Eng. 42 (1995) 777–785.
[63] N. Saito, R.R. Coifman, Local discriminant bases and their applications, J. Math. Imaging Vision 5 (1995) 337–358.
[64] A.D.C. Chan, K.B. Englehart, Continuous myoelectric control for powered prostheses using hidden Markov model, IEEE Trans. Biomed. Eng. 52 (2005) 121–124.

[65] Y. Huang, K.B. Englehart, B. Hudgins, A.D.C. Chan, A Gaussian mixture model based classification scheme for myoelectric control of powered upper limb prostheses, IEEE Trans. Biomed. Eng. 52 (2005) 1801–1811.

[66] V.N. Vapnik, Statistical Learning Theory, Wiley, New York, 1998.

[67] M.A. Oskoei, H. Hu, Support vector machine-based classification scheme for myoelectric control applied to upper limb, IEEE Trans. Biomed. Eng. 55 (2008) 1956–1965.

[68] M.F. Kelly, P. Parker, R.N. Scott, The application of neural networks to myoelectric signal analysis: a preliminary study, IEEE Trans. Biomed. Eng. 37 (1990) 221–230.

[69] T. Tsuji, O. Fukuda, M. Kaneko, K. Ito, Pattern classification of time-series EMG signals using neural networks, Int. J. Adapt. Control Signal Process. 14 (2000) 829–848.

[70] T. Tsuji, O. Fukuda, H. Ichinobu, M. Kaneko, A log-linearized Gaussian mixture network and its application to EEG pattern classification, IEEE Trans. Syst., Man, Cybern. C, Appl. Rev. 29 (1999) 60–72.

[71] T. Tsuji, N. Bu, O. Fukuda, M. Kaneko, A recurrent log-linearized Gaussian mixture network, IEEE Trans. Neural Netw. 14 (2003) 304–316.

[72] H. Hayashi, T. Shibanoki, K. Shima, Y. Kurita, T. Tsuji, A recurrent probabilistic neural network with dimensionality reduction based on time-series discriminant component analysis, IEEE Trans. Neural Netw. Learn. Syst. 26 (2015) 3021–3033.

[73] B. Karlik, M.O. Tokhi, M. Alci, A fuzzy clustering neural network architecture for multifunction upper-limb prosthesis, IEEE Trans. Biomed. Eng. 50 (2003) 1255–1261.

[74] M. Khezri, M. Jahed, A neuro-fuzzy inference system for sEMG-based identification of hand motion commands, IEEE Trans. Ind. Electron. 58 (2011) 1952–1960.

[75] Y. Rong, D. Hao, X. Han, et al., Classification of surface EMGs using wavelet packet energy analysis and a genetic algorithm-based support vector machine, Neurophysiology 45 (2013) 39–48.

[76] J.J. Darragh, I.H. Witten, M.L. James, The reactive keyboard: a predictive typing aid, IEEE Comput. 23 (1990) 41–49.

[77] E. Smirnova, S.M. Watt, Context sensitive mathematical character recognition, in: Proceedings IAPR International Conference Frontiers in Handwriting Recognition, 2008, pp. 604–610.

[78] G. Hoffman, C. Breazeal, Cost-based anticipatory action selection for human-robot fluency, IEEE Trans. Robot. 23 (2007) 952–961.

[79] W. Sheng, A. Thobbi, Y. Gu, An integrated framework for human-robot collaborative manipulation, IEEE Trans. Cybern. 45 (2015) 2030–2041.

[80] W. Reisig, Petri Nets: An Introduction, Springer-Verlag, New York, 1988.

[81] I. Zukerman, D. Albrecht, Predictive statistical models for user modeling, User Model. User-Adap. 11 (2001) 5–18.

[82] G. Chang, D. Kulic, Robot task learning from demonstration using Petri nets, in: Proceedings of the 22nd IEEE International Symposium on Robot and Human Interactive Communication (RO-MAN), 2013, pp. 31–36.

[83] T. Cao, A.C. Sanderson, Task sequence planning using fuzzy Petri nets, IEEE Trans. Syst., Man, Cybern., Syst. 25 (1995) 755–768.

[84] J. Rosell, Assembly and task planning using Petri nets: a survey, J. Eng. Manuf. 218 (2004) 987–994.

[85] J. Pearl, Probabilistic Reasoning in Intelligent Systems: Networks of Plausible Inference, Morgan Kaufmann, San Mateo, CA, 1988.

[86] E. Horvitz, J. Breese, D. Heckerman, D. Hovel, K. Rommelse, The Lumière project: Bayesian user modeling for inferring the goals and needs of software users, in: Proceedings of the 14th Conference on Uncertainty in Artificial Intelligence, 1998, pp. 256–265.

[87] J. Hirayama, M. Nakatomi, T. Takenouchi, S. Ishii, Collaborative prediction by multiple Bayesian networks and its application to printer usage modeling, Behaviormetrika 35 (2008) 99–114.

[88] Q. Ding, X. Zhao, J. Han, A hybrid EMG model for the estimation of multijoint move-
      ment in activities of daily living, in: Proceedings of 2014 International Conference on
      Multisensor Fusion and Information integration for Intelligent Systems, 2014, pp. 1–6.
[89] L. Xu, A. Krzyzak, C.Y. Suen, Methods of combining multiple classifiers and their
      applications to handwriting recognition, IEEE Trans. Syst., Man, Cybern. 22 (1992)
      418–435.
[90] J. Kittler, M. Hatef, R.P.W. Duin, J. Matas, On combining classifiers, IEEE Trans. Pat-
      tern Anal. Mach. Intell. 20 (1998) 226–239.
[91] O. Fukuda, T. Tsuji, M. Kaneko, An EMG controlled robotic manipulator using neural
      network, in: Proceedings of the 6th IEEE International Workshop on Robot and
      Human Communication, 1997, pp. 442–447.
[92] O. Fukuda, T. Tsuji, M. Kaneko, A human supporting manipulator based on manual
      control using EMG signals, J. Robot. Soc. Jpn. 18 (2000) 387–394.
[93] R. Daly, Q. Shen, S. Aitken, Learning Bayesian networks: approaches and issues,
      Knowl. Eng. Rev. 26 (2011) 99–157.

CHAPTER FOUR

Personalized Modeling for Home-Based Postural Balance Rehabilitation

M. Hayashibe, A. González, P. Fraisse
INRIA, University of Montpellier, Montpellier, France

1. INTRODUCTION

The need for in-home care, monitoring, and rehabilitation of the elderly and impaired is increasing as the world population continues to age. The United Nations expects that 16% of the population will reach 65 years or older by 2050. This tendency continues to expand all over the world, especially for the developed countries. Patients in need of rehabilitation, prosthetics and assistance are usually supported by social welfare and have difficulties in reestablishing a normal life after their injury. The social cost in terms of work force and medical care will thus become more significant every year if no action is taken. In addition, the present situation of hospital capacity, rehabilitation support and nursing is also worsening with a lack of human resources. In the current scenario, the cost for those medical treatments would be drastically increased in the future. Together with the fact that falls are a leading cause of injury in the elderly [1], the aging social environment has prompted the need to remotely monitor the condition of older adults and the motor impaired, such as poststroke subjects, and to provide them with an efficient rehabilitation tool for balance assessment.

Following a constant physical training program can improve physical rehabilitation outcomes and will help a motor-impaired subject to achieve the best motor performance possible [2]. Important improvements in motor function have been observed when subjects continue training during everyday life [2]. In a similar way, a personalized balance training and monitoring system for the elderly for outpatient or at-home use can help to prevent falls by maximizing balance training.

Standing balance has been generally used to predict fall risks in elderly populations. For example, the Berg Balance Scale (BBS) and the Timed Up and Go (TUG) are commonly used by clinicians to determine static and dynamic balance, respectively [3]. A posturography, or the analysis of center of pressure (CoP) trajectory, has also shown that human balancing strategies change with age [4].

In order to prevent falls, balance training is common after orthopedic surgery or cardiovascular accidents [2,5]. Some research teams have focused on creating preventive systems and promoting fitness training at home. For example, Kinect-based video games have been developed to encourage players to move while performing tasks designed to improve functional reach [6,7]. Similarly, a Wii balance board (WBB) has been used to increase the range of CoP excursion during rehabilitation [5]. As with BBS and TUG, in practice, it is preferable to determine balance without the use of force platforms or force sensors in order to increase the patient's range of motion without requiring him/her to stand on top of the device. We propose to use video based motion capture to reduce the number of sensors attached to the subject for home-based applications. This would be useful for the analysis of unconstrained motion such as walking, sit-to-stand (StS), or general training specified by their therapists. However, the currently available system for such purpose does not consider subject-specific body dynamics, which can differ from standard values, especially for patient populations that need neurorehabilitation.

2. HOME-BASED POSTURAL BALANCE REHABILITATION

To provide a patient with the best chances of motor recovery, special interest is being taken in the development of home rehabilitation devices that will guide exercise sessions in the absence of a therapist. This aspect is important also for socioeconomic reasons to minimize the rehabilitation cost for our society. Such systems are needed to achieve the ideal repetition intensity by having a patient continue to exercise at home. That is, home rehabilitation devices should improve the effectiveness of rehabilitation exercises without increasing their costs. A number of systems have already been developed to track and record a subject's movements outside of the laboratory environment through the use of wearable sensors. For example, sensors already present in cellular phones can determine activity levels and detect falls [8,9]. When this data is aggregated over a period of time, it can be used to detect changes in motion patterns in the patient's lifestyle and help to evaluate his motor function.

Similarly, in a current trend, video game controllers are utilized for the subject to perform a series of full body movements such as reaching or shifting his weight to access in–game functions. This idea has intrigued the rehabilitation community and they have used it to create a series of games meant to guide a patient during a series of exercises. For example, Lange et al. [6] created one where the subject is asked to move the arms in order to collect gems; Borghese et al. ask subjects to change their stance in order to collect moving fruit [10]. Both of these examples aim to encourage the subject by adapting the difficulty in response to the subject's performance.

Multiple research teams work towards the creation of nonintrusive surveillance tools to be installed inside a subject's home that could be capable of recording gait parameters over time [11], or of monitoring other everyday activities [12]. These systems are meant to support unassisted living [13,14] by helping the subject maintain high-level motor functions without the need for constant human assistance. Additionally, they may be used to accurately predict important events, such as a fall, so that action can be taken accordingly. This prediction would require constant evaluation of a subject's posture.

Furthermore, the role that sensory feedback has on balance training has been studied using virtual reality (VR) based rehabilitation. For example, walking [15] and cycling [16] simulators were developed and results suggest an improvement in the subject's performance when sensory feedback was introduced. The elderly and balance impaired can also benefit from VR rehabilitation systems [7,17,18] inside their homes. Other studies, such as [19], focus on the role that visual feedback has on balance training. Their subjects were asked to change their CoP position while on-screen feedback was given; the CoP sway was used as an indicator of balance. A similar system was implemented by Kennedy et al. [5] with a WBB. A multilevel electrotherapy paradigm is proposed by Dutta et al. [20] where a virtual reality-based adaptive response technology is proposed for poststroke balance rehabilitation.

Among the previously listed systems, it is rare to employ personalized modeling for the subject balance estimate, as it is challenging to take into consideration the different body's dynamic characteristics. We wish to contribute to home rehabilitation systems by establishing a balance measure based on subject-specific postural modeling that can be easily implemented even in the home environment. We believe that the personalized CoM can

Fig. 1 A scene of home-based postural balance rehabilitation. The data management with the personalized performance improvement through a tablet would be useful for tracking of training history for the user.

be used to accurately determine balance instabilities [21] and contributes to a personalized evaluation of the subject's motor performance while following rehabilitation protocols. A scene of home-based postural balance rehabilitation is presented in Fig. 1. The data management with the personalized performance improvement through a tablet would be useful for tracking of training history for the user.

3. BODY SEGMENT PARAMETERS

The human body can be divided into several segments, each one with its unique mass and geometry and whose three-dimensional position and orientation contributes to the overall CoM position. The first attempts to determine the relevant parameters were recorded by Harless [22] who used cadavers to determine the mass and CoM position of the individual body segments. This technique was later improved by Braune and Fischer [23] who froze the cadavers before carefully measuring them in order to avoid the loss of fluids. Since then several other anthropometric studies have been performed and reported [24–26]. The tables proposed by Zatsiorsky, Winter and De Leva have been used quite popularly in the biomechanics domain.

In general, anthropometric parameters will change with age, somatotype and even fitness level. We have to take care to use the appropriate segment information for all calculations; data gathered from infants, the elderly, and from young and otherwise healthy subjects should not be used interchangeably. However, simple scaling by using the subject height and weight is normally employed for the usage of these anthropometric tables with regression equations [24,27].

Other studies [25,28] show how to obtain either the CoM position of a living subject's limbs, or its mass. This is achieved by having the subject lie down on a balance board and measuring the reaction moments for a variety of postures. When all segments are to be measured, this method can be very time consuming since every segment should be moved individually. Additionally, either CoM position or the segment's mass is assumed to be known. This introduces uncertainty in the method [25,28,29]. A similar idea is used by Venture et al. [30] and Jaffrey [29] to find the best inertial parameters that reproduce the forces and moments measured during a dynamic motion. Ayusawa et al. [31] have estimated center of mass and inertial parameters of both humans and humanoid robots. The same group has also developed an on-line parameter estimation tool capable of real-time feedback. The method was also applied to determine subject-specific muscle strength [32]. Even if these are powerful methods for the identification of inertial parameters, they assume that the lengths of the segments are known. Additionally, the numerical differentiation involved may introduce inaccuracies, and then these approaches need to rely on a high-end motion capture system and sensitive force plate measurements available only in a lab environment.

Finally, another approach is to make use of imaging techniques to determine the volume and shape of an individual segment [24,33,34]. In this way its inertial properties may be estimated. For simplicity these methods often assume homogeneous rigid limbs [29].

In summary, the methods that estimate CoM position from mass parameters of the individual body segments have different degrees of accuracy, depending on the quality of the segment parameters they use. It is possible to determine a subject-specific set of parameters but special treatment and laboratory grade equipment are often required to obtain reliable parameters.

4. ESTIMATING CENTER OF MASS POSITION FOR HUMAN SUBJECTS

The CoM is the point where the concentrated mass of any real object can be represented. The position of the CoM is defined as the centroid of all mass elements [35], also described as the mean position for all masses. It can be divided into n segments each with its own mass (m_i) and a segmental CoM position (c_i). The overall CoM position (c) can be found as:

$$c = \frac{1}{M} \sum c_i m_i \tag{1}$$

where M is the total mass. When the segments are small enough, the CoM can be expressed as the following integral:

$$c = \frac{1}{M} \int p \, dm \tag{2}$$

where p is the position of the differential mass.

The CoM position is completely dependent on the mass distribution and position of the objects. For certain configurations it is possible that the CoM is located outside of the system. Placing the arms and legs far from the body could result in a stable posture where the CoM is located outside the body, as long as the reaction force from the ground could go through the supporting polygon.

The study of an individual's CoM kinematics is useful to understand gait and balance, and has been used extensively in the study of human locomotion [36]. For example, Braun and Fischer [23] in 1889 published a study on the determination of CoM position under different loading conditions. This was for better understanding of the effects of heavy loading on soldiers and to optimize the weight distribution of their equipment. A recent work with similar goals studies a subject's response to load changes while walking [37]. For medical purposes, the analysis of CoM is directly concerned with the diagnosis of motor deficiencies and to determine the best course of action to correct them [38]. Additionally, the amplitude of the oscillations of the CoM, also known as sway, during quiet standing is generally considered an accurate measurement of balance [39–41]. Finally, the analysis of the CoM's acceleration, for example while running, can determine the forces applied on the body [42] and by extension, the muscle forces required to support it. In its essence, if the mass distribution of an object and its position and orientation are known, then the position of its CoM is also known.

Caron et al. [43] estimated the ground projection of a subject's CoM by removing high frequency components of the measured CoP. They found an invariant relationship between the amplitudes of CoM and CoP oscillations [40]. This relationship was obtained from the dynamic analysis of an inverted pendulum-like model as measured forces can be related to the acceleration of a mass oscillating above the ankles. This method was validated for quiet standing and stepping in place; however, it is applicable only for motions where the overall CoM remains stable, at a constant height and with its motion described as a sum of sinusoidal functions.

Some have used the fact that ground reaction forces are directly related to the acceleration of the CoM to estimate its position [40,44–46]. After double integration of the acceleration, the CoM position can be written as:

$$c = \frac{1}{2}\left(g + \frac{f}{M}\right)t^2 + \dot{c}_0 t + c_0 \tag{3}$$

where $\dot{c}_0$ and c_0 are the initial velocity and position of the CoM and appear as integration constants. Although this approach is rigorous because it considers all of the body's dynamics and can be used to determine CoM position for all types of motions, its implementation is not straightforward: low-frequency noise can be amplified by the integration and will cause drift [45,46]; also a good knowledge of the initial conditions is required. Additionally, force measurements are needed for the duration of the motion, meaning that the subject must remain inside the force-sensing area or use special instrumented shoes as in [44,47].

Current research involving this method focuses on removing drift and obtaining good initial conditions. For example, the zero-point-to-zero-point (ZPZP) method presented by Zatsiorky and King [46] constantly corrects the current CoM estimate by observing the measured forces. Instead of performing the double integral over a large period of time, something prone to drift [29], the method segments the recording by finding measurements from static postures and uses them to obtain the integration constants. Thus, the ZPZP method cannot be used to estimate the CoM trajectory unless two static postures have been found. Jaffrey [29] extensively studied the ZPZP method and determined that in its original form it is inappropriate for CoM estimation outside quiet standing. He applied a modified version to other movements such as jumping. Another approach is given by Schepers et al. [44], who combine the double integral and low-pass filtering methods. To do this, they take the high-frequency component of the double integral, and add the low-frequency component of the CoP. In this way they remove the drift while maintaining an accurate CoM estimation for quiet standing. In summary, the double integral methods are difficult to apply in real-time due to noisy measurements that require filtering and the determination of unknown integration constants.

4.1 Statically Equivalent Serial Chain

Cotton et al. [48,49] proposed the statically equivalent serial chain (SESC) method to identify a set of subject-specific parameters relevant to the CoM position. Their work is based on that of Espiau and Boulic [50] and was originally developed to improve the design and control of mechanisms, specifically humanoid robots. The SESC models the CoM as the end position of an open chain whose link lengths are determined by the mass and geometric parameters of the original segments. Cotton et al. [49] were able to successfully identify the relevant parameters and predict CoM position for several subjects in the sagittal plane. Although they developed the method for three-dimensional motion, their experiments involving human subjects only covered planar motion. The parameter identification requires measuring a number of static postures, much like the method detailed by Pataky et al. [28], but makes no assumptions about the mass or geometry of the individual segments. It also allows for several segments to be moved at the same time. After a full model is identified, the SESC method allows three-dimensional estimation of the CoM position, as long as the segments' orientations can be measured without needing any other ground reaction force measurement.

Due to its simple application and subject-specific capabilities, we use the SESC method to estimate CoM position and elaborate upon it. The validation of SESC for three-dimensional motions was performed in [51]. However, the conventional SESC approach has been to use high-end motion capture systems to estimate the CoM position [44,49] which could not be used inside a patient's home. Works like [52] propose to use an inertial measurement unit (IMU) to track the CoM position. This low-cost sensor approach is suitable for a home, but the method is only accurate for walking or quiet standing. This prompted us to design a portable, accurate, and *real-time* CoM estimation and visualization system to be used inside unstructured environments including in-home situations.

5. METHOD

5.1 SESC Computation

A linked chain's CoM position can be expressed in terms of its link orientations and a set of parameters determined by the link geometry and mass. This representation is equivalent to the geometric description of an open, virtual serial chain whose end-effector is the original chain's CoM. This virtual chain is known as the SESC [48].

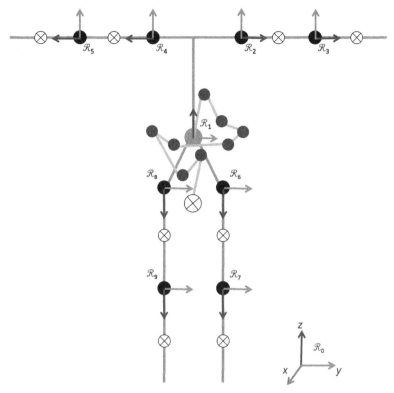

Fig. 2 The skeleton model is composed of nine rigid segments. Shoulders and hips are represented with spherical joints, while elbows and knees are treated as hinge joints. The CoM position is indicated by the end-point of the SESC.

In order to estimate the CoM position for a human subject, we use a skeleton model composed of nine links connected by four spherical and four hinge joints (see Fig. 2). We associate a frame $\mathcal{R}_i$ to each link with the SESC parameters expressed with respect to those frames. The orientation between two frames is expressed using the rotation matrix $\mathbf{A}_i^j$. Additionally, the translational vector between two frames is written as $\boldsymbol{p}_i^j$. The CoM of a system with n number of links, each with a mass m_i, can be represented by performing the matrix multiplication:

$$c = \frac{1}{M} \sum_{i=1}^{n} m_i (\mathbf{A}_i c_i + \boldsymbol{p}_i) = \begin{bmatrix} \mathbf{E} & \mathbf{A}_1 & \dots & \mathbf{A}_9 \end{bmatrix} \begin{bmatrix} \boldsymbol{p}_1 \\ \boldsymbol{r}_1 \\ \vdots \\ \boldsymbol{r}_9 \end{bmatrix} \quad (4)$$

where $\mathbf{E}$ is an identity matrix and $\mathbf{A}_i$ is the 3-by-3 orientation matrix of link i with respect to the global frame. The values in r_i can be explicitly determined as a function of the link masses (m_i), CoM positions in their local reference frames (c_i), and p_i^j. M is the total mass. The superindex 0 used to denote a position or orientation measured from the global reference frame has been omitted for convenience.

For our model, r_i is a constant 3-by-1 vector. This is due to the fact that only spherical and hinge joints were considered. Moreover, the CoM can be referenced to a floating frame attached to the skeleton at $\mathcal{R}_1$. Using the torso as a base for the SESC, (4) can be rewritten as follows:

$$c^1 = \begin{bmatrix} \mathbf{A}_1 & \dots & \mathbf{A}_9 \end{bmatrix} \begin{bmatrix} r_1 \\ \vdots \\ r_9 \end{bmatrix} = \mathbf{B}r \qquad (5)$$

where c^1 is the CoM position measured from the origin of $\mathcal{R}_1$, $\mathbf{B}$ contains the orientation matrices for all of the chain's links, and r represents a 27-by-1 vector of subject-specific parameters. The explicit expressions of r can be found in [53].

It was noted in [48] that due to the SESC redundancy, there are several solutions to the identification problem. To constrain the solution, while hoping to find the one closest to our model, we put forward the following assumptions:

(a) The floating frame $\mathcal{R}_1$ is attached to the skeleton at the torso.
(b) As depicted in Fig. 2, one axis of the associated reference frame $\mathcal{R}_i$ lies on the line segment connecting two joints.
(c) The CoM position for all limb segments, c_i^i, also lies on this straight line. The last two assumptions reduce the size of r once the rows which are known to equal zero are removed. Accordingly, $\mathbf{B}$ is reduced by removing the corresponding columns. In summary, r_1 remains a 3-by-1 vector while $r_{2\dots9}$ are represented by one scalar each. The number of unknown SESC parameters was reduced from 27 to 11 constants, which can then be estimated using recursive techniques.

5.2 SESC Parameter Identification and Visual Feedback Using a Kalman Filter

To determine the geometric parameters of a serial chain such as the one described in Eq. (5), we use a recursive approach where each new measurement improves our knowledge of $\hat{r}$ [54]. The Kalman filter can be used to determine a set of constant values from a group of noisy measurements. Consider the linear system described by:

$$x_k = \mathbf{F}_{k-1} x_{k-1} + w_{k-1} \tag{6}$$

where the subindex k denotes the time step. The evolution of the state vector x_k is determined by $\mathbf{F}_{k-1}$ and a zero mean process noise w_k with covariance $\mathbf{Q}_k$. An estimate of vector x_k can be found from a number of y_k measurements. Each measurement is a linear combination of the states and a zero mean measurement noise v_k with covariance $\mathbf{R}_k$. Each measurement can be expressed in the form:

$$y_k = \mathbf{H}_k x_k + v_k \tag{7}$$

where $\mathbf{H}_k$ is known as the configuration matrix.

The Kalman filter provides an optimal linear solution for systems defined by Eqs. (6) and (7). When estimating a constant vector, the linear Kalman filter may be written as follows [55]:

$$\mathbf{P}_k^- = \mathbf{P}_{k-1}^+ + \mathbf{Q}_{k-1} \tag{8}$$

$$\mathbf{K}_k = \mathbf{P}_k^- \mathbf{H}_k{}^T \left(\mathbf{H}_k \mathbf{P}_k^- \mathbf{H}_k^T + \mathbf{R} \right)^{-1} \tag{9}$$

$$\hat{x}_k = \hat{x}_{k-1} + \mathbf{K}_k \left(y_k - \mathbf{H}_k \hat{x}_{k-1} \right) \tag{10}$$

$$\mathbf{P}_k^+ = (\mathbf{E} - \mathbf{K}_k \mathbf{H}_k) \mathbf{P}_k^- \tag{11}$$

where $\mathbf{P}_k^-$ and $\mathbf{P}_k^+$ are the covariance matrices before and after the state update, $\mathbf{E}$ is an identity matrix of suitable size, and $\mathbf{K}_k$ is the optimal filter gain for minimizing the estimation error. Convergence of the estimation is reflected by the eventual decrease in the covariance matrix ($\mathbf{P}$) magnitude, as the confidence on the current estimate increases. That is, the estimation improves as $\mathbf{P}$ approaches zero.

In order to determine the effect that the adaptive visual feedback has on the identification, the color of each of the skeleton segments is varied to represent the convergence of the corresponding SESC parameters. $\mathbf{P}$ can thus be used to determine the color of each skeleton segment. $\mathbf{P}$ has a large value at the beginning of the identification: the skeleton is drawn completely in red. As $\mathbf{P}$ decreases the corresponding segments turn to green.

To perform a correct identification, a large number of measurements with an appropriate set of configuration matrices should be obtained. The orientation of each limb is directly measurable, but this is not the case for the subject's CoM. CoP offers a good approximation of the ground projection of CoM during quiet standing. This is due to the small CoM accelerations [48] that occur in these postures. To determine if a posture is stable enough, we observe the CoP position and the limb roll-pitch-yaw angles during a 1 s window. We look at the standard deviation of both the angles and CoP values to determine if the pose was stable enough during this window to be used as identification data.

5.3 Zero Rate of Change of Angular Momentum

The zero moment point (ZMP) was developed by Vukobratović [56] to control walking gaits for legged robots by assigning the dynamics of the robot's limbs. The ZMP is defined as the position in the ground plane where the moments perpendicular to the horizontal axis are equal to zero; that is, the robot's foot can lie flat on the ground with no rotation. This definition is equivalent to that of the CoP as we have expressed it previously. This equivalence has been recognized by Sardain and Bessonnet [57], and Goswami [58]. This criteria guarantees balance as long as the ZMP point is contained within the area of support.

The use of the zero rate of change of angular momentum (ZRAM) to determine the stability of humanoid robots during the single and double support phases was developed by Goswami [59] and originates from the same metric as ZMP control [60]. The main difference between ZMP/CoP and the ZRAM point is that the former are confined to the support polygon, while the latter can be used to determine a point on the ground where the reaction forces should be positioned in order to maintain balance.

For a walking robot on level ground, a sum of moments on its CoM can be written as:

$$\dot{\mathcal{L}}_c = \left(c_p - c\right) \times f \tag{12}$$

where $\dot{\mathcal{L}}_c$ is the rate of change of angular momentum, f is the vector of reaction forces, and c_p gives the point of its application. The ZRAM point is defined as the position on the ground where $\dot{\mathcal{L}}_c = 0$, ie, the action line of the ground reaction force passes through the CoM. Let p_0 represent a point on the ground plane and n a vector normal to the plane; then the ZRAM point can be determined as follows:

$$p_f = \frac{(p_0 - c) \cdot n}{f \cdot n} f + c \tag{13}$$

Unlike ZMP and CoP, the ZRAM point may exit the support polygon. This excursion can be seen as an unstable situation and used to provide a dynamic measure of balance but does not predict a fall. Falling may still be avoided if a step is taken or the limbs are used to balance. A full dynamic analysis is necessary to determine if a fall cannot be avoided.

Kajita et al. [60] present an equivalence between the sum of forces acting on a linked chain and the acceleration of its CoM.

$$f = M(\ddot{c} + g) \tag{14}$$

where M is total mass of the chain and g the acceleration of gravity. Using Eqs. (13) and (14), the ZRAM point can be determined from the trajectory of CoM alone, without the need for external force sensors.

5.3.1 Measuring Balance

Whenever the CoP is measured (by means of a force platform), $\dot{\mathcal{L}}_c$ can be used to evaluate balance. It is obtained from Eq. (12) and it should remain close to zero for stable motions. On the other hand, the ZRAM point can be obtained and used to evaluate balance in the absence of CoP measurements. We propose to use the position of the ZRAM point (p_f) with respect to the center of the support polygon to determine stability [59]. We distinguish two cases.

(i) A single support phase where the distance of p_f to the ankle is found as

$$d = \| p_{l,r} - p_f \| \tag{15}$$

where $p_{l,r}$ is the three-dimensional position of the support ankle (left or right).

(ii) A double support phase where we determine the distance of p_f to the line defined by both ankles. When this line is expressed as a function of the left ankle position and of the unit vector $v = (p_r - p_l) / \| p_r - p_l \|$, the distance between p_f and the line can be found as

$$d = \left\| \left(p_f - p_l \right) - \left(\left(p_f - p_l \right) \cdot v \right) v \right\| \tag{16}$$

5.4 Experiment

Eight healthy volunteers (two females and six males, age $= 27 \pm 3.0$ years, weight $= 76.4 \pm 20.9$ kg, height $= 1.76 \pm 0.06$ m) were asked to participate in this study and gave their informed consent. The subjects had never participated in a SESC parameter identification before. The subjects were asked to stand on top of a WBB and to maintain a series of static postures. They were also asked to hold each posture for approximately 2 s before changing to a different one. No instructions were given regarding the number or type of postures to perform, but they were asked to move all joints during the experiment.

The procedure was performed three times:

(i) once without visual feedback of the identification status;

(ii) once with visual feedback (where the skeleton color indicates the SESC parameter covariance); and

(iii) a final recording without feedback to be used as cross-validation for the identified SESC model.

That is, in (a) and (iii) the subject can observe the skeleton in real-time but without a color change, and in (b) the subject can observe the skeleton changing colors in real-time as the covariance of each SESC parameter changes during the identification.

The Kinect sensor was placed 3 m away from the subject on the WBB. Both the depth information from the Kinect and the force and CoP data from the WBB were recorded synchronously using a custom application capable of reading and storing data. A custom program was created in C++ to:

(1) Collect and save the data.

(2) Display the information in 3D.

(3) Allow user input for navigation in a 3D environment.

(4) Identify the SESC parameters on-line.

(5) Give visual feedback to the user/subject regarding the SESC parameter identification status.

Handling of the Kinect (Microsoft, Redmond, WA, USA) and WBB sensor (Wii balance board by Nintendo Co., Ltd., Kyoto, Japan) was managed using open-source code from *OpenNI* (PrimeSense, Ltd., Tel Aviv, Israel) [61] and the *WiiUse* project [62], respectively. The visual interface was created using the OpenGL GLUT library, while OpenCV was used to perform the necessary matrix operations. The created software allows us to draw in 3D: the skeleton provided by *OpenNI*, the time-varying color information representing the Kalman filter's state convergence, the SESC, and the resultant CoM position estimated with the current model.

6. RESULTS

6.1 Comparison: High-End vs. Portable Sensors

In this work, we proposed a new method for estimating the whole body center of mass that can be used outside of the laboratory by utilizing the

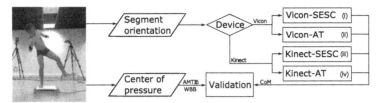

Fig. 3 Using the same static postures, the volunteer's CoM is estimated from laboratory-grade measurements with: (i) an SESC and (ii) anthropometric table (AT). To validate the use of low-cost sensors the CoM estimation is repeated using (iii) an SESC and (iv) anthropometric table (AT).

SESC and a Kinect. We evaluated the differences between the SESC's CoM estimate obtained from Vicon-AMTI6 data and one created using the Kinect-WBB. For this purpose, the SESC parameters were identified twice over the same static postures, using low-cost and high-end equipment. For completeness, the CoM was also estimated using anthropometric table data [25] to compare these results to the SESC estimate. Fig. 3 summarizes this process. An example of such a comparison between the measured CoP the estimated CoP using SESC and the estimated CoP using anthropometric table (AT) from Winter [25] is shown in Fig. 4. We can see that the experimentally identified personalized model can, in general, provide stable estimate performance for different postures in contrast to the anthropometric table approach. Here, we show the digest version of the result. The detailed results and analysis are presented in [53,63].

Fig. 5 shows the averaged root mean square error (RMSE) of the estimation for all subjects and all postures. The best performance was the (i) case with a 12.89 ± 9.11 mm average error. The (ii) and (iii) cases performed similarly to each other. The largest estimation error came from case (iv) using literature values with Kinect measurements. Table 1 shows the mean and standard deviation of the RMSE for the (i)–(iv) estimations. This table also shows the error in the subjects' anterior-posterior (AP) and medio-lateral (ML) directions and the coefficient of determination of each case compared to the CoP measurements of the validation posture set.

The validation of human CoM estimation methods is an open problem, as this quantity cannot be directly measured. The segmentation method, using laboratory instruments and anthropometric tables, is considered as the standard for whole body CoM estimation [40]. However, no study to our knowledge has yet investigated the possibility of using low-cost instruments to provide a subject-specific CoM estimation. We evaluated the accuracy of portable sensors (the Kinect and the WBB) to estimate CoM by

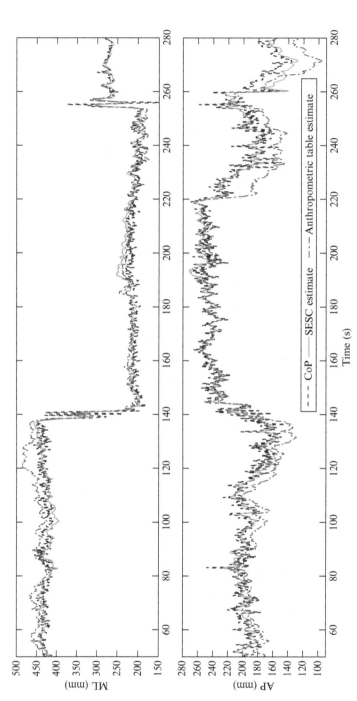

Fig. 4 An example of the comparison between the measured CoP, and estimates using SESC and anthropometric table data from Winter is plotted for Vicon measurement case. It is shown for the axis of anterior-posterior (AP) and medio-lateral (ML) directions, respectively. We can see that the experimentally identified-personalized model can provide stable estimate performance in general for different postures in contrast to the case of the anthropometric generic table approach.

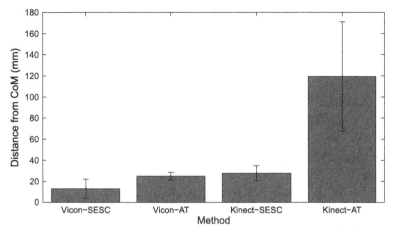

Fig. 5 Summary of the performance of each CoM estimation method. The bars correspond to the RMSE $\pm$ STD averaged for all subjects. We observe an increase in the accuracy of the identified SESCs with respect to the literature estimates (AT). In addition, the performance of the Kinect-SESC was found to be equivalent to that of the literature-based estimate using high-end sensors.

Table 1 The Root Mean Square Error for Each CoM Estimation Method in Cross-Validation

CoM Estimation	Error (mm)	AP (mm)	ML (mm)	R^2
Vicon-SESC	12.8 $\pm$ 9.1	10.4 $\pm$ 6.6	10.2 $\pm$ 6.9	0.9 $\pm$ 0.1
Vicon-AT	24.9 $\pm$ 3.7	23.1 $\pm$ 5.9	13.9 $\pm$ 7.3	0.8 $\pm$ 0.1
Kinect-SESC	26.6 $\pm$ 6.0	23.4 $\pm$ 6.8	17.1 $\pm$ 8.0	0.8 $\pm$ 0.2
Kinect-AT	118.4 $\pm$ 50.0	98.82 $\pm$ 70.2	51.8 $\pm$ 24.2	NA

The error is measured in the world reference frame, averaged for all subjects $\pm$ std. Anterior-posterior (AP) and medio-lateral (ML) give the AP and ML direction errors, respectively. R^2 is the coefficient of determination of the cross-validation set.

comparing it with that obtained with conventional sensors (Vicon and an AMTI6 force platform). With the Vicon system, the estimation error of the literature-based CoM estimate was found to be 24.9 ± 3.7 mm; this error was reduced to 12.8 ± 9.1 mm using the SESC method. With the Kinect, the literature-based estimate had an error of 118.4 ± 50.0 mm, while the subject-specific SESC error was 26.6 ± 6.0 mm. We find that the subject-specific SESC estimation with low-cost sensors performed as well as a literature-based one with high-end sensors.

Regarding CoM estimation accuracy, we improve on the literature due to the SESC's subject-specific nature. Fig. 5 focuses on this. A lower RMSE was observed with the Vicon-SESC method than in the literature sources [25]. Similarly, using literature values with the Kinect (iv) results in large estimation errors (see Table 1). The SESC estimate for the Vicon-SESC has the same error magnitude for both the AP and ML directions, whereas the Kinect-SESC has a larger mean error on the AP (depth) direction than the ML one. This could be due to the noisy joint positions given by the Kinect skeleton, as only one camera is available to reconstruct the kinematics. In contrast, the Vicon skeleton offers a better-defined joint obtained from the markers' positions. Finally, the performance of the Kinect-SESC estimate gets closer to the Kinect's known measurement error. The Vicon-SESC estimate error is larger than that of the Vicon's measurement. This is probably due to the simplified skeletal model.

A two-way ANOVA test was performed to determine the influence of the sensor (Vicon or Kinect) and of the origin of the parameters (AT or SESC) on the RMSE. A significant effect was found from both factors ($p < 0.01$). Additionally, a strong interaction of both factors was found ($p < 0.01$), suggesting that the low RMSE found for the Vicon-SESC case was due to both the SESC method and the high-quality measurements of the Vicon system.

6.2 Convergence-Skeleton Coloring Feedback vs. No Feedback

SESC identification is performed and compared between: (a) without color visual feedback and (b) with color feedback. The color indicates the status of the ongoing identification. For (a), the skeleton was visible but its color remained red. In Fig. 6 we show the adaptive interface for the identification used for (b). The current CoM estimate is also visible. The skeleton color gradually changes from red to green indicating the convergence of the SESC model. The color change is determined from the magnitude of the SESC parameter covariance. We observe a high estimation error for the first frames and the CoM gradually entering the subject's body as the identification proceeds. In this figure, we see the gradual change in limb color. This happens as the confidence on the limb parameters increases as a larger amount of significant information is gathered. In Fig. 6, we see that the arms are the first segments to turn green; the subject presented them in many different orientations before moving the legs or thorax. With this interface, the subject understands which segments should still be reoriented. Fig. 7 represents the mean covariance value among the SESC parameters for all eight subjects. We show both cases: (a) seen in red, and (b) in blue. The center line

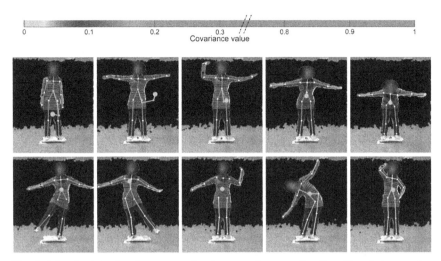

Fig. 6 Adaptive interface for personalized center of mass estimation: The skeleton color changes from red to green indicating the convergence of the SESC model adaptively identified by the Kalman filter. Each segment color is determined by its respective covariance, where the highest value corresponds to red. The current CoM is also visible. We observe a high estimation error for the first frames and the CoM gradually entering the subject's body as the parameter identification proceeds. A video showing the identification procedure is available at http://youtu.be/J-yqOzRK5Ts. For interpretation of the references to color in this figure legend, see the online version of the book.

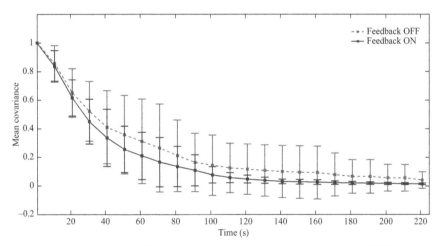

Fig. 7 Mean and standard deviation of the maximum covariance among SESC parameters for all eight subjects during on-line identification. The red line shows the results without the adaptive interface while the blue line shows the results with the interface. The error bars show the *STD* transition for each case. The effects of the visual feedback on identification are visible in the speed of the convergence. For interpretation of the references to color in this figure legend, see the online version of the book.

represents the mean value while the surrounding error bars show the *STD* range. The feedback effect is visible. It reduces the overall parameter covariance, as well as their *STD*. As a result, we observed a faster convergence of the SESC parameter identification in (b) than in (a). The result of this adaptive identification interface to accelerate the SESC model establishment is a digest version. For detailed results and discussions, refer to [64,65].

6.3 Cross-Validation With a New Motion Set

Cross-validation with a new motion set was done to evaluate the identified SESC model in terms of CoM estimation. The final SESC parameters from the (a) feedback off and (b) feedback on recordings were used to estimate the subject's CoM during a third trial and compared to the measured CoP. Table 2 presents the average RMSE of the CoM estimation along with the mean covariance as a function of time spent in identification. SESC parameters obtained at 15 s intervals were used to estimate the CoM position of the cross-validation trial. The average RMSE value decreases over time. That is, the overall identification improves as the time spent on identification increases. We observed faster convergence of the CoM estimate when the adaptive interface was used. These results may serve as a guideline for the identification phase. To obtain an RMSE of 27 mm, the identification time could be halved, ie, dropping from 180 s without visual feedback to 90 s with it.

6.4 Postural Stability Index

As an application example, we monitor the stability of a dynamic movement using the personalized CoM trajectory. The ZRAM point distance to the

Table 2 Summary of Results Obtained for the (a) Feedback Off and (b) Feedback On Cases as a Function of the Identification Time and Averaged for All Eight Subjects

		Mean(P)		RMSE (mm)	
		Off	**On**	**Off**	**On**
Time (s)	30	$0.91 \pm (0.08)$	$0.82 \pm (0.15)$	$69.14 \pm (20.57)$	$51.87 \pm (16.54)$
	60	$0.59 \pm (0.28)$	$0.46 \pm (0.21)$	$44.15 \pm (21.74)$	$38.64 \pm (18.17)$
	90	$0.35 \pm (0.25)$	$0.26 \pm (0.21)$	$37.90 \pm (23.95)$	$27.74 \pm (10.99)$
	120	$0.25 \pm (0.27)$	$0.12 \pm (0.07)$	$33.52 \pm (20.99)$	$24.87 \pm (9.89)$
	150	$0.20 \pm (0.29)$	$0.07 \pm (0.05)$	$32.33 \pm (21.97)$	$24.82 \pm (9.33)$
	180	$0.16 \pm (0.24)$	$0.05 \pm (0.03)$	$26.53 \pm (8.33)$	$23.09 \pm (7.87)$

Root mean square error of the CoM estimation and standard deviations are presented against the time progress. Convergence of CoM estimate appeared faster with the adaptive interface case. The CoM accuracy is also improved via the interface. For example, for the same RMSE error, almost half the time is required when the adaptive interface is provided.

supporting area is used for this purpose. We consider a motion stable when the ZRAM point remains inside the support polygon. If the ZRAM point gets closer to the edge of the supporting area or gets out of the range, we regard this situation as an unstable dynamic posture. This personalized balance assessment for home-based rehabilitation was first proposed and reported in [66].

In Fig. 8, we show the ZRAM point (p_f) transitions against the support polygon for dynamic motions during which the subject did not stand on top of the WBB, ie, CoP position is not available. With the proposed method, we can see the balance measure is still available even if the subject is not stepping on the WBB. The body mass distribution doesn't change every day, thus the personalized model is established from a one-time session with WBB. The personalized balance assessment is always available as long as we have the kinematic information from Kinect. The training without WBB is also safe for the users, especially for the elderly, to avoid presenting an

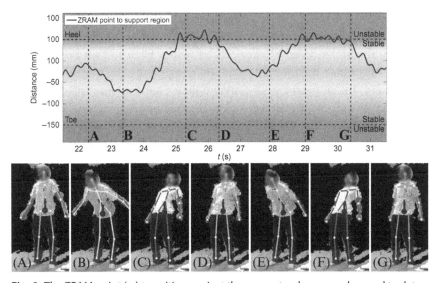

Fig. 8 The ZRAM point (p_f) transition against the support polygon can be used to determine the current balance during dynamic motions. The smaller distance situation between p_f and the edge of the support area indicates an unstable situation. This stability metric can be obtained in real time and shown to the subject. Starting from a stable position (A) the subject bends the trunk forward as in (B), where ZRAM is moved to toe side. Then, the color entered into yellow to show some instability. Then, bending the trunk backward like (C) and (F), the posture reaches high instability as shown in red as the ZRAM point gets closer to the heel side of the foot. It also distinguishes well from the modest level of instability during the backward situation of (D) as shown in orange. For interpretation of the references to color in this figure legend, see the online version of the book.

uneven floor during rehabilitation training. This aspect is quite important in rehabilitation practice. The ZRAM point is shown as a pink sphere in the ground plane. Starting from a stable position (A) the subject bends the trunk forward as in (B), where ZRAM is moved toward the toe. There, the color is yellow to show the situation is a little unstable. Then, bending the trunk backward like (C) and (F), the posture is highly unstable as shown by the color red as the ZRAM point gets closer to the heel side of the foot. This high instability area is distinguishable from the modest level of instability in the backward situation (D) as shown in orange. The metric ZRAM can reflect the effect of the personalized CoM's accelerations in dynamic motions. The faster backward action gives us lower stability information than the case of slower backward action; it is also sensitive to the stability level changes by different speeds.

7. DISCUSSION

In this work, we proposed a new method to perform the personalized modeling for home-based balance rehabilitation. Each subject body's mass distribution cannot be assumed in advance, especially for motor-impaired populations. In biomechanics, conventionally anthropometric tables are employed to compute the subject's center of mass by applying simple scaling. Such information is helpful and makes certain level's estimation if his body characteristics is close to the averaged body characteristics. However, the elderly or the motor-impaired can easily have a difference from the standard. Sometimes such a body difference can be a cause of motor-coordination problems. In this work, we tried to provide an efficient CoM identification interface. It makes it possible to identify personalized CoM with portable sensors for home usage, and in a self-directed manner to be done by the subject himself. We tested the hypothesis that an adaptive visual feedback during the SESC parameter identification may reduce the total time needed to establish a personalized CoM estimation. When they performed the identification procedure with the help of the adaptive interface, each subject was also able to see the color change and was asked to make the skeleton green. The addition of an adaptive feedback interface reduced the average RMSE values across all subjects and across all trials. It also decreased the average parameter covariance. In this way, the model could be established faster. To obtain less than RMSE of 30 mm, the identification time could be halved, going from 180 s without visual feedback to 90 s with it. Conversely, a lower error was found for the same identification time. For an identification session of 120 s in length, RMSE was of 24.8 mm when feedback was provided, and 33.5 mm when it was not provided.

We believe that the improvement in the identification accuracy is due to the increased amount of information that the subjects were able to present to the identification algorithm. The subjects were able to interpret the color-based feedback as an indication that he/she should perform a different posture to continue changing the color of the skeleton. In this way, the subject can use the adaptive interface tool to perform the parameter identification in a self-directed manner and in the home environment.

8. CONCLUSION

As the world population continues to age, there is an increased need for efficient systems that can improve rehabilitation standards for the elderly, poststroke patients, and other motor-impaired subjects. A self-training system that takes into account subject-specific differences would contribute in a positive way to personalized home rehabilitation protocols. After the SESC model is identified, a personalized CoM estimate can be obtained just by using Kinect information. The CoM estimated with the SESC model is also valid for dynamic motions. Thus, a personalized balance stability evaluation is available through the ZRAM point. We validated the use of our model to perform the personalized balance visualization. This chapter developed a personalized measure of balance that considers subject-specific variations. This is a step towards the creation of versatile and portable self-balance assessment tools. Fall risks and subject balance improvement during training would be accurately evaluated when considering subject-specificity. Balance evaluation studies with clinical application using the proposed method are scheduled as part of our future work.

REFERENCES

[1] J.A. Painter, S.J. Elliott, S. Hudson, Falls in community-dwelling adults aged 50 years and older: prevalence and contributing factors, J. Allied Health 38 (4) (2009) 201–207.
[2] J.E. Deutsch, D. Robbins, J. Morrison, P. Guarrera Bowlby, Wii-based compared to standard of care balance and mobility rehabilitation for two individuals post-stroke, in: Proceedings of the International Conference on Virtual Rehabilitation, Haifa, Israel, 2009, pp. 117–120.
[3] F.A. Langley, S.F.H. Mackintosh, Functional balance assessment of older community dwelling adults: a systematic review of the literature, Internet J. Allied Health Sci. Pract. 5 (4) (2007).

[4] T. Fujita, S. Nakamura, M. Ohue, Y. Fujii, A. Miyauchi, Y. Takagi, H. Tsugeno, Effect of age on body sway assessed by computerized posturography, J. Bone Mine. Metab. 23 (2) (2005) 152–156.

[5] M.W. Kennedy, J.P. Schmiedeler, C.R. Crowell, M. Villano, A.D. Striegel, J. Kuitse, Enhanced feedback in balance rehabilitation using the Nintendo Wii balance board, in: Proceedings of the IEEE International Conference on e-Health Networking, Applications and Services, Columbia, MO, USA, 2011, pp. 162–168.

[6] B. Lange, C.-Y. Chang, E. Suma, B. Newman, A.S. Rizzo, M. Bolas, Development and evaluation of low cost game-based balance rehabilitation tool using the Microsoft Kinect sensor, in: Proceedings of the Annual International Conference IEEE Engineering in Medicine and Biology Society (IEEE/EMBC), MA, USA, Boston, 2011, pp. 1831–1834.

[7] A.V. Dowling, O. Barzilay, Y. Lombrozo, A. Wolf, An adaptive home-use robotic rehabilitation system for the upper body, IEEE J. Trans. Eng. Health Med. 2 (2014) 1–10.

[8] B.-C. Lee, J. Kim, S. Chen, K.H. Sienko, Cell phone based balance trainer, J., Neuroeng. Rehabil. 9 (2012) 10.

[9] G. Mastorakis, D. Makris, Fall detection system using Kinect's infrared sensor, 2012.

[10] N.A. Borghese, M. Pirovano, R. Mainetti, P.L. Lanzi, An integrated low-cost system for at-home rehabilitation, in: Proceedings of the 2012 18th International Conference on Virtual Systems and Multimedia, VSMM 2012: Virtual Systems in the Information Society, 2012, pp. 553–556.

[11] E.E. Stone, M. Skubic, Evaluation of an inexpensive depth camera for passive in-home fall risk assessment, in: Proceedings of the International Conference on Pervasive Computing Technologies for Healthcare (PervasiveHealth), Dublin, Ireland, 2011, pp. 71–77.

[12] S. Patel, H. Park, P. Bonato, L. Chan, M. Rodgers, A review of wearable sensors and systems with application in rehabilitation, J. NeuroEng. Rehabil. 9 (2012) 21:1—21:17.

[13] R. Igual, C. Medrano, I. Plaza, Challenges, issues and trends in fall detection systems, Biomed. Eng. Online 12 (1) (2013) 66.

[14] M. van Diest, C.J.C. Lamoth, J. Stegenga, G.J. Verkerke, K. Postema, Exergaming for balance training of elderly: state of the art and future developments, J. Neuroeng. Rehabil. 10 (1) (2013) 101. http://dx.doi.org/10.1186/1743-0003-10-101

[15] J. Fung, C.F. Perez, Sensorimotor enhancement with a mixed reality system for balance and mobility rehabilitation, in: Proceedings of the Annual International Conference IEEE Engineering in Medicine and Biology Society (IEEE/EMBC), vol. 2011, MA, Boston, 2011, pp. 6753–6757.

[16] C.G. Song, J.Y. Kim, N.G. Kim, A new postural balance control system for rehabilitation training based on virtual cycling, IEEE Trans. Inform. Technol. Biomed. 8 (2) (2004) 200–207.

[17] A.A. Rendon, E.B. Lohman, D. Thorpe, E.G. Johnson, E. Medina, B. Bradley, The effect of virtual reality gaming on dynamic balance in older adults, Age Ageing 41 (4) (2012) 549–552.

[18] F. Kamieth, P. Dähne, R. Wichert, J.L. Villalar, V. Jimenez-Mixco, A. Arca, M.T. Arredondo, Exploring the Potential of Virtual Reality for the Elderly and People With Disabilities, InTech, Croatia, 2010.

[19] L. Czerwosz, J. Blaszczyk, M. Mraz, M. Curzytek, Application of virtual reality in postural stability rehabilitation, in: Proceedings of the Virtual Rehabilitation International Conference, Haifa, Israel, 2009, p. 214.

[20] A. Dutta, U. Lahiri, A. Das, M.A. Nitsche, D. Guiraud, Post-stroke balance rehabilitation under multi-level electrotherapy: a conceptual review, Front. Neurosci. 8 (403) (2014). http://dx.doi.org/10.3389/fnins.2014.00403.

[21] M.E. Hahn, L.S. Chou, Can motion of individual body segments identify dynamic instability in the elderly? Clin. Biomech. 18 (8) (2003) 737–744.

[22] E. Harless, The static moments of human limbs [Die statischen Momente der menschlichen Gliedmassen], Abhandlungen der Bayerischen Akademie der Wissenschaften, Mathematisch-Physikalische Klasse 8 (2) (1857).

[23] W. Braune, O. Fischer, On the Centre of Gravity of the Human Body, Springer, Berlin, Heidelberg, Germany, 1985.

[24] V.M. Zatsiorsky, Kinetics of Human Motion, Human Kinetics, Champaign, IL, 2002.

[25] D.A. Winter, Biomechanics and Motor Control of Human Movement, Processing 2nd (Book, Whole), 1990. p. 277. http://doi.wiley.com/10.1002/9780470549148.

[26] P. de Leva, Adjustments to Zatsiorsky-Seluyanov's segment inertia parameters, J. Biomech. 29 (9) (1996) 1223–1230.

[27] M.J. Pavol, T.M. Owings, M.D. Grabiner, Body segment inertial parameter estimation for the general population of older adults, J. Biomech. 35 (5) (2002) 707–712.

[28] T.C. Pataky, V.M. Zatsiorsky, J.H. Challis, A simple method to determine body segment masses in vivo: reliability, accuracy and sensitivity analysis, Clin. Biomech. 18 (4) (2003) 364–368.

[29] M.A. Jaffrey, Estimating centre of mass trajectory and subject-specific body segment parameters using optimisation approaches, PhD thesis, Victoria University, Melbourne, Australia, 2008.

[30] G. Venture, K. Ayusawa, Y. Nakamura, Real-time identification and visualization of human segment parameters, in: Proceedings of the Annual International Conference IEEE Engineering in Medicine and Biology Society (IEEE/EMBC), Minneapolis, MN, 2009, pp. 3983–3986.

[31] K. Ayusawa, G. Venture, Y. Nakamura, Identifiability and identification of inertial parameters using the underactuated base-link dynamics for legged multibody systems, Int. J. Robot. Res. 33 (3) (2013) 446–468.

[32] M. Hayashibe, G. Venture, K. Ayusawa, Y. Nakamura, Muscle strength and mass distribution identification toward subject-specific musculoskeletal modeling, in: Proceedings of the IEEE/RSJ International Conference on Intelligent Robots and Systems (IROS), CA, San Francisco, 2011, pp. 3701–3707.

[33] M.T.G. Pain, J.H. Challis, High resolution determination of body segment inertial parameters and their variation due to soft tissue motion, J. Appl. Biomech. 17 (4) (2001) 326–334.

[34] J. Norton, N. Donaldson, L. Dekker, 3D whole body scanning to determine mass properties of legs, J. Biomech. 35 (1) (2002) 81–86.

[35] B.M. Nigg, W. Herzog, Biomechanics of the Musculo-Skeletal System, second ed., John Wiley & Sons, Inc., Chichester, England, 2002.

[36] J.B. Saunders, V.T. Inman, H.D. Eberhart, The major determinants in normal and pathological gait, J. Bone Joint Surg. Am. 35 (3) (1953) 543–558.

[37] R.R. Caron, R.C. Wagenaar, C.L. Lewis, E. Saltzman, K.G. Holt, Center of mass trajectory and orientation to ankle and knee in sagittal plane is maintained with forward lean when backpack load changes during treadmill walking, J. Biomech. 46 (1) (2013) 70–76.

[38] R. Baker, Gait analysis methods in rehabilitation, J. NeuroEng. Rehabil. 3 (2006) 4:1–4:10.

[39] M. Goffredo, M. Schmid, S. Conforto, T. D'Alessio, A markerless sub-pixel motion estimation technique to reconstruct kinematics and estimate the centre of mass in posturography, Med. Eng. Phys. 28 (7) (2006) 719–726.

[40] D. Lafond, M. Duarte, F. Prince, Comparison of three methods to estimate the center of mass during balance assessment. J. Biomech. 37 (9) (2004) 1421–1426, http://dx.doi.org/10.1016/S0021-9290(03)00251-3.

[41] D.A. Winter, Human balance and posture standing and walking control during, Gait Posture 3 (1995) 193–214.

[42] D.P. Ferris, M. Louie, C.T. Farley, Running in the real world: adjusting leg stiffness for different surfaces, Proc. R. Soc. Biol. Sci. 265 (1400) (1998) 989–994.

[43] O. Caron, B. Faure, Y. Brenière, Estimating the centre of gravity of the body on the basis of the centre of pressure in standing posture, J. Biomech. 30 (11/12) (1997) 1169–1171.

[44] H.M. Schepers, E.H.F. van Asseldonk, J.H. Buurke, P.H. Veltink, Ambulatory estimation of center of mass displacement during walking, IEEE Trans. Biomed. Eng. 56 (4) (2009) 1189–1195.

[45] M. Zok, C. Mazzà, U. Della Croce, Total body centre of mass displacement estimated using ground reactions during transitory motor tasks: application to step ascent, Med. Eng. Phys. 26 (9) (2004) 791–798.

[46] V.M. Zatsiorsky, D.L. King, An algorithm for determining gravity line location from posturographic recordings, J. Biomech. 31 (2) (1998) 161–164.

[47] P. Sardain, G. Bessonnet, Zero moment point—measurements from a human walker wearing robot feet as shoes, IEEE Trans. Syst. Man Cybern. A Syst. Hum. 34 (5) (2004) 638–648.

[48] S. Cotton, A.P. Murray, P. Fraisse, Estimation of the center of mass: from humanoid robots to human beings, IEEE/ASME Trans. Mechatronics 14 (6) (2009) 707–712.

[49] S. Cotton, M. Vanoncini, P. Fraisse, N. Ramdani, E. Demircan, A.P. Murray, T. Keller, Estimation of the centre of mass from motion capture and force plate recordings: a study on the elderly, Appl. Bionics Biomech. 8 (1) (2011) 67–84.

[50] B. Espiau, R. Boulic, On the computation and control of the mass center of articulated chains, 1998. Tech. Rep. 3479, Inria, Grenoble, France.

[51] V. Bonnet, A. González, C. Azevedo-Coste, M. Hayashibe, S. Cotton, P. Fraisse, Determination of subject specific whole-body centre of mass using the 3D statically equivalent serial chain, Gait Posture 41 (1) (2015) 70–75, http://dx.doi.org/10.1016/j.gaitpost.2014.08.017.

[52] M.J. Floor-Westerdijk, H.M. Schepers, P.H. Veltink, E.H.F. van Asseldonk, J.H. Buurke, Use of inertial sensors for ambulatory assessment of center of mass displacements during walking, IEEE Trans. Biomed. Eng. 59 (7) (2012) 2080–2084.

[53] A. González, M. Hayashibe, V. Bonnet, P. Fraisse, Whole body center of mass estimation with portable sensors: using the statically equivalent serial chain and a kinect, Sensors 14 (9) (2014) 16955–16971.

[54] B.W. Mooring, Z.S. Roth, M.R. Driels, Fundamentals of Manipulator Calibration, John Wiley & Sons, Inc., New York, NY, 1991.

[55] D. Simon, Optimal State Estimation: Kalman H Infinity, and Non Linear Approaches, John Wiley & Sons, Inc., New York, NY, 2006.

[56] M. Vukobratovic, B. Borovac, Zero-moment point—thirty five years of its life, Int. J. Hum. Robot. 1 (1) (2004) 157–173.

[57] P. Sardain, G. Bessonnet, Forces acting on a biped robot. Center of pressure—zero moment point, IEEE Trans. Syst. Man Cybern. A Syst. Hum. 34 (5) (2004) 630–637.

[58] A. Goswami, Postural stability of biped robots and the foot-rotation indicator (FRI) point, Int. J. Robot. Res. 18 (6) (1999) 523–533.

[59] A. Goswami, V. Kallem, Rate of change of angular momentum and balance maintenance of biped robots, in: Proceedings of the IEEE International Conference on Robotics and Automation (ICRA), New Orleans, LA, 2004, pp. 3785–3790.

[60] S. Kajita, H. Hirukawa, K. Harada, K. Yokoi, Introduction à la Commande des Robots Humanoïdes, Springer-Verlag France, Paris, France, 2009.

[61] PrimeSense, Inc., Open-source SDK for 3D sensors—OpenNI, http://www.openni. org/.

[62] R. Pavlik, WiiUse: main page, http://www.vrac.iastate.edu/vancegroup/docs/ wiiuse/.

[63] A. González, M. Hayashibe, P. Fraisse, Estimation of the center of mass with kinect and Wii balance board, in: Proceedings of the IEEE/RSJ International Conference on Intelligent Robots and Systems (IROS), Vilamoura, Alagarve, Portugal, 2012, pp. 1023–1028.

[64] A. González, M. Hayashibe, P. Fraisse, Online identification and visualization of the statically equivalent serial chain via constrained Kalman filter, in: Proceedings of the IEEE International Conference on Robotics and Automation (ICRA), Karlsruhe, Germany, 2013, pp. 5303–5308.

[65] A. Gonzalez, P. Fraisse, M. Hayashibe, Adaptive interface for personalized center of mass self-identification in home rehabilitation, IEEE Sens. J. 15 (5) (2015) 2814–2823, http://dx.doi.org/10.1109/JSEN.2014.2379431.

[66] A. Gonzalez, P. Fraisse, M. Hayashibe, A personalized balance measurement for home-based rehabilitation, in: Proceedings of the 2015 Seventh International IEEE/EMBS Conference on Neural Engineering (NER), 2015, pp. 711–714, http://dx.doi.org/ 10.1109/NER.2015.7146722.

CHAPTER FIVE

Modeling and Dynamic Optimization of a Hybrid Neuroprosthesis for Gait Restoration

N. Sharma, N. Kirsch
University of Pittsburgh, Pittsburgh, PA, United States

1. INTRODUCTION

Spinal cord injury (SCI) or other neurological disorders such as stroke and traumatic brain injury often impair lower extremity function such as standing and walking. Potentially, functional electrical stimulation (FES) can be used to restore lower limb function in individuals with paraplegia, allowing them to regain the ability to walk again using their own muscles [1–7]. The Parastep system (Sigmedics, Inc., Fairborn, OH), an FDA approved FES device, is one of the examples that has been successfully shown to restore standing and walking function in individuals with an SCI [8]. Among many challenges, the rapid onset of muscle fatigue during FES [9] remains the main challenge. Although build up in muscle fatigue during FES can be reduced by training the muscles via neuromuscular electrical stimulation [10], or by using appropriate stimulation waveforms [11], the current achievable duration of walking is still not sufficient for activities of daily living. Therefore, interventions must be developed that are more resistant to muscle fatigue.

One strategy for compensating for muscle fatigue is to combine FES with orthosis, such as a reciprocating gait orthosis [12–14], controlled-brake orthosis [15], variable coupling hip reciprocal orthosis [16], or joint couple orthosis [17]. An orthotic brace can lower metabolic fatigue by transferring load from the arms to the exoskeleton, which supports a user's body weight, and reducing FES-induced muscle fatigue by supporting the standing phase, which removes the requirement to stimulate the quadriceps muscle. It can also aid by restricting the degrees of freedom of the lower limbs and increasing the stability of a user. Another way to reduce muscle fatigue is to design joint-angle trajectories that require less muscle stimulation. Joint-angle trajectories of an able-bodied person can also be employed for closed-loop tracking [14,18–20] but imposing these trajectories on a subject using an FES/orthosis system may not be optimal. This is due to the fact that joint torques or motion elicited via FES in persons with an SCI, when compared to volitional contractions in able-bodied persons, is restricted or reduced [10]. A walking system that combines a passive orthosis with FES has the disadvantage that it cannot add power, especially when the muscle fatigue caused by FES sets in and leads to reduction in available muscle joint torque.

Powered exoskeleton technology uses an electric motor [21–23] or a pneumatic or hydraulic power source to move the lower-limb joints [24,25]. Recently, portable exoskeletons that use an electric motor at each lower-limb joint are increasingly being explored for walking restoration [21–23,26] and gait rehabilitation [27]. However, the ability of an electric motor-based powered exoskeleton to sustain walking for a longer time period depends on its power source's capacity; ie, an extended use will need to house a larger, heavier battery.

A hybrid technology that combines an FES system with a powered exoskeleton [16,28–31] seems most promising for use as an assistive device. With such a system, the limitations of FES such as limited and unreliable muscle torque generation can be overcome by using an electric motor while using FES intermittently to get either its physiological benefits or to conserve the battery charge required for an electric motor, thus allowing smaller size and weight of the battery. Among existing electric motor-based powered exoskeletons, namely, the Vanderbilt exoskeleton [21], ReWalk exoskeleton [32], Mina exoskeleton [22], and EKSO (Ekso Bionics, Richmond, CA) [23]. To this point, FES has been combined with the Vanderbilt exoskeleton [30,31].

This chapter discusses the use of a forward dynamic optimization, which was performed on a three-link dynamic walking model, to calculate joint angle trajectories that require minimum electrical stimulation and/or motor torque inputs. Importantly, these trajectories were computed without tracking any nominal trajectory or given able-bodied trajectories. In the subsequent section, the dynamic walking model that represents walking using a powered or unpowered orthosis, FES, and an assistive device such as a walker is presented. The key feature of the dynamic model is that it is a low degree of freedom (DOF) walking model which eases computational time required to perform dynamic optimization. Furthermore, optimization results of different actuation strategies that utilize FES or a combination of

FES and an electric motor to generate a walking step are depicted. The computed optimal results of these stimulation strategies are compared, and the feasibility of these strategies under various scenarios are discussed. Moreover, the optimal joint angles computed for each strategy were compared with joint angles of able-bodied persons [33] to illustrate how they differ.

2. DYNAMIC MODEL

In [34], a low DOF planar model that represents walking generated with a knee ankle foot orthosis (KAFO), walker, and FES (see Fig. 1) was developed. The model consists of four phases: initial double support phase (DSP), single support phase (SSP), impact, and final DSP. Because a user will use a walker as an aid, a support force was added at the hip joint. Also, as the user's arms can stabilize the upper body during walking, the dynamics of the trunk were neglected in the model, ie, only the lower extremities were modeled. However, the dynamics of the upper body were modeled as a point mass at the hip joint. The following subsections briefly describe or derive the complete model and the resulting equations of motion for each phase. A more detailed derivation of the model is given in [34].

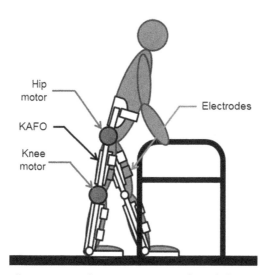

Fig. 1 Assistive walking generated via an FES, powered exoskeleton, and a walker. This is considered a hybrid neuroprosthesis, because it uses a combination of FES and electric motors to produce the gait motion.

2.1 Initial Double Support Phase

It is assumed that the user begins walking with both feet flat on the ground, as shown in Fig. 2. This phase is defined as the initial DSP. In this phase a KAFO is assumed to lock both knee joints, allowing both legs to be considered a single segment. The model can rotate about the hip joint, ankle joints, and at the toe of the trailing leg. Because walking during this phase can be represented as a closed-link chain, the system has a single DOF and was parameterized by the stance leg angle q_{d_1}. To parameterize the model in one angle, the following kinematic loop-closure equations were utilized:

$$
\begin{aligned}
l_f \cos\left(q_{d_f}\right) + l\cos\left(q_{d_2}\right) &= l_{0x} + l\cos\left(q_{d_1}\right) \\
l_f \sin\left(q_{d_f}\right) + l\sin\left(q_{d_2}\right) &= l_{0y} + l\sin\left(q_{d_1}\right),
\end{aligned}
\tag{1}
$$

where q_{d_f} and q_{d_2} are the angles of the foot and ankle of the swing leg, respectively. The parameters $l_f, l, l_{0x}, l_{0y} \in \mathbb{R}^+$ are the length of the foot, length of the leg from ankle joint to hip joint, horizontal distance between ankle joints, and vertical distance between ankle joints, respectively. After differentiating Eq. (1) with respect to time, the following angular velocity expression was obtained:

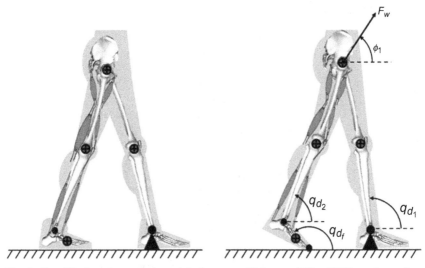

Fig. 2 Simplified mathematical model of a user utilizing an FES, a KAFO, and a walker during the initial DSP. The left image shows the initial position of the model at the beginning of a step, and the right image shows the degrees of freedom and inputs to the system.

$$\begin{bmatrix} l_f \sin(q_f) & l \sin(q_{d_2}) \\ l_f \cos(q_{d_f}) & l \cos(q_{d_2}) \end{bmatrix} \begin{bmatrix} \dot{q}_{d_f} \\ \dot{q}_{d_2} \end{bmatrix} = \begin{bmatrix} l \sin(q_{d_1}) \\ l \cos(q_{d_1}) \end{bmatrix} \dot{q}_{d_1}. \tag{2}$$

Using the Lagrangian formulation, Eqs. (1) and (2), the equation of motion in the initial DSP was derived as

$$J\ddot{q}_{d_1} = C(q_{d_1}, \dot{q}_{d_1}) + G(q_{d_1}) + M_w, \tag{3}$$

where J is the inertial term, $C(q_{d_1}, \dot{q}_{d_1})$ is the Coriolis term, $G(q_{d_1})$ is the gravitational term, and M_w, which is expressed as

$$M_w = F_w l \sin(\phi_1 - q_{d_1}),$$

is the moment generated by the arm reaction force F_w that acts at the hip. The details of the inertial, Coriolis, and gravitational terms are given in the Appendix.

2.2 Single Support Phase

A three-link model was developed to describe motion during the SSP (see Fig. 3). In this phase, the stance leg was assumed to be locked by a KAFO, which means it was considered as the single segment link. The stance leg was modeled to rotate at its ankle joint, while the corresponding

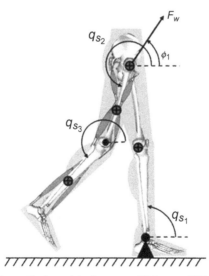

Fig. 3 Simplified mathematical model of a user utilizing an FES, a KAFO, and a walker during the SSP.

foot was assumed to be fixed to the ground. A KAFO on the swing leg was assumed to be unlocked, which means that the swing leg was modeled as two segments. The ankle of the swing leg was modeled as fixed, while knee and hip joints of the swing leg were modeled as rotating. During the SSP torque generation for knee flexion/extension and hip flexion/extension of the swing leg were modeled as an FES actuation. The Newton-Euler method was used to derive the following equations of motion:

$$\mathbf{I}\ddot{\Theta} = \mathbf{C}(\Theta,\dot{\Theta})\dot{\Theta}^2 + \mathbf{G}(\Theta) + \mathbf{T}. \tag{4}$$

In Eq. (4) the vectors Θ, $\dot{\Theta}^2$, and $\ddot{\Theta}$ are defined as

$$\Theta = \begin{bmatrix} q_{s_1} \\ q_{s_2} \\ q_{s_3} \end{bmatrix}, \quad \dot{\Theta}^2 = \begin{bmatrix} \dot{q}_{s_1}^2 \\ \dot{q}_{s_2}^2 \\ \dot{q}_{s_3}^2 \end{bmatrix}, \quad \text{and} \quad \ddot{\Theta} = \begin{bmatrix} \ddot{q}_{s_1} \\ \ddot{q}_{s_2} \\ \ddot{q}_{s_3} \end{bmatrix},$$

where q_{s_1}, q_{s_2}, and q_{s_3} are the stance leg angle, thigh angle, and shank angle, respectively. The matrix $\mathbf{I} \in \mathbb{R}^{3\times3}$ is the symmetric inertia matrix whose scalar components are defined as

$$I_{11} = J_1 + m_1 l_{1c}^2 + (m_2 + m_3 + m_h)l_1^2$$
$$I_{22} = J_2 + m_2 l_{2c}^2 + m_3 l_2^2$$
$$I_{33} = J_3 + m_3 l_{3c}^2$$
$$I_{12} = I_{21} = -(m_2 l_1 l_{2c} + m_3 l_2 l_1)\cos(q_{s_1} - q_{s_2})$$
$$I_{13} = I_{31} = -m_3 l_{3c} l_1 \cos(q_{s_1} - q_{s_3})$$
$$I_{23} = I_{32} = m_3 l_2 l_{3c} \cos(q_{s_2} - q_{s_3}).$$

In the inertia matrix, the terms $J_1, J_2, J_3 \in \mathbb{R}^+$ are defined in the Appendix. The mass parameters $m_1, m_2, m_3, m_f, m, m_h \in \mathbb{R}^+$ are the mass of the stance leg, mass of the thigh, mass of the shank, mass of the foot, mass of both legs, and the point mass representing the head, arms, and trunk. The length parameters $l_{1c}, l_{2c}, l_{3c} \in \mathbb{R}^+$ are the distances to the center of mass of the stance leg from the ankle, the thigh from the hip, and the shank from the knee.

The matrix $\mathbf{C}(\Theta) \in \mathbb{R}^{3\times3}$ in Eq. (4) is the Coriolis matrix, whose primary diagonal components C_{11}, C_{22}, and C_{33} are each zero. The off-diagonal components of the Coriolis matrix are defined as

$$C_{12} = C_{21} = (m_2 l_1 l_{2c} + m_3 l_2 l_1)\sin(q_{s_1} - q_{s_2})$$
$$C_{23} = -C_{32} = -m_3 l_2 l_{3c}\sin(q_{s_2} - q_{s_3})$$
$$C_{13} = m_3 l_{3c} l_1 \sin(q_{s_1} - q_{s_3})$$
$$C_{31} = m_3 l_{3c} l_1 \sin(q_{s_1} - q_{s_2}).$$

The vector $\mathbf{G}(\Theta) \in \mathbb{R}^3$ in Eq. (4) contains the gravitational components of the dynamics, and is defined as

$$\mathbf{G}(\Theta) = \begin{bmatrix} (m_1 l_{1c} + m_h l_1 + m_3 l_1 + m_2 l_1)g\cos(q_{s_1}) \\ -m_2 g l_{2c}\cos(q_{s_2}) - m_3 g l_2 \cos(q_{s_3}) \\ -m_3 g l_{3c}\cos(q_{s_3}) \end{bmatrix}.$$

Finally, $\mathbf{T} \in \mathbb{R}^3$ in Eq. (4) is the joint torque vector that is defined as

$$\mathbf{T} = \begin{bmatrix} M_h - M_w \\ M_h - M_k \\ M_k \end{bmatrix}, \tag{5}$$

where M_w is the moment created by the arm reaction force F_w, M_h is the moment acting on the hip joint, and M_k is the moment acting on the knee joint. The terms M_h and M_k in Eq. (5) are the moments due to the joint torque generated by external electrical stimulation of hip flexors/extensors and knee extensors/flexors, respectively.

The moment acting on the hip joint and the moment acting on the knee joint are defined as

$$M_h = M_{hf} - M_{he} - M_{hr}, \quad M_k = M_{ke} - M_{kf} - M_{kr}, \tag{6}$$

where M_{hr} and M_{kr} are the passive/resistive moments generated by the muscles and are defined as

$$M_{hr} = \Gamma_{hr}(q_{s_2}, \dot{q}_{s_2}), M_{kr} = \Gamma_{kr}(q_{s_2}, q_{s_3}, \dot{q}_{s_2}, \dot{q}_{s_3}). \tag{7}$$

In Eq. (7), the functions Γ_{hr} and Γ_{kr} are functions that model the stiffness and damping at the hip and knee joints, respectively. For more details on these functions see [20]. M_{hf}, M_{he}, M_{kf}, and M_{ke} in Eq. (6) are active hip flexion/extension torques and knee flexion/extension torques, respectively, and are defined as

$$M_i = \Omega_i(q_{s_2})\Psi_i(\dot{q}_{s_2})a_i \quad \text{for } i = he, hf \tag{8}$$
$$M_j = \Omega_j(q_{s_2}, q_{s_3})\Psi_j(\dot{q}_{s_2}, \dot{q}_{s_3})a_j \quad \text{for } j = ke, kf, \tag{9}$$

where $\Omega_{i,j}(\cdot), \Psi_{i,j}(\cdot) \in \mathbb{R}$ are the torque equivalents of the muscle force-length and muscle force-velocity relationships, respectively. In Eqs. (8) and (9), $a_{i,j} \in \mathbb{R}$ are the first-order muscle activation dynamics of the hip flexors/extensors and the knee flexors/extensors, which were modeled as

$$
\dot{a}_k = \begin{cases} (u_k - a_k)(\tau_1 u_k + \tau_2(1 - u_k)), & u_k \geq a_k \\ (u_k - a_k)\tau_2, & u_k < a_k \end{cases}, \quad k = hf, he, kf, ke, \quad (10)
$$

where τ_1 and τ_2 are activation and deactivation constants, and u_k are the normalized stimulation levels. For more details on Eqs. (8)–(10) see [20].

The normalized stimulation levels u_k in Eq. (10) are defined with the following piecewise function:

$$
u_k = \begin{cases} 0, & V_k < V_t \\ \dfrac{V_k - V_t}{V_{sat} - V_t}, & V_{sat} \geq V_k \geq V_t, \\ 1, & V_k > V_{sat} \end{cases} \quad (11)
$$

where V_k is the muscle stimulation intensity, V_t is the muscle stimulation threshold level, and V_{sat} is the muscle stimulation saturation level. For an applied voltage less than V_t the muscle will not produce any force, and for an applied voltage greater than V_{sat} the muscle force will not increase. The manner in which the stimulation is selected depends on the stimulation strategy used and is explained in the next section.

2.3 Final Double Support and Impact Phase

After the completion of the swing phase and just before the start of the final DSP, the impact of the swing leg with the ground was accounted for in the model. Since the final DSP model had one degree of freedom, q_{d_1} was the only unknown variable (q_{d_1} was computed from the model geometry). With the assumption that the angular momentum about the impact point is conserved [35]), $q_{d_1}^+$ was computed from the following algebraic equation:

$$
\rho q_{d_1}^+ = H^-, \quad (12)
$$

where $\rho q_{d_1}^+$ is the angular momentum after the impact and H^- is the angular momentum before impact which is defined as

$$
\begin{aligned}
H^- = &(l_{1c}\bar{l}_2 m_1 + l_1 \bar{l}_{2c}\bar{m}_1 + l_1 \bar{l}_2 m_h) \cos(q_{s_1}^- - q_{s_2}^-)\dot{q}_{s_1}^- \\
&- l_{1c} l_{1c}^* m_1 - \bar{l}_{2c} l_{2c}^* m_1 - \bar{l}_{2c} l_{2c}^* \bar{m}_1 \dot{q}_{s_2}^-,
\end{aligned} \quad (13)
$$

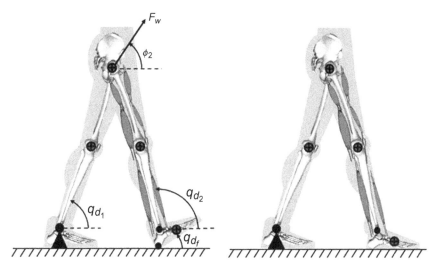

Fig. 4 Simplified mathematical model of a user utilizing an FES, a KAFO, and a walker during the final DSP. The left image shows the degrees of freedom and inputs to the system, and the right image shows the terminal position of the model.

where $\bar{l}_2 = l_2 + l_3$, $\bar{l}_{2c} = \bar{l}_2 - l_{2c}^*$, $l_{1c}^* = l_1 - l_{1c}$. The details of ρ are provided in the Appendix.

The final DSP was modeled as shown in Fig. 4. The model had joints at the hip, both ankles, and heel and its equation of motion was derived as in Eq. (4). In Fig. 4, q_{d_2} is the angle of the swing leg after the heel strike, q_{d_f} is the angle of the foot in the swing leg. The equation of motion was parameterized by only one angle q_{d_1}, which is the angle of the stance leg. In the final DSP, the arm reaction force F_w was applied at an angle ϕ_2 on the hip.

3. DYNAMIC OPTIMIZATION

To compute optimal joint-angle trajectories and their corresponding control inputs for different actuation strategies (explained in the next section), two different cost functionals, defined as

$$\Pi_1 = \int_{t_0}^{t_1} w_1 \bar{F}_w^2 dt + \int_{t_1}^{t_2} \left(w_1 \bar{F}_w^2 + \sum_i w_i u_i^2 \right) dt + \int_{t_2}^{t_f} w_1 \bar{F}_w^2 dt, \qquad (14)$$

$$\Pi_2 = \int_{t_0}^{t_1} w_1 \bar{F}_w^2 dt + \int_{t_1}^{t_2} \left(w_1 \bar{F}_w^2 + \sum_i w_i u_i^2 + \sum_k w_k M_k^2 \right) dt$$
$$+ \int_{t_2}^{t_f} w_1 \bar{F}_w^2 dt, \tag{15}$$

were used. Note that no tracking type optimal control was set up to compute the joint angle trajectories. In Eqs. (14) and (15), u_i, $(i = h_f, h_e, k_f, k_e)$ are the normalized muscle stimulation levels of hip flexors, hip extensors, knee flexors, and knee extensors. $\bar{F}_w$ is the normalized arm reaction force profile, M_i denotes either a joint moment generated by stimulation of a muscle group or motor torque, t_0 is the initial time of the step, t_1 is the end time of the initial DSP, t_2 is the end time of the SSP, and t_f is the time taken to complete one step. The optimization problem was subject to dynamic constraints in Eqs. (3), (4), and (12), a dynamic equation for final DSP (similar to Eq. 3), and the constraints explained in the following subsection. Note that t_1, t_2, and t_f were not fixed; $w_{i,k}$ in Eqs. (14) and (15) are the weights for the control inputs.

3.1 Constraints

To obtain a realistic result, the optimizations were subject to the following endpoint constraints

$$q_d(t_0) = \alpha_1, \quad q_d(t_f) = \alpha_2,$$
$$\dot{q}_d(t_0) = \alpha_3, \quad \dot{q}_d(t_f) = \alpha_4, \tag{16}$$

where $\alpha_i (i = 1-4)$ are known arbitrary constants. These constraints specified the segment angle and segment angular velocities of the stance leg at the start and end of a step. The following constraints provided the upper and lower bounds on the time spent in each walking phase and control inputs to the model:

$$0 < [t_1 \ t_2 \ t_f]^T < \delta,$$
$$\gamma_1 < [F_w \ \bar{V}_h \ \bar{V}_k]^T < \gamma_2, \tag{17}$$

where $\gamma_1, \gamma_2 \in \mathbb{R}^3$ are known constant vectors, and $\delta \in \mathbb{R}^{3 \times 1}$ is a known constant vector. The constant vectors in Eq. (17) were chosen as $\delta = [2s \ 2s \ 2s]^T$, $\gamma_1 = -\gamma_2, \gamma_2 = [860N \ 2000 \ 2000]^T$. These bounds were chosen to prevent the optimization from providing solutions that are not realistic, eg, bounds on walking phase time should not exceed 2 s. Similarly, the bound on the arm reaction force was chosen such that the solution could not exceed the maximum weight of a person (eg, 860 N). The muscle stimulation intensity (arbitrary units) was chosen as 2000. During optimization, this bound on the stimulation intensity influences only its slope between the grid points. Moreover, the stimulation intensity is limited by the saturation value of a muscle model, as modeled in Eq. (11), which can be chosen as per the physiological limit (eg, maximum pulse width or stimulation voltage that can be given during FES). We chose this value arbitrarily, so a lower value can also be chosen. The optimization was also subject to the following constraints:

$$
\begin{aligned}
q_{s_2}(t) &< q_{s_3}(t) \quad \forall t \in (t_1, t_2), \\
\gamma_1(t), \gamma_2(t) &> 0 \quad \forall t \in (t_0, t_f),
\end{aligned}
\tag{18}
$$

where γ_1, γ_2 are the positions of the heel and toe of the swing leg relative to the ground. The first constraint in Eq. (18) prevented knee hyperextension during the swing phase, and the second constraint prevented the heel and the toe of the swing leg from going below the ground. Additional assumptions and constraints that are specific to each strategy are expressed in their corresponding subsection.

4. SIMULATIONS AND RESULTS

Different FES-based or hybrid (FES and electric motor-based) actuation strategies for generating a walking step were considered: (1) switching between antagonist pairs, (2) coactivation of antagonist pairs, and (3) hybrid joint actuation. In the following subsections, we explain these different actuation scenarios.

4.1 Switching Between Antagonist Pairs

In this switching strategy, optimization selects the stimulation levels for hip/knee flexion/extension such that antagonistic muscles are never stimulated simultaneously. For example, if the desired torque at the hip is in the direction of hip flexion, then only the hip flexion is stimulated, and vice versa. This muscle stimulation strategy during the swing phase was modeled as

$$
\begin{aligned}
V_{h_f} &= \bar{V}_h, \quad V_{h_e} = -\bar{V}_h, \\
V_{k_e} &= \bar{V}_k, \quad V_{k_f} = -\bar{V}_k,
\end{aligned}
\tag{19}
$$

where V_{h_f}, V_{h_e}, V_{k_e}, and V_{k_f} are the muscle stimulation intensities as defined in Eq. (11), and $\bar{V}_h$ and $\bar{V}_k$ are the computed (through optimization) muscle stimulation intensity profiles for the hip and knee muscles. For computing joint angle trajectories which use this strategy, the cost functional defined in Eq. (14) was considered.

4.2 Coactivation of Antagonist Pairs

In this strategy, antagonist pairs can be stimulated simultaneously. Using co-contractions of an antagonist pair for generating movement may not seem optimal (as more muscles are being activated to produce the movement), but there are some advantages to using co-contractions. For example, able-bodied subjects use co-contractions to increase joint stability and control [36,37]. Therefore, it is possible that using a co-contraction strategy for an FES aided gait could improve performance of an automatic control, which means more stability during walking for a user. However, the disadvantages of this strategy are apparent as well, as coactivation uses antagonistic muscles simultaneously so the amount of stimulation necessary to produce a gait motion would increase. This means that the muscles would be stimulated more in each walking step, thus increasing the rate of fatigue, which further decreases the walking duration. For computing joint angle trajectories that use this strategy, the cost functional defined in Eq. (14) was considered.

4.3 Hybrid Joint Actuation

Aforementioned strategies are difficult to implement with transcutaneous FES because hip muscles, which are harder to access through surface electrodes, would require stimulation using implanted FES electrodes. A gait can also be restored by stimulating only the peroneal nerve (a synergistic sensory nerve that evokes hip/knee flexion and dorsiflexion) and a quadriceps for knee extension [2,4,6,38,39]. One benefit of this method is simplified control as the number of stimulations are reduced, and the second benefit is that both of these stimulations can be elicited via surface FES electrodes. Therefore, we attempted simulations using only stimulation of hip flexors and knee extensors to generate walking. Since we are unaware of a mathematical model to represent peroneal nerve stimulation, we approximated this stimulation by hip flexor torque alone. Our simulation results were unable to meet the ground clearance constraints, and thus did not converge to an optimized solution. Therefore, to assist the FES of knee extensors and hip flexors, a powered actuation was incorporated to the hip and knee joints. This added actuation could be thought of as a motor that could be added to the joints, eg, adding a powered orthosis to assist FES. This strategy is similar to the aforementioned strategies, except the stimulation of the hip extensors and knee flexors are removed from the dynamics and the cost function. For computing joint angle trajectories that use this strategy, the cost functional defined in Eq. (15) was considered.

4.4 Results

Each optimization was performed using MATLAB's (version R2011a, MathWorks, Inc., USA) parallel computation toolbox, fmincon optimization routine, and ODE45 function for numerical solution of ordinary differential equations. The optimization code was run on a computer with a quad-core processor (Intel Corporation, USA). Simulations were run for each stimulation strategy. The hip and knee joint angles were calculated from the model, and the results are shown in Fig. 5A and B, respectively. Knee and hip joint angles of able-bodied individuals during walking were obtained from [33], and are compared with the computed optimal hip and knee angles of the simulations. Toe-off, for the leg whose joint angles are shown, occurs at 60% of the gait cycle for the able-bodied data and the simulations. A positive hip angle corresponds to hip flexion, while a negative hip angle corresponds to hip extension. Similarly for the knee, a positive knee angle corresponds to knee flexion, and a negative knee angle corresponds to knee extension. In the plots of the joint angles, as well as the following plots, when referring to the cost function used in coactivation, cost 1 will refer to the cost function that minimizes the stimulation and cost 2 will refer to the cost function that minimizes the moment.

From Fig. 5 it can be observed how the simulation joint angles differ from the able-bodied joint angles. The minimum hip angles that occur just before toe-off are larger for the simulations than they are for the able-bodied data. The knee of the able-bodied subject flexes to approximately 40 degrees before toe-off, which reduces the amount of hip extension necessary before toe-off. Also, since the torso is not modeled, the hip joint angles in the simulation are actually absolute angles, and not relative angles as are used in the able-bodied data. During gait an able-bodied subject would perform plantar flexion and then lean forward slightly before taking the step. This combination of plantar flexion and forward leaning would make the relative joint angle smaller than the absolute angle.

It can also be observed that the peak hip and knee angles are larger for all of the simulations than they are for the able-bodied data. This is a result of

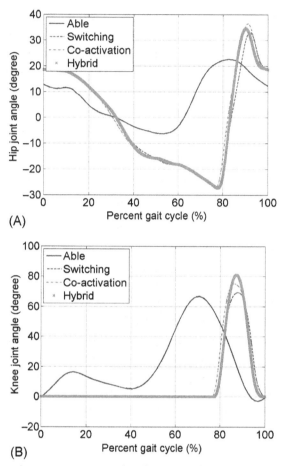

Fig. 5 (A) Hip angle vs. percent gait cycle. (B) Knee angle vs. percent gait cycle. The gait cycle begins and ends with heel strike.

the ankle of the swing leg in the model being fixed during swing, and being unable to perform dorsiflexion. Because the model cannot dorsiflex, the hip and knee flexions need to be larger to avoid ground contact. The orthoses that are used with FES aided gait typically have fixed ankle joints, and therefore would also not be able to dorsiflex. This makes it difficult for the foot of their swing leg to clear the ground. Using an able-bodied trajectory on an FES/orthosis aided gait would likely result in the foot of the swing leg dragging on the ground, due to the hip and knee flexion not being large enough.

The normalized stimulation levels and joint actuation moments that were calculated as a result of the simulations are shown in Figs. 6 and 7,

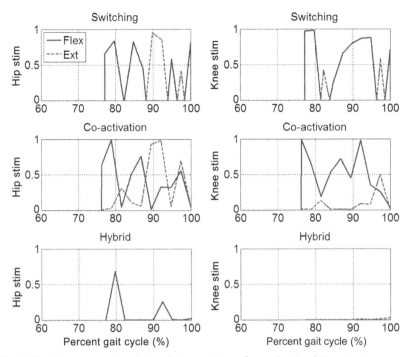

Fig. 6 Optimized normalized stimulation patterns for each stimulation strategy.

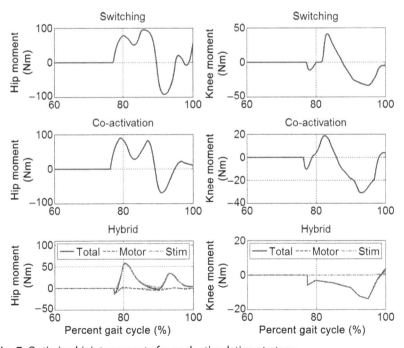

Fig. 7 Optimized joint moments for each stimulation strategy.

respectively. For the normalized stimulation levels shown in Fig. 6 the solid line indicates a stimulation to the flexor, while the dashed line indicates a stimulation to the extensor. For the plots of the moments for the switching and both coactivation strategies the moment that is plotted is the total muscle moment, which includes both the active and passive muscle elements. For the plots of the moments for the hybrid joint actuation strategy, the total moment that is shown is the sum of the muscle moment, both passive and active, and the ideal actuator moment. The total moment is broken down into the muscle and actuator moments for the hybrid strategy to better illustrate the contribution of each individual component.

On comparing the moments of the coactivation strategy to the moments of the switching strategy, it can be observed that the coactivation moments, at the knee and at the hip, are approximately half of the switching strategy moments. Therefore it can be concluded that using a coactivation strategy would decrease the required moments at the joints to achieve gait.

For the hybrid joint actuation strategy, the resulting stimulation intensities show that stimulation of the hip flexors is only required during the swing phase and just before heel strike. It can also be observed that the knee extension stimulation is required for a very brief period, towards the end of the gait phase, for this strategy. The electric motor at the knee primarily generates flexion torque, which allows the foot to clear the ground. Note that when the model was optimized using stimulation of only hip flexor and knee extensor muscles, a feasible gait was not achieved due to lack of foot clearance. The amount of work that is done by the electric motors, compared to the muscles, can be adjusted by changing the cost function. Increasing the weights on the stimulation would reduce the amount of stimulation used, and increase the amount of torque from electric motors to achieve gait. Proper tuning of the cost function weights could result in a hybrid joint actuation FES that achieves considerably longer walking durations.

5. CONCLUSION AND FUTURE WORK

In this chapter, we discussed optimal joint angle trajectories under different scenarios of joint actuation, such as using an FES combined with a passive orthosis or a combination of a powered orthosis and an

FES. These optimal trajectories were computed using a low DOF dynamic walking model, which eased the computational effort required to perform the dynamic optimization. Our optimization results depict that the optimally computed joint angle trajectories of the considered different stimulation strategies vary significantly from an able-bodied person's joint angle trajectories. This may imply that imposing able-bodied trajectories on subjects using FES/orthosis devices is not optimal. Also, typically a KAFO has fixed ankle joints, and therefore a user cannot dorsiflex his/her foot to clear ground contact. In this case, using an able-bodied trajectory on an FES/orthosis aided gait would likely result in foot-drag along the ground during the swing phase.

Our results also explored two different ways to stimulate antagonist pairs, namely switching stimulation between antagonist pairs and costimulating antagonist pairs. The switching and coactivation strategies resulted in similar joint angle trajectories; however, their corresponding stimulation and joint moment profiles were different. Although the coactivation strategy used more stimulation than the switching strategy, it resulted in lower joint moments. A disadvantage of coactivation strategy is excessive muscle stimulation, which could result in rapid muscle fatigue. However, it may result in a smoother and more natural gait. To explore this strategy, further experiments using implanted electrodes for FES aided walking can be conducted. Our results also show that a hybrid joint actuation can potentially achieve longer walking durations, as results depicted lower joint moments and muscle stimulation. Moreover, optimization simulations with hip flexion only and knee extension only could not converge to a solution. Only adding an electric motor torque to hip flexion and knee extension stimulation resulted in the optimal solution.

Overall, the dynamic optimization allowed us to explore different actuation strategies that can be potentially harnessed in walking rehabilitation research. Further, it was shown that the dynamic optimization can be applied under different actuation scenarios and can potentially provide insight to achieve longer walking durations and improved control.

APPENDIX

In Eq. (3) J is the inertial term, $C(q_{d_1})$ is the Coriolis term, and $G(q_{d_1})$ is the gravitational term. These are defined by the following equations:

$$J = 2(J_f S_f^2 + J_2 S_2^2 + J_1 + PS_f S_2)$$

$$C = -\Omega \dot{q}_{d_1}^2$$

$$G = G_\theta S_f + S_2 G_\alpha + G_\phi$$

$$J_f = \frac{1}{2}(m_f l_{fc}^2 + I_f + m_2 l_f^2)$$

$$J_2 = \frac{1}{2}(m l_c^2 + I)$$

$$J_1 = \frac{1}{2}(m l_c^2 + I + m_h l^2)$$

$$P = m l_f l_c \cos\left(q_{d_f} - 1_{d_2}\right)$$

$$G_\alpha = -mgl_c \cos\left(q_{d_2}\right)$$

$$G_\phi = -m_f gl_c \cos\left(q_{d_1}\right) - m_h gl \cos\left(q_{d_1}\right)$$

$$G_\theta = -(m_f l_{fc} + m l_f)g \cos\left(q_{d_1}\right)$$

$$\Omega = 2J_f S_f Z_f + 2J_2 S_2 Z_2 + P(Z_f S_2 + Z_2 S_f) + PS_f S_2 Z_1$$

$$Z_f = S_{f\theta} S_f + S_2 S_{fa} + S_{f\phi}$$

$$Z_2 = S_{2\theta} S_f + S_2 S_{2\alpha} + S_{2\phi}$$

$$Z_1 = \frac{l}{l_f} \sin\left(q_{d_2} - q_{d_1}\right) - \sin\left(q_{d_1} - q_{d_f}\right)$$

$$S_f = \frac{l \sin\left(q_{d_2} - q_{d_1}\right)}{l_f \sin\left(q_{d_2} - q_{d_f}\right)}$$

$$S_{f\theta} = \frac{l}{l_f} \cot\left(q_{d_f} - q_{d_2} \csc\left(q_{d_f} - q_{d_2}\right) \sin\left(q_{d_2} - q_{d_1}\right)\right)$$

$$S_{fa} = -\frac{l}{l_f} \csc\left(q_{d_f} - q_{d_2}\right)\left(\cos\left(q_{d_2} - q_{d_1}\right)\right.$$
$$\left. + \cot\left(q_{d_1} - q_{d_2}\right) \sin\left(q_{d_2} - q_{d_1}\right)\right)$$

$$S_{f\phi} = \frac{l}{l_f} \cos\left(q_{d_2} - q_{d_1}\right) \csc\left(q_{d_f} - q_{d_2}\right)$$

$$S_{2\theta} = \cot\left(q_{d_f} - q_{d_2}\right) \csc\left(q_{d_f} - q_{d_2}\right) \sin\left(q_{d_2} - q_{d_1}\right)$$

$$S_{2\alpha} = -\csc\left(q_{d_f} - q_{d_2}\right)\left(\cos\left(q_{d_2} - q_{d_1}\right)\right.$$
$$\left. + \cot\left(q_{d_f} - q_{d_2}\right) \sin\left(q_{d_2} - q_{d_1}\right)\right)$$

$$S_{2\phi} = \frac{\cos\left(q_{d_2} - q_{d_1}\right)}{\sin\left(q_{d_f} - q_{d_2}\right)}.$$

In the impact phase of the model it is assumed that angular momentum is conserved. Using this assumption the next state is solved for using the equation $\dot{q}_{d_1}^+ = H^-/\rho$, where ρ is expressed as

$$\rho = \frac{1}{2}\left(-2l_{fc}^2 m_f \frac{l_1 \sin\left(q_{d_2}^+ - q_{d_1}^+\right)}{l_f \sin\left(q_{d_2}^+ - q_{d_f}^+\right)}\right.$$

$$+ \left(-2(l_{1c}l_f m_1 + l_1 l_{fc} m_f)\cos\left(q_{d_f}^+\right)\cos\left(q_{d_2}^+\right)\right.$$

$$- l_{1c} m_1 \left(l_{1c} + l_{1c}^* + (l_{1c} - l_{1c}^*)\cos\left(2q_{d_2}^+\right)\right) - 2\left(l_{1c}^* l_f m_1\right.$$

$$+ l_1 l_{fc} m_f\right)\sin\left(q_{d_f}^+\right)\sin\left(q_{d_2}^+\right)\right)\frac{l_1 \sin\left(q_{d_1}^+ - q_{d_f}^+\right)}{l_f \sin\left(q_{d_2}^+ - q_{d_f}^+\right)}$$

$$+ 2\left(-l_{1c}l_1 m_1 + (l_{1c}l_f m_1\right.$$

$$+ l_1(l_{fc} m_f + l_f(m_1 + m_h)))\cos\left(q_{d_f}^+ - q_{d_1}^+\right)2(-l_{1c}l_{1c}^* m_1$$

$$+ (l_{1c}l_f m_1 + l_1(l_{fc} m_f + l_f(m_1 + M_h)))\cos\left(q_{d_f}^+ - q_{d_1}^+\right)$$

$$+ l_1(2l_{1c} m_1 + l_1 m_h)\cos\left(q_{d_2}^+ - q_{d_1}^+\right))).$$

REFERENCES

[1] A. Kralj, T. Bajd, Functional Electrical Stimulation: Standing and Walking After Spinal Cord Injury, CRC, Boca Raton, FL, 1989.

[2] R. Kobetic, R. Triolo, E. Marsolais, Muscle selection and walking performance of multichannel FES systems for ambulation in paraplegia, IEEE Trans. Rehabil. Eng. 5 (1) (1997) 23–29.

[3] T. Bajd, A. Kralj, R. Turk, H. Benko, J. Šega, The use of a four-channel electrical stimulator as an ambulatory aid for paraplegic patients, Phys. Ther. 63 (1983) 1116–1120.

[4] E. Marsolais, R. Kobetic, Functional electrical stimulation for walking in paraplegia, J. Bone Joint Surg. 69 (5) (1987) 728–733.

[5] E. Hardin, R. Kobetic, L. Murray, M. Corado-Ahmed, G. Pinault, J. Sakai, S. Bailey, C. Ho, R. Triolo, Walking after incomplete spinal cord injury using an implanted FES system: a case report, J. Rehabil. Res. Dev. 44 (3) (2007) 333–346.

[6] M. Granat, A. Ferguson, B. Andrews, M. Delargy, The role of functional electrical stimulation in the rehabilitation of patients with incomplete spinal cord injury—observed benefits during gait studies, Spinal Cord 31 (4) (1993) 207–215.

[7] C. Ho, R. Triolo, A. Elias, K. Kilgore, A. DiMarco, K. Bogie, A. Vette, M. Audu, R. Kobetic, S. Chang, M. Chan, S. Kukelow, D. Bourbeau, S. Brose, K. Gustafson, Z. Kiss, V. Mushahwar, Functional electrical stimulation and spinal cord injury, Phys. Med. Rehabil. Clin. 25 (3) (2014) 631–654.

[8] K. Klose, P. Jacobs, J. Broton, R. Guest, B. Needham-Shropshire, N. Lebwohl, M. Nash, B. Green, Evaluation of a training program for persons with SCI paraplegia using the Parastep 1 ambulation system: part 1. Ambulation performance and anthropometric measures, Arch. Phys. Med. Rehabil. 78 (8) (1997) 789–793.

[9] S.A. Binder-Macleod, L. Snyder-Mackler, Muscle fatigue: clinical implications for fatigue assessment and neuromuscular electrical stimulation, Phys. Ther. 73 (12) (1993) 902–910.

[10] E. Rabischong, F. Ohanna, Effects of functional electrical stimulation (FES) on evoked muscular output in paraplegic quadriceps muscle, Paraplegia 30 (1992) 467–473.

[11] Y. Laufer, J. Ries, P. Leininger, G. Alon, Quadriceps femoris muscle torques and fatigue generated by neuromuscular electrical stimulation with three different waveforms, Phys. Ther. 81 (7) (2001) 1307–1316.

[12] M. Solomonow, R. Baratta, H. Shoji, M. Ichie, S. Hwang, N. Rightor, W. Walker, R. Douglas, R. D'Ambrosia, FES powered locomotion of paraplegics fitted with the LSU reciprocating gait orthoses (RGO), in: Proc, IEEE EMBS 4 (1988) 1672.

[13] M. Solomonow, R. Baratta, S. Hirokawa, N. Rightor, W. Walker, P. Beaudette, H. Shoji, R. D'Ambrosia, The RGO generation II: muscle stimulation powered orthosis as a practical walking system for thoracic paraplegics, Orthopedics 12 (10) (1989) 13099–21315.

[14] D.B. Popović, M. Radulović, L. Schwirtlich, N. Jauković, Automatic vs hand-controlled walking of paraplegics, Med. Eng. Phys. 25 (1) (2003) 63–73.

[15] M. Goldfarb, W. Durfee, Design of a controlled-brake orthosis for FES-aided gait, IEEE Trans. Rehabil. Eng. 4 (1) (1996) 13–24.

[16] R. Kobetic, C. To, J. Schnellenberger, M. Audu, T. Bulea, R. Gaudio, G. Pinault, S. Tashman, R. Triolo, Development of hybrid orthosis for standing, walking, and stair climbing after spinal cord injury, J. Rehabil. Res. Dev. 46 (3) (2009) 447–462.

[17] R. Farris, H. Quintero, T. Withrow, M. Goldfarb, Design and simulation of a joint-coupled orthosis for regulating FES-aided gait, in: IEEE ICRA, 2009, pp. 1916–1922.

[18] S. Dosen, D.B. Popovic, Accelerometers and force sensing resistors for optimal control of walking of a hemiplegic, IEEE Trans. Biomed. Eng. 55 (8) (2008) 1973–1984.

[19] S. Dosen, D.B. Popovic, Moving-window dynamic optimization: design of stimulation profiles for walking, IEEE Trans. Biomed. Eng. 56 (5) (2009) 1298–1309.

[20] D. Popović, R. Stein, M. Oğuztöreli, M. Lebiedowska, S. Jonić, Optimal control of walking with functional electrical stimulation: a computer simulation study, IEEE Trans. Rehabil. Eng. 7 (1) (1999) 69–79.

[21] R. Farris, H. Quintero, M. Goldfarb, Preliminary evaluation of a powered lower limb orthosis to aid walking in paraplegic individuals, IEEE Trans. Neural Syst. Rehabil. Eng. 19 (6) (2011) 652–659.

[22] P. Neuhaus, J. Noorden, T. Craig, T. Torres, J. Kirschbaum, J. Pratt, Design and evaluation of mina: a robotic orthosis for paraplegics, in: IEEE ICORR, 2011, pp. 1–8.

[23] K. Strausser, H. Kazerooni, The development and testing of a Human Machine Interface for a mobile medical exoskeleton, in: IEEE/RSJ IROS, 2011, pp. 4911–4916.

[24] C.S. To, R. Kobetic, T.C. Bulea, M.L. Audu, J.R. Schnellenberger, G. Pinault, R. J. Triolo, Stance control knee mechanism for lower-limb support in hybrid neuroprosthesis, J. Rehabil. Res. Dev. 48 (7) (2011) 839.

[25] K.A. Shorter, G.F. Kogler, E. Loth, W.K. Durfee, E.T. Hsiao-Wecksler, A portable powered ankle-foot orthosis for rehabilitation, J. Rehabil. Res. Dev. 48 (4) (2011) 459–472.

[26] R. Farris, H. Quintero, S. Murray, K. Ha, C. Hartigan, M. Goldfarb, A preliminary assessment of legged mobility provided by a lower limb exoskeleton for persons with paraplegia, IEEE Trans. Neural Syst. Rehabil. Eng. 22 (3) (2014) 482–490.

[27] S.K. Banala, S.H. Kim, S.K. Agrawal, J.P. Scholz, Robot assisted gait training with active leg exoskeleton (ALEX), IEEE Trans. Neural Syst. Rehabil. Eng. 17 (1) (2009) 2–8.

[28] T.C. Bulea, R. Kobetic, M.L. Audu, J.R. Schnellenberger, R.J. Triolo, Finite state control of a variable impedance hybrid neuroprosthesis for locomotion after paralysis, IEEE Trans. Neural Syst. Rehabil. Eng. 21 (1) (2013) 141–151.

[29] A.J. del Ama, A.D. Koutsou, J.C. Moreno, A. de-los Reyes, A. Gil-Agudo, J. L. Pons, Review of hybrid exoskeletons to restore gait following spinal cord injury, J. Rehabil. Res. Dev. 49 (4) (2012) 497–514.

[30] K. Ha, S. Murray, M. Goldfarb, An approach for the cooperative control of FES with a powered exoskeleton during level walking for persons with paraplegia, IEEE Trans. Neural Syst. Rehabil. Eng. 24 (4) (2015) 455–466.

[31] K. Ha, H. Quintero, R. Farris, M. Goldfarb, Enhancing stance phase propulsion during level walking by combining FES with a powered exoskeleton for persons with paraplegia, in: IEEE EMBC, 2012, pp. 344–347.

[32] A. Esquenazi, M. Talaty, A. Packel, M. Saulino, The ReWalk powered exoskeleton to restore ambulatory function to individuals with thoracic-level motor-complete spinal cord injury, Am. J. Phys. Med. Rehabil. 91 (11) (2012) 911–921.

[33] D. Winter, Biomechanics and Motor Control of Human Movement, Wiley, New York, 2009.

[34] N. Sharma, V. Mushahwar, R. Stein, Dynamic optimization of FES and orthosis-based walking using simple models, IEEE Trans. Neural Syst. Rehabil. Eng. 22 (1) (2014) 114–126.

[35] M. Garcia, A. Chatterjee, A. Ruina, M. Coleman, The simplest walking model: stability, complexity, and scaling, J. Biomech. Eng. 120 (1998) 281–288.

[36] P. Gribble, L. Mullin, N. Cothros, A. Mattar, Role of cocontraction in arm movement accuracy, J. Neurophysiol. 89 (2003) 2396–2405.

[37] M. Gardner-Morse, I. Stokes, The effects of abdominal muscle coactivation on lumbar spine stability, Spine 23 (1) (1998) 86–91.

[38] M. Goldfarb, K. Korkowski, B. Harrold, W. Durfee, Preliminary evaluation of a controlled-brake orthosis for FES-aided gait, IEEE Trans. Neural Syst. Rehabil. Eng. 11 (3) (2003) 241–248.

[39] G. Braz, M. Russold, G. Davis, Functional electrical stimulation control of standing and stepping after spinal cord injury: a review of technical characteristics, Neuromodulation 12 (3) (2009) 180–190.

CHAPTER SIX

Soft Wearable Robotics Technologies for Body Motion Sensing

Y.-L. Park*,†
*Robotics Institute, Carnegie Mellon University, Pittsburgh, PA, United States
†Seoul National University, Seoul, Republic of Korea

1. BODY MOTION SENSING

Body motion sensing has been one of the long-standing fundamental questions in the area of assistive and rehabilitation technologies, since information on motion could be very useful not only for verifying biomechanical models but also for understanding the dynamics of the body. Furthermore, the emergence of wearable robots, such as exoskeletons [1–4] and robotic orthotics [5–8], has caused a need for information on body motion, such as orientation of the limbs and joint angle changes, in real time. Figs. 1 and 2 show examples of exoskeletons and robotic orthotic devices. However, the human body is composed of many rotational joints with one or more degrees of freedom (DOFs) as well as covered by complex three-dimensional (3D) curved surfaces, which makes it very difficult to integrate physical sensors into the body. For this reason, traditional rigid sensing mechanisms, such as manual, electronic, or optical goniometers [9–12], have been used with a very limited number of joints and only in lab-constrained environments in most cases. One of the major advantages of this type of device is the mechanical simplicity that consists of linkages and a rotational joint. Depending on mounting methods, they can be used for different joints. However, they can measure the angle change only with a single DOF per joint, and multiple devices are required for simultaneous multijoint measurement. Moreover, the rigid structure and the mechanical pin joint mechanism not only physically constrain the natural motion of the human body but also do not fully represent complex motions of the human joints, introducing disturbances in the measurement (Fig. 3).

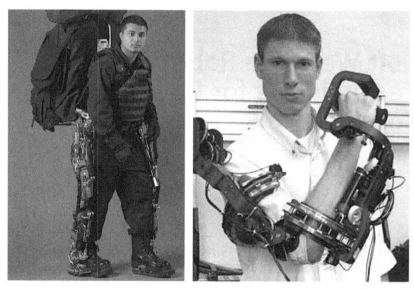

Fig. 1 Examples of exoskeletons for lower [1] (*left*) and upper [2] (*right*) body assistances.

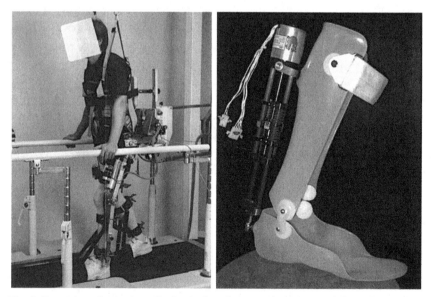

Fig. 2 Examples of robotic orthotic devices for paraplegic [5,64] (*left*) and drop-foot [7,65] (*right*) gait assistances.

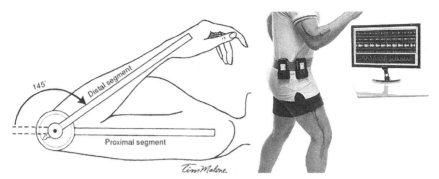

Fig. 3 Body motion sensing using manual (*left*) and electronic (*right*) goniometers. *(Left image: From C.C. Norkin, D.J. White, Measurement of Joint Motion: A Guide to Goniometry, F.A. Davis Company, Philadelphia, PA, 2009 and right image: Twin Axis Goniometers, Biometrics Ltd.)*

To address this issue, extensive research and development has been carried out on the use of optical motion capture systems. Optical systems utilize two-dimensional (2D) image data captured from different angles using more than one camera to triangulate the 3D position of a human subject [13,14]. The positions can be obtained using special markers attached to the subject body. Based on the kind of marker, optical systems can be divided into two types: passive systems and active systems [15]. Passive systems use optically reflective markers that reflect the light emitted from the camera and process the positions of the markers. In this case, since the markers are passive, they can be relatively small and have minimal constraints on the number and the locations. Also, the threshold of the cameras can be adjusted and the background objects can be easily ignored, not affecting the results. Fig. 4 shows examples of commercially

 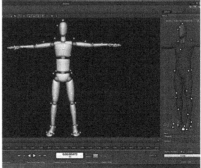

Fig. 4 Examples of commercially available passive-marker motion capture systems from Vicon (*left*) and OptiTrack (*right*).

available passive-marker systems. Active systems use small electronic devices that actively emit light, such as light-emitting diodes (LEDs), as markers. The markers could flicker one by one very quickly to identify themselves. Multiple markers could also illuminate simultaneously to provide relative distances between them. Since the lights from the active markers are much brighter than the reflections from passive markers, active systems could provide a higher signal-to-noise ratio and a higher resolution than passive systems. However, the markers require both a power source and a control circuit, making the system quite complex. Fig. 5 shows examples of commercially available active-marker systems. In general, optical systems can measure 3D body motions with high accuracy, few physical constraints, and a relatively large workspace compared to goniometers. However, one of the main limitations is that they can be used only indoors due to the requirement of multiple cameras. Therefore, it is not suitable for continuous monitoring of body motions for outdoor or large-space activities. Moreover, the entire system could be massive and very expensive depending on the size and the applications.

To overcome the challenge of a large workspace or an outdoor use, inertial measurement units (IMUs) have been proposed and widely used. An IMU is a small electronic device that can measure linear and angular motions using accelerometers and gyroscopes integrated in the device, providing six-DOF measurements. When a magnetometer is added, the device can have three additional DOFs [16,17]. For body motion sensing, multiple IMUs can be attached to different segments of the body of a subject and measure their orientations and motions [18–20]. By combining the data collected from the IMUs, the motions of the subject can be reconstructed in 3D. One of the biggest advantages of IMU motion sensing is outdoor use with

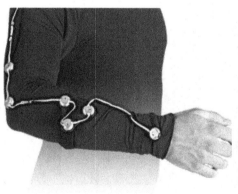
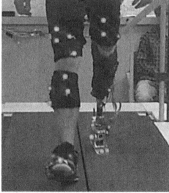

Fig. 5 Examples of commercially available active-marker motion capture systems from NDI Measurement Sciences (*left*) and PhaseSpace Motion Capture (*right*).

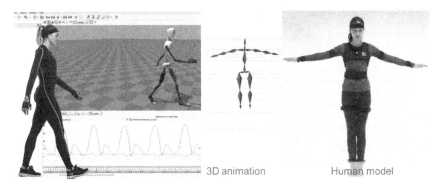

Fig. 6 Examples of commercially available IMU motion capture systems from XSense (*left*) and LpMocap (*right*).

little limitation of workspace. Since this type of system does not use optical detection, it not only is immune to ambient light but also requires no external devices, such as cameras, allowing for more freedom for physical activities. Also, the system is relatively cheap compared to optical systems. Fig. 6 shows examples of commercially available IMU systems for body motion sensing. However, for accurate measurement, IMUs need to be firmly attached to the designated areas of the body. Otherwise, they may slip and move, causing large errors in the estimation. Furthermore, due to signal drift and noise, the system requires techniques for filtering and processing the signals, which is computationally expensive and consequently requires relatively high electrical power consumption.

As an alternative to the previously mentioned systems, stretchable sensing materials have been recently proposed for measuring body motions in the form of wearables. Due to the extremely low elastic modulus (<100 kPa) and the high stretchability (over 100% strain), the soft sensors not only allow the natural DOFs of the body without physical constraints but also easily conform to complex shapes of different body parts. These characteristics make the soft sensors promising for motion sensing of the human body in particular. This chapter provides an overview of "soft" artificial skin technologies that mainly utilize highly stretchable and deformable sensing materials.

2. SOFT ARTIFICIAL SKIN USING EMBEDDED CONDUCTIVE LIQUIDS

This type of artificial skin sensor is made of thin hyperelastic elastomer (eg, silicone) layers with embedded microchannels. When the microchannels are filled with liquid conductors, such as room-temperature liquid metals

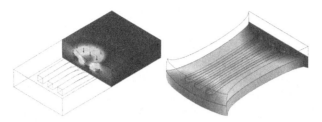

Fig. 7 Microchannel embedded sensing skin deformations due to normal pressure (*left*) and axial strain (*right*). (*From Y.-L. Park, B.-R. Chen, R.J. Wood, Design and fabrication of soft artificial skin using embedded microchannels and liquid conductors, IEEE Sens. J. 12 (8) (2012) 2711–2718.*)

including eutectic gallium-indium (EGaIn) [21,22] or Galinstan [23], and room-temperature ionic liquids (RTILs) [24,25], they form a highly flexible and stretchable electrical circuit [26,27]. Even though the elastomer is flexed or stretched, the continuity of the filled liquid is maintained, making the microchannels always conductive. However, when the material is deformed with an external stress, the geometry of the microchannels changes, which consequently changes the electrical resistance of the microchannels. Fig. 7 describes two different sensing concepts based on types of deformation. By measuring the resistance change, the degree of the deformation can be estimated. The main advantages of these sensors as wearable devices are:

(i) The devices can be highly stretchable and do not constrain natural body motions.

(ii) The devices can be easily wearable and conformable to the 3D human body.

(iii) The devices can be energy efficient since they require little electrical power.

(iv) The sensor signals are reliable since the deformation recovery is mainly governed by the host elastomer.

Fig. 8 shows various types of microchannel patterns filled with EGaIn [28]. The microchannels are usually made by casting liquid polymer poured on 3D printed plastic molds. The typical size of a microchannel is 200–500 μm for width and height in its cross-section. Instead of 3D printing for molds, soft lithography can be used for further miniaturization of microchannels, as shown in Fig. 8D.

While single-layered structures are mostly used due to the simplicity in design and fabrication, multilayered structures provide even more information, such as multimodal sensing or multilocation sensing capabilities. For pressure sensing, double-layered grid-pattern sensor arrays are typical for localizing a pressure point, as shown in Fig. 9 [29,30]. However, if sensing of a different deformation mode other than pressure, such as strain, is

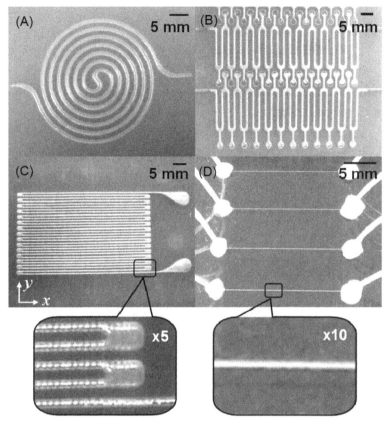

Fig. 8 Different types of microchannel patterns: (A) Spiral-shape channel for pressure sensing. (B) Serpentine-shape channel with air pockets. (C) Strain-gage shape for x-axis strain sensing. (D) Miniaturized channels with a soft lithography process; 20 μm (height) ×25, ×30, ×35, ×40 μm (width), from top to bottom [28].

involved, the design can be further modified. A soft artificial skin sensor developed by Park et al., shown in Fig. 10, utilizes a multilayered structure for multimodal sensing of multiaxis strains and a contact pressure [31]. The sensor consists of three EGaIn microchannel layers embedded in an extremely soft elastomer (modulus: 69 kPa) matrix. The bottom and middle layers have strain gage patterns configured orthogonally for multiaxis strain sensing, and the top layer has a circular pattern for contact pressure sensing. A novel manufacturing method comprised of layered molding and casting processes was used to fabricate the multilayered soft sensor circuit. Silicone rubber layers with channel patterns, cast with 3D printed molds, are bonded to create embedded microchannels, and a conductive liquid is injected into

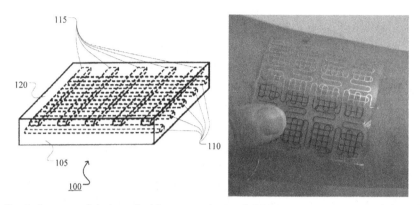

Fig. 9 Conceptual design of grid-patterned microfluidic pressure sensors [29] (*left*) and its application of soft keypad [30] (*right*).

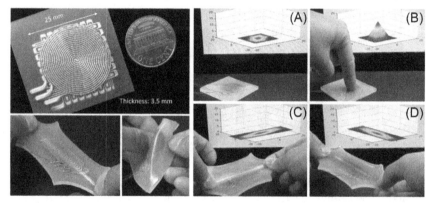

Fig. 10 Soft artificial skin for sensing multiple deformation modes: (A) no stimulus, (B) normal pressure, and (C), (D) multiaxis strain [31].

the microchannels. The channel dimensions are 200×200 μm, and the overall sensor size is 25×25 mm with a thickness of ∼3.5 mm. The skin prototype demonstrated repeatable responses to strain and pressure.

Another type of soft skin that uses EGaIn microchannels is the soft multiaxis force sensor shown in Fig. 11 [32]. While the previous soft skin sensor can detect only a normal pressure, this sensor detects shear forces in addition to a normal force. This sensor uses the same elastomer material and conductive microfluidic channels, but microchannels divided into three or four sections were embedded in a single layer, and a microscale force-post was embedded above the microchannels. When a normal force in the z-axis is applied to the force-post, all the microchannels are compressed and

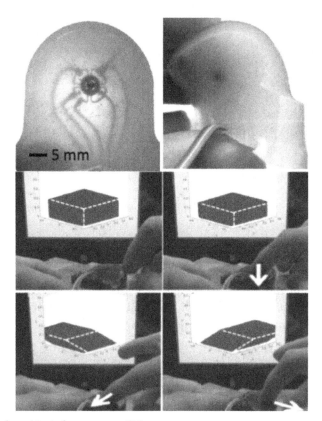

Fig. 11 Soft multiaxis force sensor [32].

increase their resistances in a similar way. However, if a shear force is applied, the force–post rotates about a horizontal axis (x- or y-axis), and only one or two microchannels are compressed, giving higher resistance changes than the other channels. The channel dimensions are $\sim$200 × 200 µm. The overall size of this sensor is 50×60 mm with a thickness of 7 mm approximately.

Although liquid metals have great advantages as a conductive medium in soft sensors and soft electronics, they are not completely safe with the human body, especially in the case of direct contact with skin or oral absorption when they accidentally leak out of the device. Ionic solutions (eg, saline solutions) and RTILs have been proposed as biocompatible alternatives. A soft strain sensor, shown in Fig. 12 (left), embedded with saline-solution-filled microchannels has been proposed [33]. This sensor demonstrated a linear strain response with higher sensitivity than that of a liquid metal strain sensor.

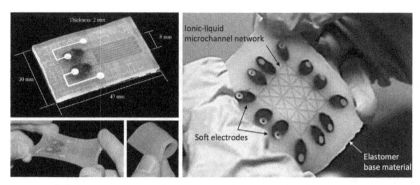

Fig. 12 Soft strain sensor using ionic-solution filled microchannels [33] (*left*) and soft tactile skin with an ionic-liquid filled microchannel network [34] (*right*).

However, the ionic solution was not stable in a silicone matrix. In other words, the liquid started to be absorbed by the silicone matrix, and the sensor lost its functionality after a while. Therefore, RTILs were used as a replacement due to their relatively high chemical stability compared to "water-based" ionic solutions. RTILs also have lower surface tensions than liquid metals, making the injection process easier. The soft tactile skin, shown in Fig. 12 (right), has an RTIL microchannel network for detecting both the magnitudes and the locations of surface contacts using electrical impedance tomography [34]. Carbon grease can be another option as a conductive medium in microfluidic soft sensors. Muth et al. embedded carbon grease patterns directly in an uncured silicone matrix to make similar stretchable sensing skin [35].

Using the liquid metal soft sensor technology, various wearable soft sensors for detecting body motions have been developed taking advantage of hyperelasticity of the elastomer material. While a single sensor unit can be used for detecting motions of a specific body part, multiple units can be combined and simultaneously used for motions with multiple joints.

One of the basic applications in motion sensing is a single-axis strain sensor for detecting 1-DOF finger joint angles, as shown in Fig. 13 [36]. A 1 mm × 50 µm (width and height) microchannel filled with EGaIn was embedded in a polydimethylsiloxane (PDMS) layer.

Park et al. proposed use of two single-axis soft strain sensors for detecting 2-DOF ankle motions as part of a soft active orthotic device for people with neuromuscular disorders, as shown in Fig. 14 [8]. One sensor was attached on the anterior part of the ankle for sagittal motions—dorsiflexion and plantar flexion—and the other sensor was fixed at the inside of the ankle

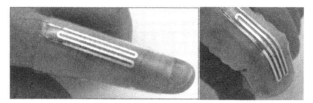

Fig. 13 Soft strain sensor for detecting finger joint motions [36].

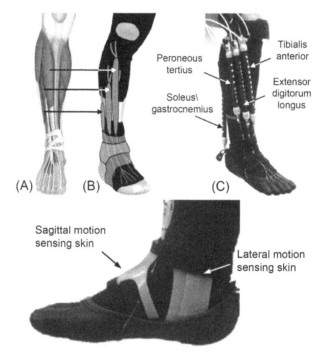

Fig. 14 Bio-inspired design of an active soft orthotic device (*top*) with 3D ankle motion sensing using multiple soft artificial skin sensors (*bottom*) [8].

for mediolateral motions—inversion and eversion. The low elastic modulus and the high stretchability of the base material provided reliable measurement without constraining the natural motions of the ankle. In this device, IMUs were used as a means of calibrating the soft sensors. The experimental results showed measurement error <3 degrees on average.

Use of multiple sensors for detecting motions combined with multiple body joints can be seen in a wearable soft sensing suit for human gait

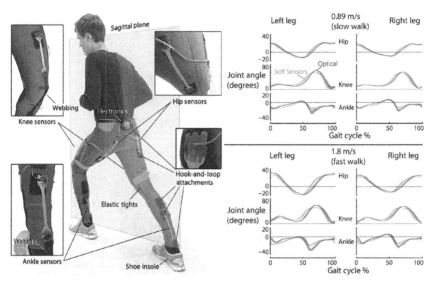

Fig. 15 Lower body wearable sensing suit (*left*) for human gait measurement and its experimental results for slow and fast walk (*right*) [37].

measurement developed by Mengüç et al. [37]. Fig. 15 shows the prototype on a human subject and its experimental results. Three soft strain sensors were attached to each leg for measuring angle changes of three different joints—ankle, knee, and hip. Since the soft sensors are responsive to pressures in addition to strains, they were located on relatively soft surfaces of the body, avoiding areas of bony prominence. Each sensor was equipped with a Bluetooth wireless communication unit, so the sensing suit became fully untethered, making the device fully mobile and suitable for outdoor use. The authors compared the motion data from the soft sensors with that of a commercial optical motion capture system (Vicon). The results showed errors of ~5 degrees and 10 degrees for walking and running, respectively. The error was smallest with the hip sensor and largest with the knee sensor. Also, the error was larger during the swing phase than during the stance phase.

Instead of liquid metals, RTILs also have been used for a wearable device. Chossat et al. developed a soft skin device, shown in Fig. 16, for detecting real-time hand motions using an array of RTIL strain sensors [38]. In this device, the strain sensors were made of RTIL microchannels, and the connections between the sensors and from the sensors to the circuit were made of liquid metal microchannels. The large difference in the

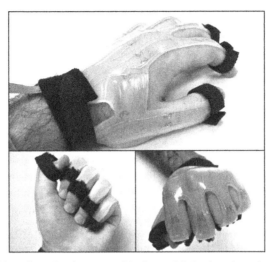

Fig. 16 Wearable soft artificial sensing skin for multijoint hand motion detection [38].

resistance changes between the two liquid channels enables rejecting the signal fluctuations from the EGaIn soft wiring, which are extremely small compared to the RTIL sensor signals.

3. STRAIN-SENSITIVE CONDUCTIVE POLYMERS

Conductive polymers are another type of soft material that can be used as soft sensors in the form of wearable skin for detecting large strains when attached to different body parts. They are typically made by mixing conductive particles called "fillers" into a liquid-state polymer (ie, prepolymer) before curing [39,40]. Types and concentrations of the conductive fillers in a polymer matrix are key factors that determine the mechanical and electrical properties of the final polymers [41,42]. Carbon nanotubes (CNT) are one of the most commonly used fillers for their relatively high conductivity. In general, the higher concentration makes the polymer more conductive. However, it also makes the polymer stiffer and less elastic. Therefore, it is important to find the right concentration that satisfies both the conductivity and the elasticity depending on the application. When a conductive polymer embedded with fillers is stretched, the distances between the fillers increase, which consequently increases the electrical resistance of the material [43,44]. Therefore, similar to microfluidic soft sensors, strain can be estimated by measuring the resistance change.

Yamada et al. developed a stretchable CNT strain sensor for motion detection [45]. Thin CNT films were first grown using water-assisted chemical vapor deposition and laid onto a flat elastomeric substrate. Then, the CNT films make a reasonably strong bond to the elastomeric substrate. Since the CNTs are aligned in the perpendicular direction to the strain axis, when stretched the CNT chains become loose and their electrical conductivity decreases. The sensor was able to measure and withstand strain up to 280% and demonstrated a linear strain response. The sensor was implemented in various wearable devices for detecting body motions, such as knee and hand motions, as shown in Fig. 17. Ryu et al. also proposed a CNT fiber strain sensor for monitoring human motion [46]. In this case, CNT fibers were directly dry spun and bonded to a hyperelastic elastomer substrate. Differently from the previous sensor, the CNT fibers were aligned in the direction of strain. When stretched, the CNT fibers lose some of the conductive paths, increasing the overall resistance. The hyperelasticity of the substrate material allowed the sensor to remain intact up to 900% strain. The sensor demonstrated a linear strain response up to 300%. Similar to the previous sensor, this sensor was implemented in wearable devices for measuring joint motions in different body parts, as shown in Fig. 18.

Instead of bonding to a stretchable substrate, conductive polymer can be made wearable in the form of textiles. One example is coating a stretchable fabric with a conductive polymer. In this sensor, nylon Lycra was coated with conductive polypyrrole [47]. The polypyrrole-coated textiles demonstrated conformation to the shape of the human body as wearable biomechanical sensors for monitoring body motions, as shown in Fig. 19. The sensor showed a reduction in electrical resistance of ∼4 and 1 kΩ up to 20% strain and from 20% to 60%,

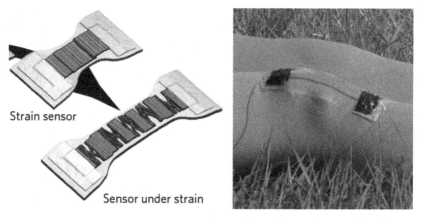

Fig. 17 Carbon nanotube film strain sensor and its application for knee angle measurement [45].

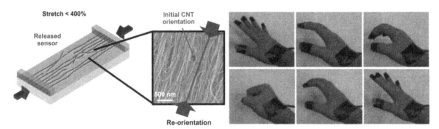

Fig. 18 Soft strain sensor made of dry spun carbon nanotube fibers and its application for hand motion sensing [46].

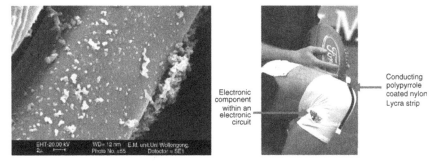

Fig. 19 Polypyrrole-coated nylon Lycra fiber and its implementation in an intelligent knee sleeve [47].

respectively. In a similar way, Lorussi et al. developed wearable sensing devices for different body parts, such as arm, knee, and hand, by patterning conductive tracks on stretchable sleeves [48]. The conductive tracks were made by directly printing a carbon/rubber mixture onto fabrics or by weaving carbon-filled rubber (CFR) coated fibers. Although the sensor response was relatively linear up to 80% strain, it showed noticeable hysteresis when released after being stretched. They also developed an algorithm for estimating body motions when the sensor arrays were used in different body parts (Fig. 20).

Another method is directly weaving conductive polymer threads into textiles. Wu et al. developed a highly stretchable strain sensor using conductive yarn [49]. They coated polyurethane yarn with conductive polymer composites made from carbon black on cellulose nanocrystals and natural rubber nanohybrid, and wove the conductive polyurethane yarn into a fabric patch. Since the sensing material was sensitive to very small strains, it was used for monitoring tiny motions in the human body, such as speech recognition in the area of vocal cords on the neck skin, on the forehead skin and the philtrum skin for facial expression recognition, and hand motion recognition, as shown in Fig. 21. Shyr et al. proposed elastic conductive

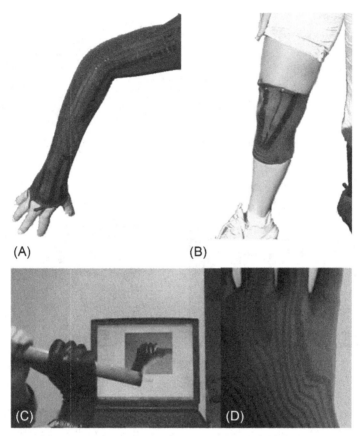

Fig. 20 Wearable sensing device prototypes using fabric-based stretchable conductive polymer sensor arrays: (A) Arm sleeve. (B) Knee pad. (C) Sensing glove for grasp motions. (D) Magnified view of conducting tracks on a sensing fabric [48].

webbing by weaving polyamide fibers coated with carbon particles [50]. They used the sensors for monitoring flexion angles of elbow and knee movements. The sensors demonstrated linear responses when stretched on the elbow and knee joints, as shown in Fig. 22. They also showed tensile stretch recovery within 30% strain.

4. FIBER OPTIC WEARABLE SENSORS FOR MOTION SENSING

Other than conductive materials, fiber optic sensing can also be used in wearable motion sensing. Although optical fibers are not stretchable in general, they can provide reasonable flexibility for body motion sensing depending on how and where they are embedded in wearable suits.

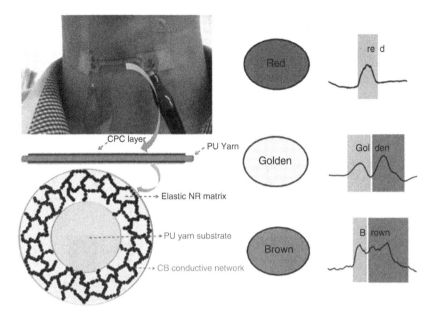

Fig. 21 Stretchable strain sensor using conductive polyurethane yarn directly attached on the neck skin for speech recognition. Different strain signal patterns are observed for different word pronunciations [49].

The fiber Bragg grating (FBG) sensor [51,52] is one of the most widely used fiber optic sensors for measuring strain. An FBG is made by inscribing multiple periodic Bragg grating patterns in the cross-sections of a short segment (typically 5–10 mm) of an optical fiber. When a light that enters from one end of the optical fiber meets the FBG, a certain wavelength of the light is reflected back to the origin, transmitting all the other wavelengths of the input light to the other end of the fiber. However, if the FBG area experiences mechanical strains, the reflection wavelength shifts. Therefore, the strain applied to the FBG area of the fiber can be estimated by measuring the change of the reflection wavelength. FBGs are usually used in rigid structures for applications such as structural health monitoring [53,54] and force sensing of robotic manipulators [55,56], due to their high accuracy in strain measurement with small strain tolerance. However, if the fiber optics can be carefully routed and located in wearable suits so they do not experience large strain, they can be used for motion sensing as well. Silva et al. sewed FBG embedded optical fibers in a fabric glove and measured finger motions in 3D, as shown in Fig. 23 [57]. FBGs were also embedded in a polyvinyl chloride (PVC) foil, and the PVC foil was attached to an elastic knee sleeve for joint motion sensing, as shown in Fig. 24 [58].

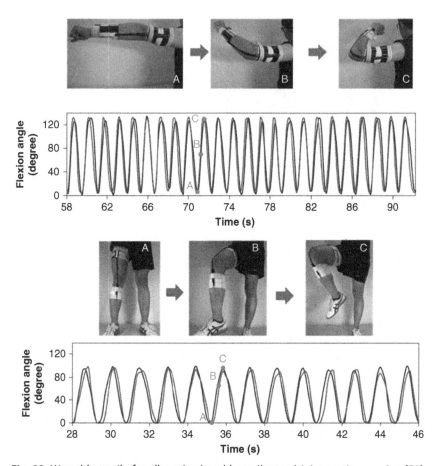

Fig. 22 Wearable textile for elbow (*top*) and knee (*bottom*) joint motion sensing [50].

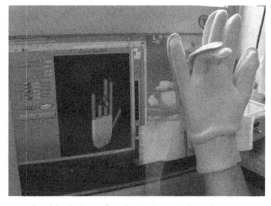

Fig. 23 FBG array embedded glove for detecting 3D hand motions [57].

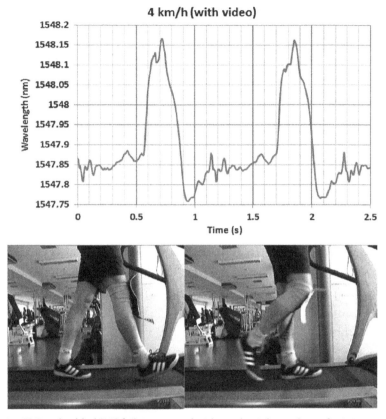

Fig. 24 FBG embedded PVC foils integrated with an elastic knee sleeve for joint motion measurement [58].

Another type of strain-sensing technique involving fiber optics uses optical power loss in a hetero-core fiber. Unlike regular optical fibers, hetero-core fibers have two different cores with different diameters. When the input light passes the hetero-core area of the fiber, part of the light escapes through the cross-section of the larger core, resulting in optical power loss in the light transmission. Depending on the strain applied to the hetero-core area, the optical power loss changes. Nishiyama et al. and Koyama et al. embedded optical fibers with hetero-core in a fabric glove [59] and an upper-body garment [60] for sensing 3D hand and trunk motions, respectively, as shown in Fig. 25.

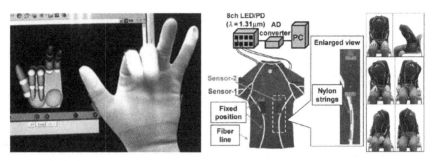

Fig. 25 Hetero-core fiber optic sensor embedded wearable devices for measuring hand [60] (*left*) and trunk [61] (*right*) motions.

5. CONCLUSIONS AND FUTURE DEVELOPMENTS

In this chapter, we discussed various types of soft (ie, stretchable and/ or flexible) wearable sensors that are promising for real-time 3D body motion sensing, one of the core elements in assistive technologies. While many of them are already close to widespread adoption, there are still some limitations in practical applications.

Microfluidic soft artificial skins demonstrated high stretchability with relatively linear and repeatable strain responses, suitable for wearable devices. However, one of the limitations is the complexity in manufacturing that requires fabrication of microchannels and injection of liquid conductors. This problem could be addressed by mixing liquid conductor particles directly into liquid-state elastomer before curing, as discussed in Refs. [61,62].

Conductive polymers have a potential as soft strain sensors since they do not contain any liquids in their sensing mechanism, making them safe and practical for wearable applications that may make direct contact with human skin. They can be easily combined, even with fabric materials that are more breathable and skin-friendly than polymer materials. However, the stretchability of the base polymer materials can be significantly reduced with increase of concentrations of the conductive filler materials for higher electrical conductivity. Hysteresis could also be amplified with a high concentration of the fillers. Therefore, it is important to identify the influence of the filler materials on both electrical and mechanical properties of the final composites, as discussed in [42].

Fiber optic sensors can also be used for strain measurement if carefully embedded and routed in fabrics. Due to their extremely low strain tolerance, they need to be located where only small strains are applied. Otherwise, they can either break or constrain the natural body motions of the wearer. One of the possible solutions to this problem could be a strain-sensitive stretchable waveguide recently proposed by To et al. [63]. Instead of using regular optical fibers, a waveguide can be made in a transparent elastomer for transmitting a light signal. When it is stretched, the reflective layer around the waveguide forms microcracks, causing optical power loss in its transmission. This optical power loss is proportional to the strain.

As we have seen, soft wearables and soft artificial skins for body motion sensing still have room for further improvement and advances compared to other technologies that are already commercially available in different forms, such as vision-based or IMU-based systems. However, once the main limitations are addressed, they will become almost ideal components in human modeling and assistive technologies.

REFERENCES

[1] A.B. Zoss, H. Kazerooni, A. Chu, Biomechanical design of the Berkeley lower extremity exoskeleton (BLEEX), IEEE Trans. Mechatron. 11 (2) (2006) 128–138.

[2] J.C. Perry, J. Rosen, S. Burns, Upper-limb powered exoskeleton design, IEEE/ASME Trans. Mechatron. 12 (4) (2007) 408–417.

[3] J.F. Veneman, R. Kruidhof, E.E.G. Hekman, R. Ekkelenkamp, E.H.F. Van Asseldonk, H. van der Kooij, Design and evaluation of the LOPES exoskeleton robot for interactive gait rehabilitation, IEEE Trans. Neural Syst. Rehabil. Eng. 15 (3) (2007) 379–386.

[4] S.K. Banala, S.H. Kim, S.K. Agrawal, J.P. Scholz, Robot assisted gait training with active leg exoskeleton (ALEX), IEEE Trans. Neural Syst. Rehabil. Eng. 17 (1) (2009) 2–8.

[5] G. Colombo, M. Joerg, R. Schreier, V. Dietz, Treadmill training of paraplegic patients using a robotic orthosis, J. Rehabil. Res. Dev. 37 (6) (2000) 693–700.

[6] S. Jezernik, G. Colombo, T. Keller, H. Frueh, M. Morari, Robotic orthosis Lokomat: a rehabilitation and research tool, Neuromodul. Technol. Neural Interface 6 (2) (2003) 108–115.

[7] J.A. Blaya, H. Herr, Adaptive control of a variable-impedance ankle-foot orthosis to assist drop-foot gait, IEEE Trans. Neural Syst. Rehabil. Eng. 12 (1) (2004) 24–31.

[8] Y.-L. Park, B.-R. Chen, N.O. Pérez-Arancibia, D. Young, L. Stirling, R.J. Wood, E.G. Goldfield, R. Nagpal, Design and control of a bio-inspired soft wearable robotic device for ankle-foot rehabilitation, Bioinspir. Biomim. 9 (1) (2014) 016007 (17 pp).

[9] E.Y.S. Chao, Justification of triaxial goniometer for the measurement of joint rotation, J. Biomech. 13 (12) (1980) 989–1006.

[10] M.P. Clapper, S.L. Wolf, Comparison of the reliability of the Orthoranger and the standard goniometer for assessing active lower extremity range of motion, Phys. Ther. 68 (2) (1988) 214–218.

[11] K. Aminian, B. Najafi, Capturing human motion using body-fixed sensors: outdoor measurement and clinical applications, Comput. Animat. Virtual Worlds 15 (2) (2004) 75–94.

[12] C.C. Norkin, D.J. White, Measurement of Joint Motion: A Guide to Goniometry, F.A. Davis Company, Philadelphia, PA, 2009.

[13] T.B. Moeslund, A. Hilton, V. Krüger, A survey of advances in vision-based human motion capture and analysis, Comput. Vis. Image Underst. 104 (2–3) (2006) 90–126.

[14] R. Poppe, Vision-based human motion analysis: an overview, Comput. Vis. Image Underst. 108 (1–2) (2007) 4–18.

[15] H. Zhou, H. Hu, Human motion tracking for rehabilitation—a survey, Biomed. Signal Process. Control 3 (1) (2008) 1–18.

[16] N. Ahmad, R.A.R. Ghazilla, N.M. Khairi, Reviews on various inertial measurement unit (IMU) sensor applications, Int. J. Signal Process. Syst. 1 (2) (2013) 256–262.

[17] F. Höflinger, J. Müller, R. Zhang, L.M. Reindl, W. Burgard, A wireless micro inertial measurement unit (IMU), IEEE Trans. Instrum. Meas. 62 (9) (2013) 2583–2595.

[18] D. Roetenberg, H. Luinge, P. Slycke, MVN: Full 6DOF Human Motion Tracking Using Miniature Inertial Sensors, XSENS Technologies, PR Enschede, 2013, pp. 1–9.

[19] T. Seel, J. Raisch, T. Schauer, IMU-based joint angle measurement for gait analysis, Sensors 14 (4) (2014). 6891, 6909.

[20] K.Y. Lim, F.Y.K. Goh, W. Dong, K.D. Nguyen, I.-M. Chen, S.H. Yeo, H.B.L. Duh, C.G. Kim, A wearable self-calibrating, wireless sensor network for body motion processing, in: Proceedings of the IEEE International Conference on Robotics and Automations, 2008, pp. 1017–1022.

[21] R. Chiechi, E.A. Weiss, M.D. Dickey, G.M. Whitesides, Eutectic gallium-indium (EGaIn): a moldable liquid metal for electrical characterization of self-assembled monolayers, Angew. Chem. 120 (1) (2007) 148–150.

[22] M.D. Dickey, R.C. Chiechi, R.J. Larsen, E.A. Weiss, D.A. Weitz, G.M. Whitesides, Eutectic gallium-indium (EGaIn): a liquid metal alloy for the formation of stable structures in microchannels at room temperature, Adv. Funct. Mater. 18 (7) (2008) 1097–1104.

[23] T. Liu, P. Sen, C.-J. Kim, Characterization of nontoxic liquid-metal alloy Galinstan for applications in microdevices, J. Microelectromech. Syst. 21 (2) (2011) 443–450.

[24] H. Ohno, M. Yoshizawa, W. Ogihara, Development of new class ion conductive polymers based on ionic liquids, Electrochim. Acta 50 (2–3) (2004) 255–261.

[25] H. Ohno, Electrochemical Aspects of Ionic Liquids, John Wiley & Sons, Hoboken, NJ, 2011.

[26] S. Cheng, A. Rydberg, K. Hjort, Z. Wu, Liquid metal stretchable unbalanced loop antenna, Appl. Phys. Lett. 94 (14) (2009). 144103 (4 pp).

[27] S. Cheng, Z. Wu, P. Hallbjorner, K. Hjort, A. Rydberg, Foldable and stretchable liquid metal planar inverted cone antenna, IEEE Trans. Antennas Propag. 57 (12) (2009) 3765–3771.

[28] Y.-L. Park, C. Majidi, P. Berard, R. Kramer, R.J. Wood, Hyperelastic pressure sensing with liquid embedded elastomer, J. Micromech. Microeng. 20 (12) (2010). 125029 (6 pp).

[29] C. Majidi, R.J. Wood, P. Berard, Y.-L. Park, Stretchable Two-Dimensional Pressure Sensor, US patent 8,316,719, 2012.

[30] R.K. Kramer, C. Majidi, R.J. Wood, Wearable tactile keypad with stretchable artificial skin, in: Proceedings of the IEEE International Conference on Robotics and Automation, 2011, pp. 1103–1107.

[31] Y.-L. Park, B.-r. Chen, R.J. Wood, Design and fabrication of soft artificial skin using embedded microchannels and liquid conductors, IEEE Sensors J. 12 (8) (2012) 2711–2718.

[32] D. Vogt, Y.-L. Park, R.J. Wood, Design and characterization of a soft multi-axis force sensor using embedded microfluidic channels, IEEE Sensors J. 13 (10) (2013) 4056–4064.

[33] J.-B. Chossat, Y.-L. Park, R.J. Wood, V. Duchaine, A soft strain sensor based on ionic and metal liquids, IEEE Sensors J. 13 (9) (2013) 3405–3414.

[34] J.-B. Chossat, H.-S. Shin, Y.-L. Park, V. Duchaine, Soft tactile skin using an embedded ionic liquid and tomographic imaging, ASME J. Mech. Robot. 7 (2) (2015) 021008 (9 pp).

[35] J.T. Muth, D.M. Vogt, R.L. Truby, Y. Mengüç, D.B. Kolesky, R.J. Wood, J.A. Lewis, Embedded 3D printing of strain sensors with highly stretchable elastomers, Adv. Mater. 26 (36) (2014) 6307–6312.

[36] R.K. Kramer, C. Majidi, R. Sahai, R.J. Wood, Soft curvature sensors for joint angle proprioception, in: Proceedings of the IEEE/RSJ International Conference on Intelligent Robots and Systems, 2011, pp. 1919–1926.

[37] Y. Mengüç, Y.-L. Park, H. Pei, D. Vogt, P. Aubin, L. Fluke, E. Winchell, L. Stirling, R.J. Wood, C.J. Walsh, Wearable soft sensing suit for human gait measurement, Int. J. Robot. Res. 33 (14) (2014) 1748–1764.

[38] J.-B. Chossat, Y. Tao, V. Duchaine, Y.-L. Park, Wearable soft artificial skin for motion detection with embedded microfluidic strain sensing, in: Proceedings of the IEEE International Conference on Robotics and Automation, 2015, pp. 2568–2576.

[39] Y.P. Mamunya, V.V. Davydenko, P. Pissis, E.V. Lebedev, Electrical and thermal conductivity of polymers filled with metal powders, Eur. Polym. J. 38 (9) (2002) 1887–1897.

[40] G. Harsanyi, Polymer films in sensor applications: a review of present uses and future possibilities, Sens. Rev. 20 (2) (2000) 98–105.

[41] W. Zhang, A.A. Dehghani-Sanij, R.S. Blackburn, Carbon based conductive polymer composites, J. Mater. Sci. 42 (10) (2007) 3408–3418.

[42] S.-H. Jang, Y.-L. Park, H. Yin, Influence of coalescence on the anisotropic mechanical and electrical properties of nickel powder/polydimethylsiloxane composites, Materials 9 (4) (2016). 239 (11 pp).

[43] N. Hu, Y. Karube, M. Arai, T. Watanabe, C. Yan, Y. Li, Y. Liu, H. Fukunaga, Investigation on sensitivity of a polymer/carbon nanotube composite strain sensor, Carbon 48 (3) (2010) 680–687.

[44] G.T. Pham, Y.-B. Park, Z. Liang, C. Zhang, B. Wang, Processing and modeling of conductive thermoplastic/carbon nanotube films for strain sensing, Compos. Part B 38 (1) (2008) 209–216.

[45] T. Yamada, Y. Hayamizu, Y. Yamamoto, Y. Yomogida, A. Izadi-Najafabadi, D.N. Futaba, K. Hata, A stretchable carbon nanotube strain sensor for human-motion detection, Nat. Nanotechnol. 6 (2011) 296–301.

[46] S. Ryu, P. Lee, J.B. Chou, R. Xu, R. Zhao, A.J. Hart, S.-G. Kim, Extremely elastic wearable carbon nanotube fiber strain sensor for monitoring of human motion, ACS Nano 9 (6) (2015) 5929–5936.

[47] J. Wu, D. Zhou, C.O. Too, G.G. Wallace, Conducting polymer coated Lycra, Synth. Mater. 155 (3) (2005) 698–701.

[48] F. Lorussi, W. Rocchia, E.P. Scilingo, A. Tognetti, D. De Rossi, Wearable, redundant fabric-based sensor arrays for reconstruction of body segment posture, IEEE Sensors J. 4 (6) (2004) 807–818.

[49] X. Wu, Y. Han, X. Zhang, C. Lu, Highly sensitive, stretchable, and wash-durable strain sensor based on ultrathin conductive layer@polyurethane yarn for tiny motion monitoring, ACS Appl. Mater. Interfaces 8 (15) (2016) 9936–9945.

[50] T.-W. Shyr, J.-W. Shie, C.-H. Jiang, J.-J. Li, A textile-based wearable sensing device designed for monitoring the flexion angle of elbow and knee movements, Sensors 14 (3) (2014) 4050–4059.

[51] K.O. Hill, G. Meltz, Fiber Bragg grating technology fundamentals and overview, J. Lightwave Technol. 15 (8) (1997) 1263–1276.

[52] R. Kashyap, Fiber Bragg Gratings, Academic Press, San Diego, CA, 1999.

[53] M. Majumder, T.K. Gangopadhyay, A.K. Chakraborty, K. Dasgupta, D.K. Bhattacharya, Fibre Bragg gratings in structural health monitoring—present status and applications, Sens. Actuators A Phys. 147 (1) (2008) 150–164.

[54] D. Balageas, C.P. Fritzen, A. Güemes, Structural Health Monitoring, ISTE Ltd., London, 2006.

[55] Y.-L. Park, S.C. Ryu, R.J. Black, K. Chau, B. Moslehi, M.R. Cutkosky, Exoskeletal force-sensing end-effectors with embedded optical fiber-Bragg-grating sensors, IEEE Trans. Robot. 25 (6) (2009) 1319–1331.

[56] L. Jiang, K. Low, J. Costa, R.J. Black, Y.-L. Park, Fiber optically sensorized multi-fingered robotic hand, in: Proceedings of the IEEE/RSJ International Conference on Robots and Systems, 2015.

[57] A.F. da Silva, A.F. Gonçalves, P.M. Mendes, J.H. Correia, FBG sensing glove for monitoring hand posture, IEEE Sensors J. 11 (10) (2011) 2442–2448.

[58] R.P. Rocha, A.F. Silva, FBG in PVC foils for monitoring the knee joint movement during the rehabilitation process, in: Proceedings of the International Conference of the IEEE Engineering in Medicine and Biology Society, 2011, pp. 458–461.

[59] M. Nishiyama, K. Watanabe, Wearable sensing glove with embedded hetero-core fiber-optic nerves for unconstrained hand motion capture, IEEE Trans. Instrum. Meas. 58 (12) (2009) 3995–4000.

[60] Y. Koyama, M. Nishiyama, K. Watanebe, A motion monitor using hetero-core fiber sensor sewed in sports wear to trace trunk motion, IEEE Trans. Instrum. Meas. 62 (4) (2013) 828–836.

[61] A. Fassler, C. Majidi, Liquid-phase metal inclusions for a conductive polymer composite, Adv. Mater. 27 (11) (2015) 1928–1932.

[62] M.D. Bartlett, A. Fassler, N. Kazem, E.J. Markvicka, P. Mandal, C. Majidi, Stretchable, high-k dielectric elastomer through liquid-metal inclusions. Adv. Mater. (2016). http://dx.doi.org/10.1002/adma.201506243.

[63] C. To, T. Hellebrekers, Y.-L. Park, Highly stretchable optical sensors for pressure, strain, and curvature measurement, in: Proceedings of the IEEE/RSJ International Conference on Intelligent Robots and Systems, 2015, pp. 5898–5903.

[64] G. Colombo, M. Wirz, V. Dietz, Driven gait orthosis for improvement of locomotor training in paraplegic patients, Spinal Cord 39 (5) (2001) 252–255.

[65] A.M. Dollar, H. Herr, Lower extremity exoskeletons and active orthoses: challenges and state-of-the-art, IEEE Trans. Robot. 24 (1) (2008) 144–158.

PART TWO

Modeling of Human Cognitive/Muscular Skills and Their Applications

Noninvasive Brain Machine Interfaces for Assistive and Rehabilitation Robotics: A Review

G. Lisi, J. Morimoto

ATR Computational Neuroscience Laboratories, Kyoto, Japan

1. INTRODUCTION

The possibility of controlling a robot by thought, as if it were part of the user's own body, has become part of mankind's imagination. Such a system would ideally allow one to naturally control all the degrees of freedom (DoF) of a robot and the associated kinematics using the brain. Great steps in this direction have been made in recent years by means of invasive brain machine interfaces (BMI) [1], while noninvasive approaches still cannot extract sufficient information for this purpose [2]. Nonetheless, noninvasive BMI has proven useful in improving the quality of life of people with severe motor disabilities, bypassing the interrupted neuromuscular pathway in order to assist an impaired user during simple daily tasks (eg, assistive robot) [3].

Moreover, it has been suggested that BMI paradigms can be used to improve volitional motor control that has been impaired by stroke. This rehabilitation procedure can be achieved, for example, by using brain activity to activate a rehabilitation robot that produces sensory input, inducing central nervous system (CNS) plasticity and leading to restoration of normal motor control [4].

In this chapter, we present a brief overview of the available noninvasive techniques, followed by a detailed description of BMI based on sensorimotor rhythms (SMR), which is extensively applied to both assistive and rehabilitation robotics. Subsequently, we describe the main challenges in building an SMR-based decoder, outline its main components, and highlight the most active branches of research in the field. Having provided the BMI background, we then discuss the applications of noninvasive BMI to nonwearable and wearable assistive robotics, with special attention to the challenges and opportunities arising from the proprioceptive feedback induced by a wearable device. Finally, we introduce CNS neuroplasticity, and review how wearable robots and BMI have been coupled in order to restore motor functions after stroke.

2. BRAIN MACHINE INTERFACES

2.1 Electroencephalography

The flow of electric currents during synaptic excitations of the dendrites in the neurons causes voltage fluctuations that can be measured by electroencephalography (EEG) electrodes placed on the scalp [5]. Among the available neuroimaging technologies—ie, functional magnetic resonance imaging (fMRI), near infrared spectroscopy (NIRS), magnetoencephalography (MEG), electrocorticography (ECoG), intracortical neuron recording—EEG represents the most suitable candidate for establishing a nonmuscular interface between the brain and a wearable robot. This is due to the fact that EEG is noninvasive, portable and has a relatively high temporal resolution. Drawbacks of this technique are the poor spatial resolution and the limited frequency range: the electrical activity on the head surface is detectable only if a large population of neurons is active, the signal is mixed and attenuated by the layers between the cortex and the scalp, and the amount of channels is limited (ie, maximum 256).

2.2 Types of Brain Activity

In order to interpret the user's intention it is possible to decode different types of brain activity from the EEG signal, including visual evoked potentials (VEPs) and steady-state VEPs (SSVEPs), P300 evoked potentials (P300s), error potentials (ErrPs), slow cortical potentials (SCPs), and SMRs. VEPs are electrical potentials that occur in the visual cortex when the subject's visual field undergoes a sensory stimulation. In SSVEP the external stimulation is provided by a flashing light and the evoked EEG signal has a frequency which is a multiple of the frequency of the flashing light, with a high signal-to-noise ratio (SNR). Multiple visual targets, flashing at different frequencies, are mapped into different commands for an external device, and SSVEPs are used to identify the target that the user is fixating [6]. Systems based on the P300 potentials make use of a matrix of letters flashing one at a time; the user selects one specific letter by concentrating his attention on it, which produces an evoked potential at approximately 300 ms after the letter flashes [7]. An error potential [8] is a positive potential, centered at the vertex, that occurs about 180 ms after a person detects a mistake. In SCP BMI, the subject is trained to decrease or increase the voltage level of a low-frequency signal (< 1 Hz) in the cortex by means of visual neurofeedback [9]. SMR consists of power modulations in the areas of the cortex directly involved in motor control (see Section 2.5). In recent years, novel BMI paradigms combine different types of brain activity (eg, VEP and SMR), or different neuroimaging technologies (eg, EEG and near-infrared spectroscopy) into systems called hybrid brain computer interfaces (BCIs), with the aim of improving decoding performance [10].

2.3 Categories of BMI

According to the nature of the different types of brain activity, BMI systems can be categorized into two broad groups, namely *exogenous* and *endogenous*. The former is based on brain activation that is elicited by an external stimulus, ie, VEP and P300. On the other hand, endogenous BMI is based on autonomous regulation of brain rhythms or potentials without external stimuli, ie, SCP and SMR. Exogenous systems do not require time-consuming setup and training, and they can achieve a high information transfer rate (ITR) (eg, 60 bits/min). Endogenous BMI typically requires neurofeedback training, in order for the user to learn to generate specific brain modulations, but it has the great advantage that, after training, it can be operated autonomously.

Another categorization of the BMI systems is based on the timing of data processing, which can be *synchronous* or *asynchronous*. In the former approach [11], subjects are instructed to start and end the mental task by means of specific cues. Therefore, the EEG signal can be analyzed only during a predefined window, while the remaining segments of the signal are ignored. On the other hand, an asynchronous system continuously processes the EEG signal and produces an output. The advantage of a synchronous BMI is that brain activity at rest, including artifacts and other unrelated cortical signals, can be ignored, simplifying the construction of the EEG decoder. Asynchronous systems are more complex, since they have to handle an additional highly variable noncontrol state; however, they allow for a self-paced communication with an external device [12].

2.4 Artifacts

Among the most critical issues when recording the EEG signal, especially during a dynamical task or during the interaction between the user and a wearable robot, there are artifacts of both physiological or non-physiological origin. The former includes electrooculography (EOG) and electromyography (EMG) activities: eye movements produce strong low-frequency (1–4 Hz) activity at frontal electrodes, while temporal muscle activations typically induce 20–60 Hz activity at temporal electrodes [13, 14]. Nonphysiological artifacts originate from outside the human body and include power-line noise and sudden impedance changes due to electrodes movement. The magnitude of muscle and mechanical artifacts is enhanced during dynamical tasks, due to head movements and shocks. Bertrand et al. [15] observed that during different types of motions, most of the artifact energy is localized in the low frequencies (< 5 Hz) of the EEG signal. However, Castermans et al. [16] found that, during locomotion, the EEG signal may be polluted up to 15 harmonics of the fundamental stepping frequency and in high γ frequency bands extending up to 15 Hz.

2.5 BMI Based on Sensorimotor Rhythms

Event-related desynchronization (ERD) is a power decrease caused by the destructive interference resulting from the desynchronization of neuronal activity. On the other hand, event-related synchronization (ERS) is a power increase due to the constructive interference ascribed to neuronal synchrony. These phenomena are time-locked to the event, but not phase-locked, so they must be analyzed by time-frequency methodologies rather than simple linear methods, such as averaging. Moreover, neuronal networks can display different states of synchrony at different frequencies, and therefore ERD and ERS are considered to be highly frequency-band specific [17]. SMRs comprise oscillations in the μ (ie, 7–13 Hz) and β (13–30 Hz) bands; and while some β rhythms are harmonics of the μ rhythms, others are independent in topography and timing [3]. Rhythms in the γ (>30 Hz) band are also modulated by specific events; however, in single-trial analysis, they are typically considered when the signal is recorded with invasive electrodes, since the SNR at high frequencies is low with noninvasive systems. ERD is thought to be the result of cortical activation during sensory or cognitive information processing, motor learning, or motor execution, while ERS may reflect a deactivated state of the corresponding networks, and especially β ERS is associated with a deactivation of the motor cortex [17].

Hand movements elicit upper μ and lower β ERD in the sensorimotor areas, that is contralateral prior to the movement onset, and becomes bilateral at the onset. The maximal ERD is reached shortly after the onset, and recovers its original level within a few seconds [5]. After the termination of a movement, a β ERS with high SNR, dominant in the contralateral primary motor cortex [17], is elicited within 1 s from the movement offset [18]. ERD/ERS patterns with high SNR are produced by relatively large and topographically independent cortical regions, such as hand, foot, and tongue areas, making them good candidates for BMI [19]. However, the timing, topography, amplitude and dominant frequencies of those rhythms are highly dependent on the task being performed and on the subject who performs it [20], making machine learning essential for the discovery of exact spatial location of the SMR, timings, and frequency bands.

SMRs elicited by motor execution share similar topographies and spectral behavior with those arising from motor imagery (MI). This means that the simple kinesthetic imagination of movements [21], after several days or weeks of training, can elicit SMR modulations in the cortical areas directly connected to the normal neuromuscular pathways. For this reason MI is considered among the most suitable strategies for the natural control of a neuroprosthesis [19], and for BCI-based rehabilitation [22]. MI has been used to elicit ERD/ERS patterns for the control of some of the most successful BCI systems, including Wadsworth [23], Berlin [24], Graz [11], and variants of Tübingen [25].

2.5.1 Challenges
Nonstationarity and Adaptive Algorithms
The EEG signal exhibits a high variability over time, called nonstationarity. This phenomenon has many concurrent causes [26] including measurement artifacts and noise, physiological artifacts (ie, electromyogram—EMG), influence of nontask-related neurophysiological processes, changes in the experimental setup (ie, stimulus appearance), changes in psychological parameters (ie, concentration), and switching of neural states during brain functioning [27]. The combination of these factors makes the performance of BCI decoders unstable over time. Given the heterogeneous nature and unpredictability of the nonstationarity sources, understanding and describing quantitatively such phenomenon is complex. Therefore, even if some

studies have tried to model the nonstationarity itself as an intrinsic characteristic of the brain functioning [27], the majority of the BCI literature has
treated it as a nuisance to be minimized. Specifically, some algorithms try to
extract features that are robust against nonstationarities, while others adapt
their underlying models accordingly (see sections "Spatial Filter Identification (ICA,CSP)" and "Adaptive Classification").

Reducing Setup Time

Most of the studies in the BMI literature have collected the EEG signal with
gel-based electrodes. The electrolyte gel minimizes the electrochemical
contact potential and increases the SNR. However, the gel must be applied
at each channel location on the scalp, requiring a long preparation;
moreover, the subjects have to wash their hair, once the experiment is
over [28]. In order to minimize the preparation time, some studies have proposed to reduce the number of channels to the minimum required [29].
Another limitation of a gel-based EEG system is that the gel tends to dry
up, leading to more pronounced nonstationarities. In order to reduce setup
time, prevent the gel from drying up and avoid hair washing, dry EEG
systems have been implemented and successfully tested in MI-based BMI
applications [30].

In addition, every time the EEG system is placed, subject-specific features must be derived and a new decoding model must be calibrated, increasing the time required to set up the BMI. In recent years, researchers have
tried to build subject-independent decoders, by exploiting a large EEG database, in order for a naive subject to start using the BMI without calibration
[31]. In this context, the adaptive and coadaptive paradigms described in the
next paragraph make it possible for the user to receive feedback as early as
possible, with minimal calibration time.

BCI Illiteracy and Coadaptive Paradigms

A significant portion of users cannot achieve BCI control by MI, due to the
inability of generating the ERD [32] and to the difficulty in understanding
the concept of kinesthetic motor imagination [33]. MI can be trained either
by simple mental rehearsal or by feedback-guided training. In the latter, after
calibrating a subject-specific brain decoder, the user is asked to perform kinesthetic motor imagination while the associated cortical activity is given as
visual [34] or auditory [35] feedback, in the form of a score or a representation of the cortical activity itself [33]. Recently, coadaptive paradigms
have been proposed to address the BCI illiteracy issue and to minimize

calibration time. From the very beginning of the training, feedback is given to the subject by means of a subject-independent decoder. Simultaneously, the incoming EEG is used to update the model in order to derive a subject-dependent decoder, for a more accurate feedback. With this approach, healthy users who previously did not have the ability to control a non-adaptive BMI were able to gain control within a 1-day session [32]. Moreover, 18 of 22 severely disabled patients were able to achieve an accuracy better than chance after 24 min of coadaptive training [36]. Indeed, while the subject improves MI skills based on visual feedback, the decoder parameters are improved based on incoming data, and adapted to tackle both the highly nonstationary EEG signal and the changes in brain activity patterns caused by the subject's learning [37].

2.5.2 Decoding Pipeline

The procedure for training a decoder [38] typically includes a preprocessing step where the EEG signal is bandpass filtered (eg, 7–30 Hz), subsampled, and cleaned from noisy chunks of data. Subsequently, spatial filters that improve SNR are identified, followed by the extraction of time-frequency features (eg, log-power, Wavelet, or Fourier transforms), and concluding with classification [39] or regression [40, 41]. In the following paragraphs, we describe the latest efforts of the BMI community to improve the main components of the decoding pipeline.

Spatial Filter Identification (ICA, CSP)

Spatial filters are useful in single-trial analysis, in order to improve the SNR [42]. Supervised methods for the identification of spatial filters, such as common spatial patterns (CSPs), are typically used for MI tasks. Unsupervised techniques, such as independent component analysis (ICA), are preferable when the experimental condition is novel or labels are not available.

ICA finds a linear transformation of non-Gaussian data, so that the resulting components are as statistically independent as possible [43]. Hence, the EEG signal is separated into independent components accounting for different neural activities, but also stereotyped nonbrain-artifact signals including eye movements, line noise, and muscle activities [44].

CSP computes the unmixing matrix W to yield features whose variances are optimal for discriminating two classes of EEG measurements [24]. The algorithm typically gives better results when executed on optimal time windows and frequency bands [42]. In order to explicitly address spectral optimization several CSP variants have been proposed [45]. Robust CSP [46] reduces the bad influence of outliers, by computing robust covariance estimates using minimum covariance determinant (MCD). Several robust algorithms have also been proposed to address the *nonstationarity issue*. Invariant CSP [47] requires extra measurements associated with external disturbances (eg, artifacts, emotions), in order to enforce invariance in the CSP model. Stationary CSP [48] is inspired by invariant CSP, but does not require extra measurements, since its objective function maximizes the variance of one class and minimizes the variance of the other class, while penalizing nonstationary directions directly using the calibration dataset. Other methods try to adapt the CSP spatial filters using incoming data: Incremental CSP [49] adapts the filters by tracking the generalized eigenvalue equation at every time step, while Tomioka et al. [50] proposed a simple but effective whitening adaptation.

Generally, once the spatial filters have been identified, it is good practice to select the ones that are strictly indispensable for decoding, in order to avoid overfitting of unrelated or noisy sources [38].

Artifact Reduction

Stereotyped artifacts, such as eye blinks and line noise, can be isolated by ICA [51] on a relatively clean EEG signal, while thresholding of higher order statistics [13] and spectral perturbation magnitude are typically used to detect artifactual epochs. Before running ICA, it is good practice to detect and remove nonstereotyped artifacts that may reduce the quality of the independent components, by visual inspection, or by thresholding higher order statistics and spectral perturbation magnitude in the raw EEG signal domain. In addition, line noise can also be reduced by filtering the EEG signal. Other methods combine stereotyped artifact-specific spatial and temporal features, to capture blinks, eye movements, and generic discontinuities [52].

With respect to artifact reduction during dynamical tasks, such as walking, Severens et al. [53] cleaned EMG artifacts by means of canonical correlation analysis. Bertrand et al. [15] proposed a multichannel linear prediction filter, based on contact impedance measurements, to remove motion artifacts. In 2010, Gwin et al. [54] employed ICA and a template subtraction method, in order to remove movement artifacts from the EEG signal of a P300 task, executed during locomotion and running. Nevertheless, this method would not work if the target of the analysis were the neural mechanisms of locomotion, since the subtraction would eliminate the information of interest. Therefore, in their study about the EEG coupling with the gait cycle, Gwin et al. [55] removed artifactual independent components, based on the inspection of their power spectra and locations of their equivalent dipoles, computed by an inverse modeling method. Castermans et al. [16] raised some concerns about the results reported by Gwin et al. [55], suggesting that researchers should take more care in cleaning the EEG signal recorded during highly dynamical tasks. For a robotic assisted treadmill walk, Wagner et al. [56, 57] followed a similar rejection approach; however, the subsequent EEG analysis was performed on a narrower band (3–40 Hz) and slower walking speed as compared to Gwin et al. [55] (1–150 Hz). According to Kline et al. [58], movement artifacts associated with highly dynamical tasks, such as sustained speed, cannot be removed by traditional signal processing methods and more sophisticated techniques are necessary. Seeber et al. [59] proposed an alternative method to running ICA in the time domain. Namely, in order to reduce muscular artifacts, they decomposed the frequency spectra into orthogonal components using principal component analysis (PCA), identified artifactual principal spectral components and ignored them in the back projection.

All the methods discussed are used offline in order to derive better models; however, during online processing little can be done other than rejecting incoming data based on the detection of abnormal statistic measures and spectral features. Recently, a promising method called artifact subspace reconstruction (ASR) [60] has been proposed, which reconstructs the clean subspace of an incoming corrupted EEG chunk, based on the properties of a clean calibration dataset.

Adaptive Classification

As previously stated, the nonstationarity of the EEG signal is a major issue in BCI. A classifier trained on the features of a given calibration dataset is likely to fail when applied to new incoming data, since the variations of the EEG properties in time cause drastic modifications of the feature space. Therefore, it is important to adapt the classifier's parameters based on incoming data, in order to keep the classifier's decision boundaries effective throughout the experiment. Most of the adaptation schemes that have been proposed rely on a linear discriminant analysis (LDA) classifier. Even though many adaptation approaches have been proposed, including supervised adaptive LDA [61], adaptive LDA [62], Kalman adaptive LDA [63], and covariate shift adaptation[64], it has been shown that one of the simplest and most effective ways to adapt a LDA classifier—called *PMean*—is to modify its bias by updating the classes' global mean using incoming unlabeled data (ie, unsupervised) [37].

2.6 Performance Evaluation

When evaluating the performance of the decoder offline it is essential to perform cross-validation, in order to avoid overoptimistic results. Additionally,
given the nonstationary nature of the EEG signal, it is good practice to keep
in the same cross-validation subset the trials recorded close in time. Typical
performance measures include accuracy and Cohen's Kappa for classification
problems, and correlation coefficient for regression problems, with the
respective confidence intervals [65].

In online synchronous BCI, the time required to accomplish a goal becomes
essential. Common metrics in BCI to measure the amount of information
sent in a given time include ITR or bit rate. It should be noted that system
delays, time required to give feedback and time required to correct errors
must also be taken into account [65, 66].

In online asynchronous BCI, ITR is often not reported since some of the
preconditions to compute this measure are typically not met. Instead, it
may be meaningful to report the time required to accomplish a sequence
of actions, such as navigating through a virtual environment [67]. However,
several recent studies have highlighted the importance of an "idle" or noncontrol class, to evaluate how well a BCI system can handle a situation in
which a user does not want to control it [36, 68]. Muralidharan et al. [69] proposed an all-or-nothing approach to evaluate trigger-events, elicited by the
transition from a relaxed (ie, noncontrol) to the attempted-movement state.

3. BMI FOR ASSISTIVE ROBOTICS

As discussed in the previous sections, noninvasive BMI does not provide sufficient information to control a high-dimensional system [2]. Moreover, the low SNR of the EEG signal and inherent uncertainty of a BMI
decoder makes the control commands not always accurate and reliable.
However, given its inherent safety and practicality, noninvasive BMI has
been widely used to control *n*-dimensional robots by means of clever strategies, such as *shared control*, that translates high-level commands into complex
actions [70]. Additionally, this approach avoids the mental burden of continuous control, allowing the user to ignore low-level problems.

3.1 Shared Control

Shared control has been widely applied in mobile robotics applications
where the human control capabilities are limited [71]. This paradigm is based

on cooperation between the human operator and an intelligent controller to achieve the final task. Depending on the application and on the quality of the human control, the autonomy of the intelligent system can be increased or decreased. For example, in some cases the user decides the final goal and the robot handles all the low-level problems (eg, obstacle avoidance) to achieve it [72]. In some other cases, where people with disabilities prefer a higher level of control, the intelligent system helps the user in solving the low-level problems [73].

3.2 BMI and Nonwearable Robots

Shared control of nonwearable robots by noninvasive BMI and an additional intelligent system have real potential to improve the quality of life of tetraplegic patients. Indeed, the possibility of regaining a limited voluntary control of simple daily-life actions would not only increase their autonomy, but also their self-esteem. A recent study [74] showed the "limited but real potential" of a robotic arm, which uses 3D eye-tracking to infer the location of the object to be grasped, and SMR-based BMI to either grab the object or put it back. This system has the additional advantage that dry electrodes are used to acquire the EEG signal, which significantly shortens the setup time. Moreover, asynchronous SMR-based BMI allows for the self-paced navigation of mobile devices, such as wheelchairs [75] and telepresence [76] robots, relying on the intelligent system for the generation of efficient trajectories. Interestingly, researchers developed a quadcopter [77] and humanoid robot [78], asynchronously steered by SMR, that may find applications in telepresence robotics.

Exogenous BMI (eg, P300, VEP, SSVEP) has also been applied for the control of telepresence [79] and humanoid robots [80], smart wheelchairs [81], robotic arms [82], and hybrid wheelchair-mounted robotic arms [83]. In this case, after the user selects the target action from several possible choices, the robot executes it in complete autonomy. This approach gives the possibility of choosing from many more target actions, compared to SMR-based BMI; however, the user has little authority after the task has been initiated.

In addition, hybrid systems, combining SMR-based and exogenous BMI, were proposed in order to overcome the respective drawbacks. A study proposed to control two DoF of an artificial upper limb by SSVEP and the grasping function by MI [84]. Another group was able to navigate a robot by SSVEP and ERD and to perform object recognition by P300 [85].

In recent years, new smart approaches have also been investigated, where ErrP-based BMI is used to guide the reinforcement learning of a robot [86]. In this framework, if the user observes a mistake during the robot automatic operation, the associated ErrP potential is detected and the probability of the corresponding action is decreased [87]. This approach has been recently used to teach motor behaviors to a robotic arm [88], and its feasibility has been studied for an autonomous wheelchair [89].

3.3 BMI and Wearable Robots

A wearable robot is a device, designed around the shape and function of the human limbs, that generates supplementary forces to overcome human physical limits [123]. Orthotic robots, such as exoskeletons, restore lost or weak limb functions, while prosthetic robots substitute the limb after amputation. Mind-controlled wearable robots are used as assistive or substitutive devices in case of permanent injuries, such as spinal cord injury (SCI), in order to bypass an impaired motor function. Additionally, recent studies have been testing the hypothesis that the haptic neurofeedback provided by the robot could promote motor recovery in stroke patients. Most of these works focused on MI, given that the reinforcement of sensorimotor cortex activation by haptic neurofeedback is the closest strategy to active training by motor execution [22, 124].

Table 1 reviews, in chronological order, the EEG-controlled wearable robots that were tested either for motor substitution or motor recovery. The main reference of each system is provided together with the BMI control modality (eg, MI and SSVEP), the type of robot, the assisted function and the condition of the targeted users in the experiments. In this section, neuroprosthesis based on functional electrical stimulation (FES) is considered to be a wearable robotic device, since human muscles are natural actuators that can be controlled by complex control strategies that activate nerves innervating paralyzed extremities [125]. From Table 1, we observe that, at the current state, the majority of the studies have been addressing the upper limbs, which is mainly due to the difficulties associated with decoding the artifact-prone EEG signal during walking, and with decoding the artifact-prone EEG signal during walking, and with the control of biped walking and balancing of a lower limb exoskeleton robot. However, given the recent advances in the field of biped robot locomotion control, studies addressing the lower limbs are increasing as part of ambitious projects aiming to design safe and stable lower-limb exoskeleton robots controlled by noninvasive BMI [126, 127].

Table 1 EEG-Based BMI and Wearable Robots

Reference	BMI Modality	Robot	Function	Condition
Pfurtscheller et al. [90]	Foot MI	Surface FES	Grasping	Tetraplegia
Müller–Putz et al. [91]	Paralyzed hand MI	Implantable FES	Grasping	SCI
Meng et al. [92]	MI	Surface FES	Wrist and hand	Stroke
Ang et al. [93]	Affected limb MI	MIT–manus	Reaching	Stroke
Daly et al. [94]	Finger MI	Surface FES	Finger extension	Stroke
Tavella et al. [95]	Hand MI	Surface FES	Grasping, writing	Healthy
Pfurtscheller et al. [96]	MI activates SSVEP	Hand orthosis	Hand open/close (o/c)	Healthy
Broetz et al. [97]	MI	Arm orthosis	Forward, backward movement	Stroke
Tan et al. [98]	MI	Surface FES	Wrist movement	Stroke
Ortner et al. [68]	SSVEP	Hand orthosis	Hand open/close (o/c)	Tetraplegia
Caria et al. [99]	Arm/hand MI	Hand orthosis	Arm movement, hand o/c	Stroke
Gomez-Rodriguez et al. [100]	Elbow MI	Elbow orthosis	Arm extension/flexion	Stroke, healthy
Shindo et al. [101]	Finger MI	Hand orthosis	Finger extension	Stroke
Tam et al. [102]	MI	Surface FES	Hand open/close (o/c)	Stroke
Ramos-Murguialday et al. [103]	Hand MI	Hand orthosis	Hand open/close (o/c)	Healthy
Frisoli et al. [104]	MI, gaze	Arm exoskeleton	Reaching	Healthy
Takahashi et al. [105]	Ankle Dorsiflexion ERD	Surface FES	Ankle dorsiflexion	Stroke
Noda et al. [106]	Left/Right MI	Legs exoskeleton	Stand up	Healthy
Ramos-Murguialday et al. [107]	Arm MI	Arm/hand orthosis	Reach and hand o/c	Stroke
Kilicarslan et al. [108]	Delta-band EEG	Legs exoskeleton	Walk, turn, etc.	Paraplegia
Rohm et al. [109]	MI, shoulder position	FES, elbow orthosis	Grasp, elbow	Tetraplegia
Sakurada et al. [110]	SSVEP	Arms exoskeleton	Grasp, etc.	SCI, healthy

Continued

Table 1 EEG-Based BMI and Wearable Robots—cont'd

Reference	BMI Modality	Robot	Function	Condition
Do et al. [111]	Walking MI	Gait orthosis	Walk	Paraplegia, healthy
Sarac et al. [112]	MI	Arm rehabilitation robot	Contour following	Healthy
Blank et al. [113]	MI	Arm exoskeleton	Elbow, wrist movement	–
Varkuti et al. [114]	MI	MIT-manus	Shoulder-elbow movements	Stroke
Witkowski et al. [115]	MI and EOG	Hand exoskeleton	Grasping	Healthy
Ang et al. [116]	Affected hand MI	Haptic knob	Wrist, hand movement	Stroke
Looned et al. [117]	Arm MI	FES, elbow orthosis	Drinking movement	Healthy
Young et al. [118]	Finger tapping SMR	Surface FES	Finger tapping	Stroke
Elnady et al. [119]	MI	FES, elbow orthosis	Reach and grasp	Stroke
Kwak et al. [120]	SSVEP	Legs exoskeleton	Walk, turn, etc.	Healthy
Naros et al. [121]	β-SMR MI	Hand orthosis	Hand o/c	Stroke
Barsotti et al. [122]	MI	Upper limb exoskeleton	Reach and grasp	Stroke
Li et al. [153]	Foot MI (dry electrodes)	Legs exoskeleton	Squat triggering	Healthy

One of the differences between a nonwearable and a wearable robot is that the latter induces movements of the user's limb. In the next two sections, we discuss how passive movements may elicit ERD/ERS interfering with the target MI, disrupting the decoder performance, and how they can be used as haptic neurofeedback for novel coadaptive paradigms.

3.3.1 Passive Movements as Nuisance

In our recent work [38], we focused on the passive movements elicited by a lower limb exoskeleton robot, and how they may interfere with the decoding of upper limb motor tasks (Fig. 1). Indeed, Müller-Putz et al. [18] suggested that passive movements of the feet produce a significant ERD/ERS not only at the vertex, but over the whole sensorimotor cortex. Our results suggest that the somatosensory afferent input, induced by the periodic leg perturbation, does not interfere with the decoding ability of an asynchronous BMI system based on the SMRs of the hand area. The results were obtained using finger tapping as a motor task, in order to reduce the uncertainty associated with MI. Therefore, future studies will have to verify whether the same conclusion can be drawn for MI, which is typically characterized by a smaller ERD/ERS magnitude [128] and requires a more intense concentration of the user compared to actual movements, especially if the user's body is perturbed simultaneously. This is an important question in a scenario where left and right hand MI is used to steer the exoskeleton robot, while the controller automatically handles biped walking and collision avoidance.

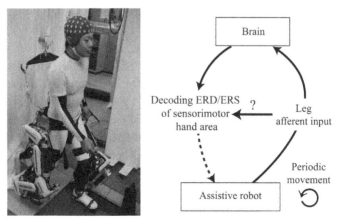

Fig. 1 Understanding how the leg afferent input, induced by a leg assistive robot, influences the decoding of the ERD/ERS of the sensorimotor hand area.

3.3.2 Haptic Feedback for Coadaptive BRI

As mentioned in Section 2.5.1, recent studies have proposed coadaptive BMI paradigms to address the "illiteracy" issue and minimize calibration time. So far, these studies have provided visual or auditory feedback, while haptic feedback has not been considered yet. Ramos-Murguialday et al. [103] showed that proprioceptive feedback improves BMI performance and enhances ERD during upper limb motor tasks and MI. However, they did not investigate advanced adaptation techniques for the BMI decoder, and performed the BMI training during four consecutive days. We believe that research should focus on combining the aforementioned topics, towards the design of a coadaptive BRI, in which the user imagines the limb movement, and SMR feedback is given in the form of proprioceptive afferent input by the movement of a wearable robot. Ideally, a haptic coadaptive BRI system would help the user to learn MI after few trials of training, by providing proprioceptive feedback that enhances the SMR desynchronization, while the underlying decoder is continuously improved.

4. BMI FOR REHABILITATION ROBOTICS

CNS plasticity involves synaptic, neuronal, and circuit modifications caused by sensory input, experience, and learning [129]. This phenomenon plays a crucial role in the reorganization of cortical areas following a brain damage after stroke. A stroke occurs when blood deprivation from a portion of the brain causes brain cell death. Following a brain lesion, activity-dependent CNS plasticity allows for motor relearning leading to the improvement of motor functions [4].

Rehabilitation paradigms based on MI [130] and SMR-mediated visual neurofeedback have been recently proposed to engage sensorimotor areas and promote activity-dependent plasticity [124, 131]. Additionally, repetitive motor practice enhances neural plasticity, leading to the improvement of motor functions [132]. Robotic devices [133], including lower limb exoskeletons [134], have been used in poststroke recovery, since they can be programmed to perform repetitive movements. This rehabilitation paradigm might be monotonous, requiring little participation and effort from the patient. A more interactive rehabilitation task would be preferable to enhance the patient's active participation [135]. Coupling BMI with a robotic orthosis engages the patients and enhances active participation [107], following a top-down rehabilitation paradigm [136]. In this section we highlight the studies in Table 1 that carried out clinical trials of motor rehabilitation using a BMI-controlled wearable robot.

4.1 Upper Limb Motor Recovery

Ang et al. [93, 137] were the first to carry out a clinical trial on stroke patients using EEG-based BMI and a rehabilitation robot (ie, MIT-manus), and to compare this paradigm with robotic-only intervention. The Fugl-Meyer Assessment (FMA) score, used to measure the motor outcome of the rehabilitation training, highlighted a significant improvement of the motor functions for both paradigms, but no significant difference was found between the two of them. The same group proposed the use of transcranial direct current stimulations (tDCS) to increase cortical excitability before MI is performed and the robot is moved [138]. Daly et al. [94], reported that a stroke patient was able to increase his finger extension after 3 weeks of rehabilitation training, where the patient performed motor execution and MI to control a FES device. Broetz et al. [97] combined daily life-oriented physical therapy with a BMI-driven robotic orthosis to treat a stroke patient with no active finger extension. After 1 year of training and motor function assessment, an increase of micro-oscillations was observed in the ipsilateral motor cortex, in conjunction with the improvement of hand and arm motor ability as well as speed and safety of gait. In Shindo et al. [101], during a period of 4 to 7 weeks, eight stroke patients controlled a hand orthosis by MI. The recovery of the electromyographic activity in the affected finger extensors was observed in four patients, while the other participants displayed an improved finger function. This was accompanied by a stronger ERD over both hemispheres during MI, and by increased cortical excitability in the affected hemisphere, which may reflect neuroplastic changes. Ramos-Murguialday et al. [107] found that combining physiotherapy with BMI-robotic intervention leads to a more prominent improvement compared to a BMI-physiotherapy control. This suggests that proprioceptive feedback rewarding ERD may prime and improve the beneficial effects of physiotherapy on motor function. Ono et al. [139] suggested a similar thesis, by showing that intervention combining BMI and proprioceptive feedback led to a greater motor gain, compared to visual feedback. Ang et al. [116] reported a higher motor gain up to 6 months for patients who underwent BMI-driven haptic knob rehabilitation, compared to patients who received standard arm therapy. Moreover, the same group [140] observed comparable motor gain with a less intensive therapy, when comparing BMI-robotic intervention with the robotic-only one, consisting of 136 and 1040 repetitions per session, respectively. Naros et al. [121] proposed to use the β-band SMR for neurofeedback, rather than the α-band, since the former mediates the disinhibition of the sensorimotor cortex and the coherent interaction with the muscles, while the latter is associated with the inhibition of task-irrelevant cortical regions. Moreover, they introduced a reinforcement learning framework, where the goal is to preserve the patient's motivation, rather than maximizing accuracy. A preliminary clinical study suggests that this approach may have potential to achieve specific motor gains; and if applied for sufficient time and appropriate intensity, it may lead to functional restoration for daily-life activities.

Other studies not only conducted clinical trials and assessed motor recovery, but also tried to evaluate the plastic cortical modifications caused by the combination of BMI and proprioceptive feedback intervention. By using diffusion tensor imaging (DTI) and fMRI, Caria et al. [99] showed for the first time that plastic changes occurred, accompanied by motor recovery, as a result of physiotherapy and BMI-driven robotic intervention. Similarly, Várkuti et al. [114] observed an increase of resting state functional connections (RS-fMRI) after BMI training with robotic feedback, compared to robotic-only rehabilitation. Results of Young et al. [118] suggest that BMI therapy with FES feedback can induce a greater involvement of the nonlesioned hemisphere during tapping of the impaired hand.

We also report two studies that did not make use of an EEG-driven robot during the intervention, but found significant evidence that BMI is effective in stroke rehabilitation. Buch et al. [141] used MEG to detect SMR associated with MI and to open/close a hand orthosis. Even though they did not observe significant motor improvement after training [141], they reported that SMR modulations during the intervention rely on "structural and functional connectivity in both ipsilesional and contralesional parietofrontal pathways involved in visuomotor information processing"; which may "serve as a future predictor of response to longitudinal therapeutic interventions" [142]. Another study by Mihara et al. [143] showed that combining MI with fNIRS neurofeedback significantly enhances the motor improvement, compared to MI with sham feedback. The fact that combining BMI with MI practice may lead to better motor functional outcomes was also supported by a recent EEG study by Pichiorri et al. [144].

4.2 Lower Limb and Gait Recovery

Lower limb and gait recovery after stroke has received little attention from the BMI community, especially from a clinical point of view. Sun et al. [145] showed that MI and BMI-mediated visual feedback could lead to an improvement of balance and walking in a 4-week training with 20 stroke patients, although they did not consider proprioceptive feedback. Takahashi et al. [105] used the ERD associated with attempted ankle dorsiflexion to activate a FES of the tibialis anterior muscle, and found on one stroke patient that, after 20 min of training, the EMG and the ankle's range of movement were significantly increased. As already mentioned, Broetz et al. [97] showed that motor recovery of the upper limb can also improve speed and safety of gait, and this may reveal some common mechanism that simultaneously influences upper and lower limb recovery [136].

Given the limited number of clinical trials for the lower limb and gait recovery, we report the latest methodological advances in this direction. Specifically, we review the major studies that found neural correlates of gait; however, we warn the reader that consensus has not been reached on the reliability of these findings, due to the strong influence of motion artifacts during walking [16, 58]. Haefeli et al. [146] found an enhanced activity during preparation and execution of obstacle stepping on a treadmill, compared to normal walking. Gwin et al. suggested that spectral power modulations are coupled with the gait cycle, and Presacco et al. decoded leg kinematics from the delta band (ie, <2 Hz); however, both studies have raised some concerns about the artifacts rejection techniques [16, 58] or the decoding model used [41]. Wagner et al. [56] found increased activity (ie, suppressed μ and β activity) over foot sensorimotor areas during active walking compared to passive walking, and over premotor and sensorimotor areas during walking as compared to a rest condition. The same research group found repeatable evidence [56, 57, 147] that power modulations in the lower γ band (25–40 Hz), over the premotor cortex and over the foot area of the primary motor cortex, are coupled with the gait cycle during active and robotic assisted treadmill walk. Seeber at al. [147] suggested that the observed suppression of μ and β rhythms throughout the gait cycle and the modulation of low γ oscillations belong to different neural networks and functions. In a more recent study, Seeber et al. [59] found that high γ components have a significantly higher power during walking compared to standing. Moreover, modulations in the high γ band are coupled with the gait cycle and negatively correlated with those in the low γ band. Sipp et al. [148] showed that the theta activity at midline cortical areas increases during walking on a balance beam as compared to treadmill walking. Wagner et al. [57] found increased activity in premotor and parietal areas during treadmill walking with adaptive virtual environment, as compared to walking in front of a mirror or with movement unrelated visual feedback. Our latest study [149] showed that μ and β rhythms in the posterior parietal cortex (PPC) are suppressed during gait speed changes on a treadmill (Fig. 2), and that a single-trial EEG classifier is able to detect them. The observed cortical activation may reflect motor planning and visuomotor transformations during online gait adaptation. Very recently, Knaepen et al. [154] found that the sensorimotor cortex is significantly more active during

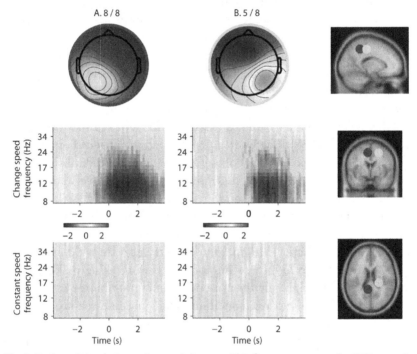

Fig. 2 Brain activity during gait speed changes. This figure compares the ERD occurring during gait speed changes (ie, acceleration and deceleration) as opposed to constant speed. In order to obtain a compact representation of the spectral perturbation across subjects, individual independent components (IC) were clustered, and time-frequency representations were averaged within the clusters. Moreover, for every IC cluster, we visualize the equivalent current dipole. The scalp maps of each cluster are visualized in the top row of the figure. Below each cluster's scalp map, we show the event related spectral perturbation (ERSP) corresponding to the *change-speed* and *constant-speed* classes. For each cluster, ERSPs are normalized using the mean log spectrum over the whole epoch length of the *constant-speed* class, and the respective color axis is visualized. On the right side, the dipoles corresponding to cluster centroids are displayed on the standard MNI brain model in sagittal, coronal, and axial view. The color of each dipole corresponds to the color of the circle surrounding the respective cluster scalp map. We observe that both clusters are characterized by ERD accompanying the change of speed. Cluster A accounts for brain activity in the posterior parietal cortex (PPC) of every subject (8/8). Cluster B, accounting for five out of eight subjects, displays a similar, but weaker, ERD perturbation compared to the other cluster, and is located more anteriorly, approximately in the primary motor cortex.

treadmill walking compared to robot-assisted treadmill walking with 100% guidance force, suggesting that a high level of guidance force should be avoided in order to increase the patient's engagement during rehabilitation. Other studies, based on fNIRS, have found evidence of cortical involvement during walking related tasks [150].

5. CONCLUSION

This chapter has reviewed the state-of-the-art of noninvasive BMI for robot control. In doing so, we described active branches of research in the BMI field, such as EEG nonstationarity, artifacts reduction, advanced spatial filters identification algorithms, adaptive classifiers and coadaptive paradigms that alleviate the problem of BMI illiteracy. Subsequently, we reviewed the studies that applied noninvasive BMI for the *shared control* of nonwearable robots. After providing the literature on BMI-controlled wearable devices, we discussed the possibilities arising from proprioceptive feedback: it induces afferent input that may interfere with motor-imagery based BMI systems, or it may be used in coadaptive paradigms to improve BMI performance. The last section described the clinical trials carried out mostly for upper limb motor recovery, providing evidence that BMI-controlled wearable robots may find application in motor neurorehabilitation, mostly as adjunctive tools for standard training. Few studies have performed clinical trials for lower limb and gait motor recovery; however, we found that great efforts have been made in order to find neural correlates of gaits, that in future may be used for rehabilitation.

Kersting et al. [151] found that motor output recovery induced by afferent stimulation is significantly larger at an approximately 100 ms delay compared to 400 ms delay from the imagination onset, suggesting that the timing of the proprioceptive feedback is crucial in order to induce cortical plasticity; most of the reviewed clinical trials used ERD in order to activate the robotic feedback, and even though some of them tried to minimize the latency [105, 107], no one has explicitly considered the timing of the feedback as a factor of the analysis. Nevertheless, efforts to reduce the decoding delay and even predict the attempted movement have been made in recent years [69].

After a comprehensive review of BMI and its applications to assistive and rehabilitation robotics, we believe that an interesting research direction to be pursued is the coadaptive BRI paradigm discussed in Section 3.3.2, Ideally, it would help a stroke patient to learn MI and simultaneously induce neuroplasticity, leading to the restoration of motor control [121, 152].

ACKNOWLEDGMENTS

This study was supported by the New Energy and Industrial Technology Development Organization (NEDO), by SRPBS from AMED, by "Development of Medical Devices and Systems for Advanced Medical Services" from AMED and by ImPACT Program of Council for Science, Technology and Innovation (Cabinet Office, Government of Japan).

REFERENCES

[1] J. L. Collinger, B. Wodlinger, J. E. Downey, W. Wang, E. C. Tyler-Kabara, D. J. Weber, A. J. McMorland, M. Velliste, M. L. Boninger, A. B. Schwartz, High-performance neuroprosthetic control by an individual with tetraplegia, Lancet 381 (9866) (2013) 557–564.

[2] M. A. Lebedev, M. A. L. Nicolelis, Brain-machine interfaces: past, present and future, Trends Neurosci. 29 (9) (2006) 536–546.

[3] J. R. Wolpaw, N. Birbaumer, D. J. McFarland, G. Pfurtscheller, T. M. Vaughan, Brain-computer interfaces for communication and control, Clin. Neurophysiol. 113 (6) (2002) 767–791.

[4] J. J. Daly, J. R. Wolpaw, Brain-computer interfaces in neurological rehabilitation, Lancet Neurol. 7 (11) (2008) 1032–1043.

[5] L. F. Nicolas-Alonso, J. Gomez-Gil, Brain computer interfaces, a review, Sensors (Basel, Switzerland), 12 (2) (2012) 1211–1279.

[6] G. Bin, X. Gao, Y. Wang, B. Hong, S. Gao, VEP-based brain-computer interfaces: time, frequency, and code modulations [Research Frontier], Comput. Intel. Mag. IEEE 4 (4) (2009) 22–26.

[7] E. Donchin, K. M. Spencer, R. Wijesinghe, The mental prosthesis: assessing the speed of a p300-based brain-computer interface, IEEE Trans. Rehabil. Eng. 8 (2000) 174–179.

[8] G. Schalk, J. R. Wolpaw, D. J. McFarland, G. Pfurtscheller, EEG-based communication: presence of an error potential, Clin. Neurophysiol. 111 (12) (2000) 2138–2144.

[9] N. Birbaumer, A. Kubler, N. Ghanayim, T. Hinterberger, J. Perelmouter, J. Kaiser, I. Iversen, B. Kotchoubey, N. Neumann, H. Flor, The thought translation device (TTD) for completely paralyzed patients, IEEE Trans. Rehabil. Eng. 8 (2) (2000) 190–193.

[10] G. Pfurtscheller, B. Z. Allison, C. Brunner, G. Bauernfeind, T. Solis-Escalante, R. Scherer, T. O. Zander, G. Mueller-Putz, C. Neuper, N. Birbaumer, The hybrid BCI, Front. Neurosci. 4 (2010) 30.

[11] G. Pfurtscheller, C. Neuper, G. R. Müller, B. Obermaier, G. Krausz, A. Schlögl, R. Scherer, B. Graimann, C. Keinrath, D. Skliris, M. Wörtz, G. Supp, C. Schrank, Graz-BCI: state of the art and clinical applications, IEEE Trans. Neural Syst. Rehabil. Eng. 11 (2) (2003) 177–180.

[12] R. Leeb, D. Friedman, G. R. Müller-Putz, R. Scherer, M. Slater, G. Pfurtscheller, Self-paced (asynchronous) BCI control of a wheelchair in virtual environments: a case study with a tetraplegic, Comput. Intell. Neurosci. 2007 (2007) 79642.

[13] A. Delorme, T. Sejnowski, S. Makeig, Enhanced detection of artifacts in EEG data using higher-order statistics and independent component analysis, Neuroimage 34 (4) (2007) 1443–1449.

[14] M. Fatourechi, A. Bashashati, R. K. Ward, G. E. Birch, EMG and EOG artifacts in brain computer interface systems: a survey, Clin. Neurophysiol. 118 (3) (2007) 480–494.

[15] A. Bertrand, V. Mihajlovic, B. Grundlehner, C. Van Hoof, M. Moonen, Motion artifact reduction in EEG recordings using multi-channel contact impedance

measurements, in: 2013 IEEE Biomedical Circuits and Systems Conference (Bio-CAS), 2013, pp. 258–261.

[16] T. Castermans, M. Duvinage, G. Cheron, T. Dutoit, About the cortical origin of the low-delta and high-gamma rhythms observed in EEG signals during treadmill walking, Neurosci. Lett. 561 (2014) 166–170.

[17] G. Pfurtscheller, F. H. Lopes da Silva, Event-related EEG/MEG synchronization and desynchronization: basic principles, Clin. Neurophysiol. 110 (11) (1999) 1842–1857.

[18] G. R. Müller-Putz, D. Zimmermann, B. Graimann, K. Nestinger, G. Korisek, G. Pfurtscheller, Event-related beta EEG-changes during passive and attempted foot movements in paraplegic patients, Brain Res. 1137 (1) (2007) 84–91.

[19] B. Graimann, B. Allison, G. Pfurtscheller, Brain-computer interfaces: a gentle intro-duction, in: B. Graimann, G. Pfurtscheller, B. Allison (Eds.), Brain-Computer Inter-faces, The Frontiers Collection, Springer, Berlin, Heidelberg, 2010, pp. 1–27.

[20] G. Pfurtscheller, C. Neuper, Future prospects of ERD/ERS in the context of brain-computer interface (BCI) developments, Prog. Brain Res. 159 (2006) 433–437.

[21] C. Neuper, R. Scherer, M. Reiner, G. Pfurtscheller, Imagery of motor actions: dif-ferential effects of kinesthetic and visual-motor mode of imagery in single-trial EEG, Brain Res. Cogn. Brain Res. 25 (3) (2005) 668–677.

[22] S. Silvoni, A. Ramos-Murguialday, M. Cavinato, C. Volpato, G. Cisotto, A. Turolla, F. Piccione, N. Birbaumer, Brain-computer interface in stroke: a review of progress, Clin. EEG Neurosci. 42 (4) (2011) 245–252.

[23] J. R. Wolpaw, D. J. McFarland, T. M. Vaughan, Brain-computer interface research at the Wadsworth Center, IEEE Trans. Rehabil. Eng. 8 (2) (2000) 222–226.

[24] B. Blankertz, F. Losch, M. Krauledat, G. Dornhege, G. Curio, K.-R. Muller, The berlin brain-computer interface: accurate performance from first-session in BCI-naive subjects, IEEE Trans. Biomed. Eng. 55 (10) (2008) 2452–2462.

[25] N. Birbaumer, L. G. Cohen, Brain-computer interfaces: communication and restora-tion of movement in paralysis, J. Physiol. 579 (Pt 3) (2007) 621–636.

[26] M. Krauledat, Analysis of nonstationarities in EEG signals for improving brain-computer interface performance, PhD thesis, Berlin Institute of Technology, 2008.

[27] A. Y. Kaplan, A. A. Fingelkurts, A. A. Fingelkurts, S. V. Borisov, B. S. Darkhovsky, Nonstationary nature of the brain activity as revealed by EEG/MEG: methodological, practical and conceptual challenges, Signal Process. 85 (11) (2005) 2190–2212.

[28] P. M. R. Reis, F. Hebenstreit, F. Gabsteiger, V. von Tscharner, M. Lochmann, Methodological aspects of EEG and body dynamics measurements during motion, Front. Hum. Neurosci. 8 (2014) 156.

[29] G. R. Müller-Putz, V. Kaiser, T. Solis-Escalante, G. Pfurtscheller, Fast set-up asyn-chronous brain-switch based on detection of foot motor imagery in 1-channel EEG, Med. Biol. Eng. Comput. 48 (3) (2010) 229–233.

[30] F. Popescu, S. Fazli, Y. Badower, B. Blankertz, K.-R. Müller, Single trial classification of motor imagination using 6 dry EEG electrodes, PLoS ONE 2 (7) (2007)e637.

[31] S. Fazli, C. Grozea, M. Danoczy, B. Blankertz, F. Popescu, K.-R. Müller, Subject independent EEG-based BCI decoding, in: Y. Bengio, D. Schuurmans, J. Lafferty, C. Williams, A. Culotta (Eds.), Advances in Neural Information Processing Systems 22, Curran Associates, Inc, Red Hook, NY, 2009, pp. 513–521.

[32] C. Vidaurre, B. Blankertz, Towards a cure for BCI illiteracy, Brain Topogr. 23 (2) (2010) 194–198.

[33] H.-J. Hwang, K. Kwon, C.-H. Im, Neurofeedback-based motor imagery training for brain-computer interface (BCI), J. Neurosci. Methods 179 (1) (2009) 150–156.

[34] B. Blankertz, G. Dornhege, M. Krauledat, K.-R. Müller, G. Curio, The non-invasive Berlin brain-computer interface: fast acquisition of effective performance in untrained subjects, Neuroimage 37 (2) (2007) 539–550.

[35] F. Nijboer, A. Furdea, I. Gunst, J. Mellinger, D. J. McFarland, N. Birbaumer, A. Kübler, An auditory brain-computer interface (BCI), J. Neurosci. Methods 167 (1) (2008) 43–50.

[36] J. Faller, R. Scherer, U. Costa, E. Opisso, J. Medina, G. R. Müller-Putz, A co-adaptive brain-computer interface for end users with severe motor impairment, PLoS ONE 9 (7) (2014)e101168.

[37] C. Vidaurre, M. Kawanabe, P. von Bünau, B. Blankertz, K. Müller, Toward unsupervised adaptation of LDA for brain-computer interfaces, IEEE Trans. Biomed. Eng. 58 (3) (2011) 587–597.

[38] G. Lisi, T. Noda, J. Morimoto, Decoding the ERD/ERS: influence of afferent input induced by a leg assistive robot, Front. Syst. Neurosci. 8 (2014) 85.

[39] F. Lotte, M. Congedo, A. Lécuyer, F. Lamarche, B. Arnaldi, A review of classification algorithms for EEG-based brain-computer interfaces, J. Neural Eng. 4 (2) (2007) R1–R13.

[40] H. Yuan, C. Perdoni, B. He, Relationship between speed and EEG activity during imagined and executed hand movements, J. Neural Eng. 7 (2) (2010) 26001.

[41] J. M. Antelis, L. Montesano, A. Ramos-Murguialday, N. Birbaumer, J. Minguez, On the usage of linear regression models to reconstruct limb kinematics from low frequency EEG signals, PLoS ONE 8 (4) (2013)e61976.

[42] B. Blankertz, R. Tomioka, S. Lemm, M. Kawanabe, K.-R. Muller, Optimizing spatial filters for robust EEG single-trial analysis, IEEE Signal Process. Mag. 25 (1) (2008) 41–56.

[43] A. Hyvärinen, E. Oja, Independent component analysis: algorithms and applications, Neural Netw. 13 (4-5) (2000) 411–430.

[44] S. Makeig, S. Debener, J. Onton, A. Delorme, Mining event-related brain dynamics, Trends Cogn Sci. 8 (5) (2004) 204–210.

[45] K. K. Ang, Z. Y. Chin, C. Wang, C. Guan, H. Zhang, Filter bank common spatial pattern algorithm on BCI competition IV datasets 2a and 2b, Front. Neurosci. 6 (2012) 39.

[46] X. Yong, R. K. Ward, G. E. Birch, Robust common spatial patterns for EEG signal preprocessing, Conf. Proc. IEEE Eng. Med. Biol. Soc. 2008 (2008) 2087–2090.

[47] B. Blankertz, M. K. R. Tomioka, F. U. Hohlefeld, V. Nikulin, K. Robert Müller, Invariant common spatial patterns: alleviating nonstationarities in brain-computer interfacing, Advances in Neural Information Processing Systems 20, MIT Press, Cambridge, MA, 2008, p. 2008.

[48] W. Samek, C. Vidaurre, K.-R. Müller, M. Kawanabe, Stationary common spatial patterns for brain-computer interfacing, J. Neural Eng. 9 (2) (2012) 026013.

[49] Q. Zhao, L. Zhang, A. Cichocki, J. Li, Incremental common spatial pattern algorithm for BCI, in: IEEE International Joint Conference on Neural Networks, IJCNN 2008 (IEEE World Congress on Computational Intelligence), 2008, pp. 2656–2659.

[50] R. Tomioka, J. Hill, B. Blankertz, K. Aihara, Adapting spatial filtering methods for nonstationary BCIs, Transformation 10 (2006) 1.

[51] C. A. Joyce, I. F. Gorodnitsky, M. Kutas, Automatic removal of eye movement and blink artifacts from EEG data using blind component separation, Psychophysiology 41 (2) (2004) 313–325.

[52] A. Mognon, J. Jovicich, L. Bruzzone, M. Buiatti, ADJUST: an automatic EEG artifact detector based on the joint use of spatial and temporal features, Psychophysiology 48 (2) (2011) 229–240.

[53] M. Severens, B. Nienhuis, P. Desain, J. Duysens, Feasibility of measuring event related desynchronization with electroencephalography during walking, Conf. Proc. IEEE Eng. Med. Biol. Soc. 2012 (2012) 2764–2767.

[54] J. T. Gwin, K. Gramann, S. Makeig, D. P. Ferris, Removal of movement artifact from high-density EEG recorded during walking and running, J. Neurophysiol. 103 (6) (2010) 3526–3534.

[55] J. T. Gwin, K. Gramann, S. Makeig, D. P. Ferris, Electrocortical activity is coupled to gait cycle phase during treadmill walking, Neuroimage 54 (2) (2011) 1289–1296.

[56] J. Wagner, T. Solis-Escalante, P. Grieshofer, C. Neuper, G. Müller-Putz, R. Scherer, Level of participation in robotic-assisted treadmill walking modulates midline sensorimotor EEG rhythms in able-bodied subjects, Neuroimage 63 (3) (2012) 1203–1211.

[57] J. Wagner, T. Solis-Escalante, R. Scherer, C. Neuper, G. Müller-Putz, It's how you get there: walking down a virtual alley activates premotor and parietal areas, Front. Hum. Neurosci. 8 (2014) 93.

[58] J. E. Kline, H. J. Huang, K. L. Snyder, D. P. Ferris, Isolating gait-related movement artifacts in electroencephalography during human walking, J. Neural Eng. 12 (4) (2015) 046022.

[59] M. Seeber, R. Scherer, J. Wagner, T. Solis-Escalante, G. R. Müller-Putz, High and low gamma EEG oscillations in central sensorimotor areas are conversely modulated during the human gait cycle, Neuroimage 112 (2015) 318–326.

[60] T. Mullen, C. Kothe, Y. M. Chi, A. Ojeda, T. Kerth, S. Makeig, G. Cauwenberghs, T.-P. Jung, Real-time modeling and 3D visualization of source dynamics and connectivity using wearable EEG, Conf. Proc. IEEE Eng. Med. Biol. Soc. 2013 (2013) 2184–2187.

[61] C. Vidaurre, A. Schlogl, R. Cabeza, R. Scherer, G. Pfurtscheller, A fully on-line adaptive BCI, IEEE Trans. Biomed. Eng. 53 (6) (2006) 1214–1219.

[62] J. Blumberg, J. Rickert, S. Waldert, A. Schulze-Bonhage, A. Aertsen, C. Mehring, Adaptive classification for brain computer interfaces, Conf. Proc. IEEE Eng. Med. Biol. Soc. 2007 (2007) 2536–2539.

[63] C. Vidaurre, A. Schlögl, R. Cabeza, R. Scherer, G. Pfurtscheller, Study of on-line adaptive discriminant analysis for EEG-based brain computer interfaces, IEEE Trans. Biomed. Eng. 54 (3) (2007) 550–556.

[64] M. Sugiyama, M. Krauledat, K.-R. Müller, Covariate shift adaptation by importance weighted cross validation, J. Mach. Learn. Res. 8 (2007) 985–1005.

[65] M. Billinger, I. Daly, V. Kaiser, J. Jin, B. Allison, G. Müller-Putz, C. Brunner, Is it significant? Guidelines for reporting BCI performance, in: B. Z. - Allison, S. Dunne, R. Leeb, J. Del, R. Millán, A. Nijholt (Eds.), Towards Practical Brain-Computer Interfaces, Biological and Medical Physics, Biomedical Engineering, Springer, Berlin, Heidelberg, 2013, pp. 333–354.

[66] P. Yuan, X. Gao, B. Allison, Y. Wang, G. Bin, S. Gao, A study of the existing problems of estimating the information transfer rate in online brain-computer interfaces, J. Neural Eng. 10 (2) (2013) 026014.

[67] R. Scherer, F. Lee, A. Schlogl, R. Leeb, H. Bischof, G. Pfurtscheller, Toward self-paced brain-computer communication: navigation through virtual worlds, IEEE Trans. Biomed. Eng. 55 (2) (2008) 675–682.

[68] R. Ortner, B. Allison, G. Korisek, H. Gaggl, G. Pfurtscheller, An SSVEP BCI to control a hand orthosis for persons with tetraplegia, IEEE Trans. Neural Syst. Rehabil. Eng. 19 (1) (2011) 1–5.

[69] A. Muralidharan, J. Chae, D. M. Taylor, Extracting attempted hand movements from EEGs in people with complete hand paralysis following stroke, Front. Neurosci. 5 (2011) 39.

[70] L. Bi, X. an Fan, Y. Liu, EEG-based brain-controlled mobile robots: a survey, IEEE Trans. Hum. Mach. Syst. 43 (2) (2013) 161–176.

[71] S. Levine, D. Bell, L. Jaros, R. Simpson, Y. Koren, J. Borenstein, The NavChair assistive wheelchair navigation system, IEEE Trans. Rehabil. Eng. 7 (4) (1999) 443–451.

[72] W. Burgard, A. B. Cremers, D. Fox, D. Hhnel, G. Lakemeyer, D. Schulz, W. Steiner, S. Thrun, Experiences with an interactive museum tour-guide robot, Artif. Intell. 114 (1999) 3–55.

[73] T. Carlson, Y. Demiris, Collaborative control for a robotic wheelchair: evaluation of performance, attention, and workload, IEEE Trans. Syst. Man Cybern. B 42 (3) (2012) 876–888.

[74] G. Onose, C. Grozea, A. Anghelescu, C. Daia, C. J. Sinescu, A. V. Ciurea, T. Spircu, A. Mirea, I. Andone, A. Spânu, C. Popescu, A.-S. Mihăescu, S. Fazli, M. Danóczy, F. Popescu, On the feasibility of using motor imagery EEG-based brain-computer interface in chronic tetraplegics for assistive robotic arm control: a clinical test and long-term post-trial follow-up, Spinal Cord 50 (8) (2012) 599–608.

[75] F. Galán, M. Nuttin, E. Lew, P. W. Ferrez, G. Vanacker, J. Philips, J. D. R. Millán, A brain-actuated wheelchair: asynchronous and non-invasive brain-computer interfaces for continuous control of robots, Clin. Neurophysiol. 119 (9) (2008) 2159–2169.

[76] R. Leeb, L. Tonin, M. Rohm, L. Desideri, T. Carlson, J. Millan, Towards independence: a BCI telepresence robot for people with severe motor disabilities, Proc. IEEE 103 (6) (2015) 969–982.

[77] K. LaFleur, K. Cassady, A. Doud, K. Shades, E. Rogin, B. He, Quadcopter control in three-dimensional space using a noninvasive motor imagery-based brain-computer interface, J. Neural Eng. 10 (4) (2013) 046003.

[78] Y. Chae, J. Jeong, S. Jo, Toward brain-actuated humanoid robots: asynchronous direct control using an EEG-based BCI, IEEE Trans. Robot. 28 (5) (2012) 1131–1144.

[79] C. Escolano, A. Ramos Murguialday, T. Matuz, N. Birbaumer, J. Minguez, A telepresence robotic system operated with a P300-based brain-computer interface: initial tests with ALS patients, Conf. Proc. IEEE Eng. Med. Biol. Soc. 2010 (2010) 4476–4480.

[80] C. J. Bell, P. Shenoy, R. Chalodhorn, R. P. N. Rao, Control of a humanoid robot by a noninvasive brain-computer interface in humans, J. Neural Eng. 5 (2) (2008) 214–220.

[81] I. Iturrate, J. Antelis, A. Kubler, J. Minguez, A noninvasive brain-actuated wheelchair based on a P300 neurophysiological protocol and automated navigation, IEEE Trans. Robot. 25 (3) (2009) 614–627.

[82] G. Muller-Putz, G. Pfurtscheller, Control of an electrical prosthesis with an SSVEP-based BCI, IEEE Trans. Biomed. Eng. 55 (1) (2008) 361–364.

[83] M. Palankar, K. De Laurentis, R. Alqasemi, E. Veras, R. Dubey, Y. Arbel, E. Donchin, Control of a 9-DoF wheelchair-mounted robotic arm system using a P300 brain computer interface: initial experiments, in: ROBIO 2008, 2009, pp. 348–353.

[84] P. Horki, T. Solis-Escalante, C. Neuper, G. Müller-Putz, Combined motor imagery and SSVEP based BCI control of a 2 DoF artificial upper limb, Med. Biol. Eng. Comput. 49 (5) (2011) 567–577.

[85] B. Choi, S. Jo, A low-cost EEG system-based hybrid brain-computer interface for humanoid robot navigation and recognition, PLoS ONE 8 (9) (2013) e74583.

[86] R. Chavarriaga, J. del Millan, Learning from EEG error-related potentials in noninvasive brain-computer interfaces, IEEE Trans. Neural Syst. Rehab. Eng. 18 (4) (2010) 381–388.

[87] I. Iturrate, L. Montesano, J. Minguez, Robot reinforcement learning using EEG-based reward signals, in: 2010 IEEE International Conference on Robotics and Automation (ICRA), 2010, pp. 4822–4829.

[88] I. Iturrate, R. Chavarriaga, L. Montesano, J. Minguez, J.D.R. Millán, Teaching brain-machine interfaces as an alternative paradigm to neuroprosthetics control, Sci. Rep. 5 (2015).

[89] X. Perrin, R. Chavarriaga, F. Colas, R. Siegwart, J. D. R. Millán, Brain-coupled interaction for semi-autonomous navigation of an assistive robot, Robot. Auton. Syst. 58 (12) (2010) 1246–1255.

[90] G. Pfurtscheller, G. R. Müller, J. Pfurtscheller, H. J. Gerner, R. Rupp, 'Thought'-control of functional electrical stimulation to restore hand grasp in a patient with tetraplegia, Neurosci. Lett. 351 (1) (2003) 33–36.

[91] G. R. Müller-Putz, R. Scherer, G. Pfurtscheller, R. Rupp, EEG-based neuroprosthesis control: a step towards clinical practice, Neurosci. Lett. 382 (1–2) (2005) 169–174.

[92] F. Meng, K.-Y. Tong, S.-T. Chan, W.-W. Wong, K.-H. Lui, K.-W. Tang, X. Gao, S. Gao, BCI-FES training system design and implementation for rehabilitation of stroke patients, in: IJCNN, 2008, pp. 4103–4106.

[93] K. K. Ang, C. Guan, K. S. G. Chua, B. T. Ang, C. Kuah, C. Wang, K. S. Phua, Z. Y. Chin, H. Zhang, A clinical study of motor imagery-based brain-computer interface for upper limb robotic rehabilitation, Conf. Proc. IEEE Eng. Med. Biol. Soc. 2009 (2009) 5981–5984.

[94] J. J. Daly, R. Cheng, J. Rogers, K. Litinas, K. Hrovat, M. Dohring, Feasibility of a new application of noninvasive brain computer interface (BCI): a case study of training for recovery of volitional motor control after stroke, J. Neurol. Phys. Ther. 33 (4) (2009) 203–211.

[95] M. Tavella, R. Leeb, R. Rupp, J. D. R. Millan, Towards natural non-invasive hand neuroprostheses for daily living, Conf. Proc. IEEE Eng. Med. Biol. Soc. 2010 (2010) 126–129.

[96] G. Pfurtscheller, T. Solis-Escalante, R. Ortner, P. Linortner, G. Muller-Putz, Self-Paced Operation of an SSVEP-based orthosis with and without an imagery-based "brain switch": a feasibility study towards a hybrid BCI, IEEE Trans. Neural Syst. Rehabil. Eng. 18 (4) (2010) 409–414.

[97] D. Broetz, C. Braun, C. Weber, S. R. Soekadar, A. Caria, N. Birbaumer, Combination of brain-computer interface training and goal-directed physical therapy in chronic stroke: a case report, Neurorehabil. Neural Repair 24 (7) (2010) 674–679.

[98] H. G. Tan, K. H. Kong, C. Y. Shee, C. C. Wang, C. T. Guan, W. T. Ang, Post-acute stroke patients use brain-computer interface to activate electrical stimulation, Conf. Proc. IEEE Eng. Med. Biol. Soc. 2010 (2010) 4234–4237.

[99] A. Caria, C. Weber, D. Brötz, A. Ramos, L. F. Ticini, A. Gharabaghi, C. Braun, N. Birbaumer, Chronic stroke recovery after combined BCI training and physiotherapy: a case report, Psychophysiology 48 (4) (2011) 578–582.

[100] M. Gomez-Rodriguez, J. Peters, J. Hill, B. Schölkopf, A. Gharabaghi, M. Grosse-Wentrup, Closing the sensorimotor loop: haptic feedback facilitates decoding of motor imagery, J. Neural Eng. 8 (3) (2011) 036005.

[101] K. Shindo, K. Kawashima, J. Ushiba, N. Ota, M. Ito, T. Ota, A. Kimura, M. Liu, Effects of neurofeedback training with an electroencephalogram-based brain-computer interface for hand paralysis in patients with chronic stroke: a preliminary case series study, J. Rehabil. Med. 43 (10) (2011) 951–957.

[102] W.-K. Tam, K.Yu. Tong, F. Meng, S. Gao, A minimal set of electrodes for motor imagery BCI to control an assistive device in chronic stroke subjects: a multi-session study, IEEE Trans. Neural Syst. Rehabil. Eng. 19 (6) (2011) 617–627.

[103] A. Ramos-Murguialday, M. Schürholz, V. Caggiano, M. Wildgruber, A. Caria, E. M. Hammer, S. Halder, N. Birbaumer, Proprioceptive feedback and brain computer interface (BCI) based neuroprostheses, PLoS ONE 7 (10) (2012)e47048.

[104] A. Frisoli, C. Loconsole, D. Leonardis, F. Banno, M. Barsotti, C. Chisari, M. Bergamasco, A new Gaze-BCI-driven control of an upper limb exoskeleton for rehabilitation in real-world tasks, IEEE Trans. Syst. Man Cybern. C 42 (6) (2012) 1169–1179.

[105] M. Takahashi, K. Takeda, Y. Otaka, R. Osu, T. Hanakawa, M. Gouko, K. Ito, Event related desynchronization-modulated functional electrical stimulation system for stroke rehabilitation: a feasibility study, J. Neuroeng. Rehabil. 9 (2012) 56.

[106] T. Noda, N. Sugimoto, J. Furukawa, M. Sato, S. Hyon, J. Morimoto, Brain-controlled exoskeleton robot for BMI rehabilitation, in: Proceedings of the IEEE-RAS International Conference on Humanoid Robots, 2012.

[107] A. Ramos-Murguialday, D. Broetz, M. Rea, L. Läer, O. Yilmaz, F. L. Brasil, G. Liberati, M. R. Curado, E. Garcia-Cossio, A. Vyziotis, W. Cho, M. Agostini, E. Soares, S. Soekadar, A. Caria, L. G. Cohen, N. Birbaumer, Brain-machine interface in chronic stroke rehabilitation: a controlled study, Ann. Neurol. 74 (1) (2013) 100–108.

[108] A. Kilicarslan, S. Prasad, R. G. Grossman, J. L. Contreras-Vidal, High accuracy decoding of user intentions using EEG to control a lower-body exoskeleton, Conf. Proc. IEEE Eng. Med. Biol. Soc. 2013 (2013) 5606–5609.

[109] M. Rohm, M. Schneiders, C. Müller, A. Kreilinger, V. Kaiser, G. R. Müller-Putz, R. Rupp, Hybrid brain-computer interfaces and hybrid neuroprostheses for restoration of upper limb functions in individuals with high-level spinal cord injury, Artif. Intell. Med. 59 (2) (2013) 133–142.

[110] T. Sakurada, T. Kawase, K. Takano, T. Komatsu, K. Kansaku, A BMI-based occupational therapy assist suit: asynchronous control by SSVEP, Front. Neurosci. 7 (2013) 172.

[111] A. H. Do, P. T. Wang, C. E. King, S. N. Chun, Z. Nenadic, Brain-computer interface controlled robotic gait orthosis, J. Neuroeng. Rehabil. 10 (2013) 111.

[112] M. Sarac, E. Koyas, A. Erdogan, M. Cetin, V. Patoglu, Brain computer interface based robotic rehabilitation with online modification of task speed, in: 2013 IEEE International Conference on Rehabilitation Robotics (ICORR), 2013, pp. 1–7.

[113] A. Blank, M. K. O'Malley, G. E. Francisco, J. L. Contreras-Vidal, A pre-clinical framework for neural control of a therapeutic upper-limb exoskeleton, Int. IEEE EMBS Conf. Neural Eng. 2013 (2013) 1159–1162.

[114] B. Várkuti, C. Guan, Y. Pan, K. S. Phua, K. K. Ang, C. W. K. Kuah, K. Chua, B. T. Ang, N. Birbaumer, R. Sitaram, Resting state changes in functional connectivity correlate with movement recovery for BCI and robot-assisted upper-extremity training after stroke, Neurorehabil. Neural Repair 27 (1) (2013) 53–62.

[115] M. Witkowski, M. Cortese, M. Cempini, J. Mellinger, N. Vitiello, S. R. Soekadar, Enhancing brain-machine interface (BMI) control of a hand exoskeleton using electrooculography (EOG), J. Neuroeng. Rehabil 11 (2014).

[116] K. K. Ang, C. Guan, K. S. Phua, C. Wang, L. Zhou, K. Y. Tang, G. J. Ephraim Joseph, C. W. K. Kuah, K. S. G. Chua, Brain-computer interface-based robotic end effector system for wrist and hand rehabilitation: results of a three-armed randomized controlled trial for chronic stroke, Front. Neuroeng. 7 (2014) 30.

[117] R. Looned, J. Webb, Z. G. Xiao, C. Menon, Assisting drinking with an affordable BCI-controlled wearable robot and electrical stimulation: a preliminary investigation, J. Neuroeng. Rehabil. 11 (2014) 51.

[118] B. M. Young, Z. Nigogosyan, L. M. Walton, J. Song, V. A. Nair, S. W. Grogan, M. E. Tyler, D. F. Edwards, K. Caldera, J. A. Sattin, J. C. Williams, V. Prabhakaran, Changes in functional brain organization and behavioral correlations after rehabilitative therapy using a brain-computer interface, Front. Neuroeng. 7 (2014) 26.

[119] A. M. Elnady, X. Zhang, Z. G. Xiao, X. Yong, B. K. Randhawa, L. Boyd, C. Menon, A single-session preliminary evaluation of an affordable BCI-controlled arm exoskeleton and motor-proprioception platform, Front. Hum. Neurosci. 9 (2015) 168.

[120] N.-S. Kwak, K.-R. Müller, S.-W. Lee, A lower limb exoskeleton control system based on steady state visual evoked potentials, J. Neural Eng. 12 (5) (2015) 056009.

[121] G. Naros, A. Gharabaghi, Reinforcement learning of self-regulated β-oscillations for motor restoration in chronic stroke, Front. Hum. Neurosci. 9 (2015) 391.

[122] M. Barsotti, D. Leonardis, C. Loconsole, M. Solazzi, E. Sotgiu, C. Procopio, C. Chisari, M. Bergamasco, A. Frisoli, A full upper limb robotic exoskeleton for reaching and grasping rehabilitation triggered by MI-BCI, in: ICORR, 2015, pp. 49–54.

[123] J. L. Pons, R. Ceres, L. Calderón, Introduction to Wearable Robotics, John Wiley & Sons Ltd, New York, 2008. pp. 1-16.

[124] K. K. Ang, C. Guan, Brain-computer interface in stroke rehabilitation, J. Comput. Sci. Eng. 7 (2) (2013) 139–146.

[125] M. Ferrarin, F. Palazzo, R. Riener, J. Quintern, Model-based control of FES-induced single joint movements, IEEE Trans. Neural Syst. Rehab. Eng. 9 (3) (2001) 245–257.

[126] J. Gancet, M. Ilzkovitz, E. Motard, Y. Nevatia, P. Letier, D. de Weerdt, G. Cheron, T. Hoellinger, K. Seetharaman, M. Petieau, Y. Ivanenko, M. Molinari, I. Pisotta, F. Tamburella, F. Labini, A. d' Avella, H. van der Kooij, L. Wang, F. van der Helm, S. Wang, F. Zanow, R. Hauffe, F. Thorsteinsson, MINDWALKER: going one step further with assistive lower limbs exoskeleton for SCI condition subjects, in: Proceedings of the Fourth IEEE RAS EMBS International Conference on Biomedical Robotics and Biomechatronics (BioRob), 2012, pp. 1794–1800.

[127] J. L. Contreras-Vidal, R. G. Grossman, NeuroRex: a clinical neural interface roadmap for EEG-based brain machine interfaces to a lower body robotic exoskeleton, Conf. Proc. IEEE Eng. Med. Biol. Soc. 2013 (2013) 1579–1582.

[128] D. J. McFarland, L. A. Miner, T. M. Vaughan, J. R. Wolpaw, Mu and beta rhythm topographies during motor imagery and actual movements, Brain Topogr. 12 (3) (2000) 177–186.

[129] B. B. Johansson, Brain plasticity and stroke rehabilitation, The Willis lecture, Stroke 31 (1) (2000) 223–230.

[130] S. J. Page, P. Levine, A. Leonard, Mental practice in chronic stroke: results of a randomized, placebo-controlled trial, Stroke 38 (4) (2007) 1293–1297.

[131] F. Cincotti, F. Pichiorri, P. Aricò, F. Aloise, F. Leotta, F. de Vico Fallani, J. D. R. Millán, M. Molinari, D. Mattia, EEG-based brain-computer interface to support post-stroke motor rehabilitation of the upper limb, Conf. Proc. IEEE Eng. Med. Biol. Soc. 2012 (2012) 4112–4115.

[132] J. W. Krakauer, Motor learning: its relevance to stroke recovery and neuro-rehabilitation, Curr. Opin. Neurol. 19 (1) (2006) 84–90.

[133] P. Lum, D. Reinkensmeyer, R. Mahoney, W. Z. Rymer, C. Burgar, Robotic devices for movement therapy after stroke: current status and challenges to clinical acceptance, Top. Stroke Rehabil. 8 (4) (2002) 40–53.

[134] J. Mehrholz, C. Werner, J. Kugler, M. Pohl, Electromechanical-assisted training for walking after stroke, Cochrane Database Syst. Rev. 7 (2007)CD006185.

[135] M. Lotze, C. Braun, N. Birbaumer, S. Anders, L. G. Cohen, Motor learning elicited by voluntary drive, Brain 126 (Pt 4) (2003) 866–872.

[136] J.-M. Belda-Lois, S. Mena-del Horno, I. Bermejo-Bosch, J. C. Moreno, J. L. Pons, D. Farina, M. Iosa, M. Molinari, F. Tamburella, A. Ramos, A. Caria, T. Solis-Escalante, C. Brunner, M. Rea, Rehabilitation of gait after stroke: a review towards a top-down approach, J. Neuroeng. Rehabil. 8 (2011) 66.

[137] K. K. Ang, C. Guan, K. S. Chua, B. T. Ang, C. Kuah, C. Wang, K. S. Phua, Z. Y. Chin, H. Zhang, Clinical study of neurorehabilitation in stroke using EEG-based motor imagery brain-computer interface with robotic feedback, Conf. Proc. IEEE. Eng. Med. Biol. Soc. 2010 (2010) 5549–5552.

[138] K. K. Ang, C. Guan, K. S. Phua, C. Wang, I. Teh, C. W. Chen, E. Chew, Transcranial direct current stimulation and EEG-based motor imagery BCI for upper limb stroke rehabilitation, Conf. Proc. IEEE Eng. Med. Biol. Soc. 2012 (2012) 4128–4131.

[139] T. Ono, K. Shindo, K. Kawashima, N. Ota, M. Ito, T. Ota, M. Mukaino, T. Fujiwara, A. Kimura, M. Liu, J. Ushiba, Brain-computer interface with somatosensory feedback

improves functional recovery from severe hemiplegia due to chronic stroke, Front. Neuroeng. 7 (2014) 19.

[140] K. K. Ang, K. S. G. Chua, K. S. Phua, C. Wang, Z. Y. Chin, C. W. K. Kuah, W. Low, C. Guan, A randomized controlled trial of EEG-based motor imagery brain-computer interface robotic rehabilitation for stroke, Clin. EEG Neurosci. 46 (4) (2015) 310–320.

[141] E. Buch, C. Weber, L. G. Cohen, C. Braun, M. A. Dimyan, T. Ard, J. Mellinger, A. Caria, S. Soekadar, A. Fourkas, N. Birbaumer, Think to move: a neuromagnetic brain-computer interface (BCI) system for chronic stroke, Stroke 39 (3) (2008) 910–917.

[142] E. R. Buch, A. Modir Shanechi, A. D. Fourkas, C. Weber, N. Birbaumer, L. G. Cohen, Parietofrontal integrity determines neural modulation associated with grasping imagery after stroke, Brain 135 (Pt 2) (2012) 596–614.

[143] M. Mihara, N. Hattori, M. Hatakenaka, H. Yagura, T. Kawano, T. Hino, I. Miyai, Near-infrared spectroscopy-mediated neurofeedback enhances efficacy of motor imagery-based training in poststroke victims: a pilot study, Stroke 44 (4) (2013) 1091–1098.

[144] F. Pichiorri, G. Morone, M. Petti, J. Toppi, I. Pisotta, M. Molinari, S. Paolucci, M. Inghilleri, L. Astolfi, F. Cincotti, D. Mattia, Brain-computer interface boosts motor imagery practice during stroke recovery, Ann. Neurol. 77 (5) (2015) 851–865.

[145] H. Sun, Y. Xiang, M. Yang, Neurological rehabilitation of stroke patients via motor imaginary-based brain-computer interface technology, Neural Regen. Res. 6 (28) (2011) 2198–2202.

[146] J. Haefeli, S. Vögeli, J. Michel, V. Dietz, Preparation and performance of obstacle steps: interaction between brain and spinal neuronal activity, Eur. J. Neurosci. 33 (2) (2011) 338–348.

[147] M. Seeber, R. Scherer, J. Wagner, T. Solis-Escalante, G. R. Müller-Putz, EEG beta suppression and low gamma modulation are different elements of human upright walking, Front. Hum. Neurosci. 8 (2014) 485.

[148] A. R. Sipp, J. T. Gwin, S. Makeig, D. P. Ferris, Loss of balance during balance beam walking elicits a multifocal theta band electrocortical response, J. Neurophysiol. 110 (9) (2013) 2050–2060.

[149] G. Lisi, J. Morimoto, EEG single-trial detection of gait speed changes during treadmill walk, PLoS ONE 10 (5) (2015)e0125479.

[150] K. L. M. Koenraadt, E. G. J. Roelofsen, J. Duysens, N. L. W. Keijsers, Cortical control of normal gait and precision stepping: an fNIRS study, Neuroimage 85 (Pt 1) (2014) 415–422.

[151] N. Mrachacz-Kersting, S. R. Kristensen, I. K. Niazi, D. Farina, Precise temporal association between cortical potentials evoked by motor imagination and afference induces cortical plasticity, J. Physiol. 590 (7) (2012) 1669–1682.

[152] W. Wang, J. L. Collinger, M. A. Perez, E. C. Tyler-Kabara, L. G. Cohen, N. Birbaumer, S. W. Brose, A. B. Schwartz, M. L. Boninger, D. J. Weber, Neural interface technology for rehabilitation: exploiting and promoting neuroplasticity, Phys. Med. Rehabil. Clin. N. Am. 21 (1) (2010) 157–178.

[153] G. Lisi, M. Hamaya, T. Noda, J. Morimoto, Dry-wireless EEG and asynchronous adaptive feature extraction towards a plug-and-play co-adaptive brain robot interface, in: Proceedings of the IEEE International Conference on Robotics and Automation (ICRA), Stockholm, Sweden, 2016, pp. 959–966.

[154] K. Knaepen, A. Mierau, E. Swinnen, H. Fernandez Tellez, M. Michielsen, E. Kerckhofs, D. Lefeber, R. Meeusen, Human-robot interaction: does robotic guidance force affect gait-related brain dynamics during robot-assisted treadmill walking? PloS one 10 (10) (2015), e0140626, http://dx.doi.org/10.1371/journal.pone.0140626. http://www.ncbi.nlm.nih.gov/pubmed/26485148.

CHAPTER EIGHT

Intention Inference for Human-Robot Collaboration in Assistive Robotics

H.C. Ravichandar, A. Dani
University of Connecticut, Storrs, CT, United States

1. BACKGROUND

Human intention inference is the first natural step for achieving safety in human–robot collaboration (HRC) [1–5]. Studies in psychology show that when two humans interact, they infer the intended actions of the other person and decide based on this inference what proactive actions could be taken for safe interaction and collaboration [6,7]. In this chapter, an inference method is presented to estimate the intentions of human actions by using an expectation–maximization (E-M) algorithm. Intentions are modeled as the goal locations of human actions. The complex dynamic motion of a human arm's joints is represented by using a state space model where a neural networks (NN) model is used to represent the state propagation [8,9,50,51]. Joint positions and velocities of the human skeletal structure are used as the states. Intention appears as parameters of the state space dynamic model. The problem of intention inference is solved as a parameter inference problem using an E-M algorithm. The intention inference problem deals with three sources of uncertainty: the uncertain system dynamics, the sensor measurement noise, and the unknown human intent. The advantage of modeling human arm motions using an NN is that the intentions can be easily included as parameters of the dynamics. Other user-specific or object-specific characteristics can be incorporated as parameters as well, but no specific results in this regard are presented, as this will be pursued as future work.

A set of demonstrations capturing human actions is collected to learn the dynamics of human arm motion. Each recorded action used for training the

NN is labeled based on the corresponding true intention, that is, the three-dimensional (3D) goal location of the reaching action. The learned model is then used in a Bayesian setting to infer the intention parameter using an approximate E-M algorithm. The proposed algorithm for intention inference extends an approximate E-M algorithm presented in Ref. [10] to the transition models learned using NNs. The expectation step of the E-M algorithm is carried out using an extended Kalman filter (EKF). The maximization step either uses numerical optimization or direct evaluation (see Section 2.2). An online learning algorithm is also developed to make the inference algorithm robust to the uncertainties in the starting arm joint positions, the trajectories taken to reach the goal location, and different human subjects. One way of updating the model online would be to use the E-M algorithm by optimizing the Q function over the model parameters along with the intention g. The first challenge in modeling arm motion comes from the high nonlinearity of human arm motion dynamics. While a closed-form expression for model update using E-M exists if the model is linear or represented using a radial basis function neural network (RBF-NN) [11,12], we do not restrict the basis function of the NN to RBFs in this case. Another challenge in modeling human arm motion stems from the fact that different subjects are likely to have different arm motion dynamics based on their personal preferences and physical characteristics. In order to make the algorithm robust to human subject-based variations, we use an identifier-based online learning technique to update the model for various subjects as new data are being observed. Experimental evidence shows that the online model-learning algorithm improves intention inference by learning a variety of arm motions from different subjects. In Fig. 1, a block diagram of the proposed algorithm is shown. The main contributions of this work can be summarized as follows:

- Modeling human arm motion dynamics using an NN and inferring human intentions (the goal location of the human hand) ahead of time using an NN-based approximate E-M algorithm.
- Online model learning for intention inference in order to adapt to the various starting arm configurations and the trajectories of different subjects using an identifier-based online learning algorithm.

1.1 Related Work

Algorithms for human intention estimation have been studied in human-computer interaction [13] and human-robot interaction [14]. Intentions have been described in many ways, for example, by using the body posture,

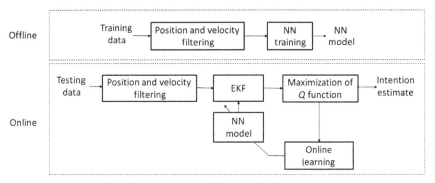

Fig. 1 Block diagram representation of the intention inference algorithm. The dotted box on top describes the training phase while the one on the bottom describes the online inference of intention.

gestures, and voice commands [15], the eye gaze [16,17], the facial expressions [18,19], or by measuring the physiological information, such as heart rate and skin response from humans [20,21]. In this chapter, we model human intention as the goal locations of the reaching motions in 3D. For safe and effective HRC, there is a technical need to model the actions of humans in order to infer the underlying intention. Human intention inference has been studied by using hidden Markov models (HMMs) [22–24], dynamic Bayesian networks [25–27], growing HMMs [28], conditional random fields [29–31], and Gaussian mixture models (GMMs) [32]. The previously mentioned studies do not consider the modeling aspects of human dynamic motion. Typically, human intentions are modeled as discrete states, whereas in the present chapter they are modeled as the parameters of a dynamic motion model. In Ref. [33], human intention is predicted by visually observing the hand-object interaction in grasping tasks. This work is specific to the grasping motions and predicts the required grasping configuration for a given task. In Ref. [22], a human intent estimation algorithm based on fuzzy inference logic is presented. In Ref. [20], the work is extended to the affective state estimation based on HMM. The affective state is represented using valence/arousal characteristics which are measured by using physiological signals, such as the heart rate and the skin response, and the valence/arousal representation is used for human intention, which only indicates the degree of approval to a given stimulus. In Ref. [34], hand-over tasks are studied and the intention to hand over an object is predicted by using key features extracted from the vision and the pose data. In Ref. [35], a latent variable model called the intention-driven dynamic model is proposed to infer intentions from the observed human movements. In Refs. [29,30], the intended activities are anticipated by modeling spatial-temporal relations that are represented by using object affordances. In Ref. [36], a robot speed control algorithm is presented that utilizes known robot joint angle values and controls the speed of the robot based upon accurate measurements of human positioning obtained from optical cameras. In Ref. [37], a vision-based algorithm is presented to infer the activity of human agents by using HMM models of a robot's experience and interaction with the environment. In Refs. [38,39], nonlinear dynamics of flying objects are learned by using nonlinear regression techniques for GMMs in order to predict the end-effector pose required to catch the objects. However, both the approaches do not explicitly model the underlying intention (goal pose in this case) as parameters of the dynamic system. In this chapter, intention is modeled as the goal location of reaching motion of the human hand. In contrast to the literature, we use an NN to represent state transitions and an adapted approximate E-M algorithm to infer the goal location ahead of time based on the measurements obtained from a Kinect sensor.

2. RESEARCH CHALLENGES AND SOLUTION APPROACH

Consider a 3D workspace with a human performing tasks, such as picking up objects placed on a table. The human operator is reaching out to different objects placed on a table and a robot is watching the human through a 3D camera sensor mounted on its head. This chapter addresses the problem of inferring the goal location, where the human hand is intending to reach. Since human motion is highly nonlinear and uncertain, an NN is used to model the human arm motion. The NN is trained by using a dataset containing multiple RGB-D demonstrations of a human reaching for predefined target locations in a given workspace. When a set of new measurements becomes available, the trained NN is used to estimate the intention (goal location) using an approximate E-M algorithm that is adapted for the dynamic models learned using the NNs. Furthermore, the weights of the NN model are updated iteratively using an identifier-based algorithm to adapt to variations in start locations and trajectories of the human arm.

2.1 System Modeling

We model the dynamics of human arm motion using a nonlinear transition function of joint positions, velocities, and intentions represented by the goal location of the human arm. Fig. 2 shows the overall structure of the model. For the previously mentioned problem scenario, we denote the intention as $g \in G$, where $G = \{g_1, g_2, \ldots, g_n\}$ and $g_i \in \mathbb{R}^3$ represents a 3D location of an object on a table. The true intention g can only represent one of the finite numbers of goal locations (target objects) on a table.

The state $x_t \in \mathbb{R}^{24}$ represents the positions and velocities of four points on the arm (shoulder, elbow, wrist, and palm) to describe the position of the arm at a given time t, and $z_t \in \mathbb{R}^{24}$ represents the measurement obtained after filtering the Kinect sensor data. All locations are specified in the 3D Carte-sian space. It should be noted that the proposed algorithm can also support g defined as a continuous variable. The state evolution depends on both the previous state and the intention g, while the measurement depends only on the current state.

2.1.1 State Transition Model

The state transition model describes the state evolution over time. The model is described by the following equation:

$$\dot{x} = f_c^*(x_t, g) + w_t \tag{1}$$

where $\{w_t\} \sim \mathcal{N}(0, Q_c) \in \mathbb{R}^{24}$ is a zero-mean Gaussian random process with a covariance matrix $Q_c \in \mathbb{R}^{24 \times 24}$, $f_c^*(x_t, g) : \mathbb{R}^{24} \times \mathbb{R}^3 \to \mathbb{R}^{24}$ is assumed to be an analytical function. The nonlinear function $f_c^*(x_t, g)$, defined in Eq. (1), is modeled using an NN as follows

$$f_c^*(x_t, g) = f_c(x_t, g) + \epsilon(s_t) = W^T \sigma(U^T s_t) + \epsilon(s_t) \tag{2}$$

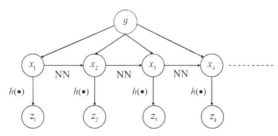

Fig. 2 Dynamic Bayesian intention estimation framework with neural network (NN) transitions.

where $s_t = \left[\left[x_t^T, g^T\right], 1\right]^T \in \mathbb{R}^{28}$ is the input vector to the NN, $\sigma\left(U^T s_t\right) =$

$$\left[\frac{1}{1 + \exp\left((-U^T s_t)_1\right)}, \frac{1}{1 + \exp\left((-U^T s_t)_2\right)}, \cdots, \frac{1}{1 + \exp\left((-U^T s_t)_i\right)}, \cdots,\right.$$

$$\left.\frac{1}{1 + \exp\left((-U^T s_t)_{n_h}\right)}\right]^T$$ is the vector-sigmoid activation function and

$(U^T s_t)_i$ is the ith element of the vector $(U^T s_t)$, $U \in \mathbb{R}^{28 \times n_h}$, and $W \in \mathbb{R}^{n_h \times 24}$ are the bounded constant weight matrices, $\epsilon(s_t) \in \mathbb{R}^{24}$ is the function reconstruction error that goes to zero after the NN is fully trained, and $n_h \in \mathbb{R}$ is the number of neurons in the hidden layer of the NN.

2.1.2 A Brief Review of Online Model Training
The NN is trained using the data consisting of the human arm's joint locations, joint velocities, and joint accelerations along with the intended target locations. The NN is trained using Bayesian regularization, which avoids overfitting of the data. Bayesian regularization is a robust algorithm when the training data sample is noisy and the sample size is small [40]. The objective function used to train an NN using Bayesian regularization is given by

$$J(U, W) = \alpha E_D + \beta E_W \tag{3}$$

where $E_D = \sum_i (y_i - a_i)^2$ is the sum of squared error, y_i and a_i represent the

target and the network's output, E_W is the sum of the squares of the NN weights, α and β are the parameters of regularization that can be used to change the emphasis between reducing the weight sizes E_W, and reducing the errors E_D. Details pertaining to gathering training data from human subjects are described in Section 3.

2.1.3 Measurement Model
The measurements of a human skeleton's joint positions are obtained using a Kinect sensor. The human skeleton is defined using 20 joints. The measurements are obtained in the Kinect's reference frame. The origins of the Kinect camera reference frame $F_c = (x_c, y_c, z_c)^T$ and the robot reference frame $F_r = (x_r, y_r, z_r)^T$ are related by

$$F_c = R_r^c F_r + T_r^c \tag{4}$$

where $R_r^c \in SO(3)$ and $T_r^c \in \mathbb{R}^3$ are the rotation matrix and the translation vector, respectively. The Kinect sensor measures the 3D locations of the skeleton's joints. The raw position measurements obtained from the Kinect sensor are fed to a local Kalman filter, such as the one in Ref. [41], to obtain the position and velocity estimates, which are used as measurements in the E-M algorithm. Design and implementation details of the Kalman filter can be found in Appendix A. The measurement model is represented in the following generic form:

$$z_t = h(x_t) + v_t \tag{5}$$

where the measurement function $h(x_t) : \mathbb{R}^{24} \to \mathbb{R}^{24}$ is given by a simple affine transformation (rotation and translation) that relates the states to the measurements as described in Eq. (5) and $\{v_t\} \sim \mathcal{N}(0, R) \in \mathbb{R}^{24}$ is a zero-mean Gaussian noise with a covariance matrix $R \in \mathbb{R}^{24 \times 24}$. The measurement noise $\{v_t\}$ is assumed to be independent of $\{w_t\}$ defined in Eq. (1).

2.2 Intention Inference

The approximate E-M algorithm presented in this chapter extends the work in Ref. [10] to the state transition models learned using NNs. Once the NN model is trained, the intention g can be inferred iteratively as new measurements arrive. The E-M algorithm requires the state transition model to be in the discrete form. The state transition model defined in Eq. (1) is discretized using a first-order Euler approximation yielding

$$x_t = f(x_{t-1}, g) + w_t T_s \tag{6}$$

where $f(x_{t-1}, g) = x_{t-1} + f_c(x_{t-1}, g) T_s$ and T_s is the sampling period. In order to infer intention, we aim to maximize the posterior probability of Z_T given the intention g using a maximum-likelihood criterion, where $Z_T = z_{1:T}$ is a set of observations[1] from time $t = 1$ to $t = T$. The covariance of the process noise $w_t T_s$ of the discretized system in Eq. (6) is given by $Q = T_s^2 Q_c$. The log-likelihood function of intention g is given by

$$l(g) = \log p(Z_T | g) \tag{7}$$

which can be obtained by marginalizing a joint distribution as shown in Eq. (8):

---

[1] T is not fixed and could be different for training and testing data.

$$l(g) = \log \int p(X_T, Z_T | g) dX_T \tag{8}$$

where $X_T = x_{1:T}$ is a collective representation of states from time $t = 1$ to $t = T$. In general, analytically evaluating this integral is extremely difficult and the E-M algorithm and other approximation techniques based on particle filtering are used to circumvent this problem. In this chapter, we use an approximate E-M algorithm with modifications for handling state transition models trained using the NN. Using the fact that $E_{X_T}\{\log[p(Z_T|g)]|Z_T, g_t\} = \log p(Z_T|g)$, the log-likelihood defined in Eq. (7) is decomposed in the following way:

$$\log p(Z_T|g) = E_{X_T}\{\log[p(Z_T, X_T|g)]|Z_T, g_t\}$$
$$- E_{X_T}\{\log[p(X_T|Z_T, g)]|Z_T, g_t\} \tag{9}$$
$$\log p(Z_T|g) = \boldsymbol{Q}(g, g_t) - \boldsymbol{H}(g, g_t) \tag{10}$$

where $E_{X_T}\{\cdot\}$ is the expectation operator, g_t is the estimate of g at time t, $\boldsymbol{Q}(g, g_t) = E_{X_T}\{\log[p(Z_T, X_T|g)]|Z_T, g_t\}$ is the expected value of the complete data log-likelihood given all the measurements and intentions, and $\boldsymbol{H}(g, g_t) = E_{X_T}\{\log[p(X_T|Z_T, g)]|Z_T, g_t\}$. It can be shown using the Jensen inequality that $\boldsymbol{H}(g, g_t) \leq \boldsymbol{H}(g_t, g_t)$ [42]. Thus, in order to iteratively increase the log-likelihood, g has to be chosen such that $\boldsymbol{Q}(g, g_t) \geq \boldsymbol{Q}(g_t, g_t)$. The E-step involves the computation of the auxiliary function $\boldsymbol{Q}(g, g_t)$ given the observations Z_T and the current estimate of the intention g_t. The M-step involves computing the next estimate of the intention g_{t+1} by finding the value of g that maximizes $\boldsymbol{Q}(g, g_t)$.

The E-step involves the evaluation of the expectation of the complete data log-likelihood given by

$$\boldsymbol{Q}(g, g_t) = E_{X_T}\{\log[p(Z_T, X_T|g)]|Z_T, g_t\}$$
$$= E_{X_T}\left\{ V_0 + \sum_{t=1}^{T} V_t(x_t, x_{t-1}, g) | Z_T, g_t \right\} \tag{11}$$

In the case of $\{v_t\}$ and $\{w_t\}$ being Gaussian, V_0 and $V_t(x_t, x_{t-1}, g)$ are given by

$$V_0 = \log[p(x_0|g)] = \log[p(x_0)]$$
$$= \text{const} - \frac{1}{2}\log[|P_0|] - \frac{1}{2}(x_0 - \mu_0)^T P_0^{-1}(x_0 - \mu_0) \tag{12}$$
$$V_t(x_t, x_{t-1}, g) = \log[p(z_t|x_t)] + \log[p(x_t|x_{t-1}, g)] \tag{13}$$

where μ_0 and P_0 are the initial state mean and covariance, $|\cdot|$ is the determinant operator,

$$\log\left[p(z_t|x_t)\right] = -\frac{1}{2}\log\left[\|R\|\right] - \frac{1}{2}\left\{(z_t - h(x_t))^T Q^{-1}(z_t - h(x_t))\right\} \quad (14)$$

and

$$\log\left[p(x_t|x_{t-1},g)\right] = -\frac{1}{2}\log\left[\|Q\|\right]$$
$$-\frac{1}{2}\left\{(x_t - f(x_{t-1},g))^T Q^{-1}(x_t - f(x_{t-1},g))\right\} \quad (15)$$

Note that in Eq. (13), $\log\left[p(z_t|x_t,g)\right]$ is replaced by $\log\left[p(z_t|x_t)\right]$. This is because for the intention inference problem the measurement z_t does not depend on the intention g. When attempting to optimize Eq. (13), the main difficulty arises because of the nonlinearity in the state transition model. The nonlinear state transition model is represented by the NN in our case. In order to compute the expectation of the log-likelihood in Eq. (15), we use the Taylor series expansion for the second line of Eq. (15) (terms inside the curly brackets) about $\bar{x}_t$ and $\bar{x}_{t-1}$. In practice, the points of linearization $\{\bar{x}_t\}$ are obtained from the measurements by ignoring the measurement noise and inverting $h(\cdot)$, the measurement mapping.[2] Let us denote $\widetilde{V}_t = (x_t - f(x_{t-1},g))^T Q^{-1}(x_t - f(x_{t-1},g))$ whose Taylor series expansion is given by

$$\widetilde{V}_t \approx \widetilde{V}_t(\bar{x}_t, \bar{x}_{t-1}, g) + \frac{\partial \widetilde{V}_t(\bar{x}_t, \bar{x}_{t-1}, g)}{\partial x_t}[x_t - \bar{x}_t]$$
$$+ \frac{\partial \widetilde{V}_t(\bar{x}_t, \bar{x}_{t-1}, g)}{\partial x_{t-1}}[x_{t-1} - \bar{x}_{t-1}]$$
$$+ \frac{1}{2}[x_t - \bar{x}_t]^T \frac{\partial^2 \widetilde{V}_t(\bar{x}_t, \bar{x}_{t-1}, g)}{\partial x_t \partial x_t}[x_t - \bar{x}_t] \quad (16)$$
$$+ \frac{1}{2}[x_{t-1} - \bar{x}_{t-1}]^T \frac{\partial^2 \widetilde{V}_t(\bar{x}_t, \bar{x}_{t-1}, g)}{\partial x_{t-1} \partial x_{t-1}}[x_{t-1} - \bar{x}_{t-1}]$$
$$+ \frac{1}{2}[x_t - \bar{x}_t]^T \frac{\partial^2 \widetilde{V}_t(\bar{x}_t, \bar{x}_{t-1}, g)}{\partial x_t \partial x_{t-1}}[x_{t-1} - \bar{x}_{t-1}]$$

---

[2] The mapping $h(\cdot)$ is a simple affine transformation and could be reversed to obtain the points of linearization from the measurements.

Since $f(x_t, g)$ is represented by an NN in Eq. (2), the derivatives of $\tilde{V}_t$ are given by the following equations:

$$\frac{\partial \tilde{V}_t}{\partial x_t} = \left(Q^{-1} + Q^{-T}\right)\left(x_t - f(x_{t-1}, g)\right) \tag{17}$$

$$\frac{\partial \tilde{V}_t}{\partial (x_{t-1})_i} = \left[\frac{\partial \tilde{V}_t}{\partial f}\right]^T \frac{\partial f}{\partial (x_{t-1})_i} \tag{18}$$

$$\frac{\partial^2 \tilde{V}_t}{\partial x_t^2} = \left(Q^{-1} + Q^{-T}\right) \tag{19}$$

$$\frac{\partial^2 \tilde{V}_t}{\partial x_t x_{t-1}} = -\left(Q^{-1} + Q^{-T}\right)\left[\frac{\partial f}{\partial x_{t-1}}\right] \tag{20}$$

$$\frac{\partial^2 \tilde{V}_t}{\partial (x_{t-1})_i \partial (x_{t-1})_j} = \left[\frac{\partial^2 \tilde{V}_t}{\partial f (\partial x_{t-1})_i}\right]^T \frac{\partial f}{\partial (x_{t-1})_j} + \frac{\partial^2 f}{\partial (x_{t-1})_i \partial (x_{t-1})_j}\left[\frac{\partial \tilde{V}_t}{\partial f}\right] \tag{21}$$

where

$$\left[\frac{\partial \tilde{V}_t}{\partial f}\right] = -\left(Q^{-1} + Q^{-T}\right)\left(x_t - f(x_{t-1}, g)\right) \tag{22}$$

$$\left[\frac{\partial^2 \tilde{V}_t}{\partial f (\partial x_{t-1})_i}\right] = \left(Q^{-1} + Q^{-T}\right)^T \frac{\partial f}{\partial (x_{t-1})_i} \tag{23}$$

Note that $\dfrac{\partial f}{\partial x_{t-1}}$ is the submatrix of the Jacobian of the NN that can be obtained by ignoring the rows pertaining to $\dfrac{\partial f}{\partial g}$. Thus, the Jacobian $\dfrac{\partial f}{\partial x_{t-1}}$ can be easily derived by taking the first n columns of $\dfrac{\partial f}{\partial s}$, where n is the number of states. The Hessian can be derived in a similar fashion. The analytical expressions for the Jacobian and Hessian are provided in Appendix B. Using Eqs. (16)–(23), the expectation in Eq. (11) can be written as

$$Q(g,\hat{g}) = -\frac{1}{2}\log[|P_0|] - \frac{1}{2}\mathrm{tr}\left\{P_0\left(\hat{P}_0 + (\hat{x}_0 - \mu_0)(\hat{x}_0 - \mu_0)^T\right)\right\}$$

$$-\frac{T}{2}\log[|R|] - \frac{T}{2}\log[|Q|]$$

$$-\frac{1}{2}\sum_{t=1}^{T}\mathrm{tr}\left\{R^{-1}\left((z_t - H\hat{x}_t)(z_t - H\hat{x}_t)^T + H\hat{P}_t H^T\right)\right\}$$

$$-\frac{1}{2}\sum_{t=1}^{T}\tilde{V}_t(\bar{x}_t, \bar{x}_{t-1}, g) - \frac{1}{2}\sum_{t=1}^{T}\left[\frac{\partial\tilde{V}_t(\bar{x}_t, \bar{x}_{t-1}, g)}{\partial x_t}\right]^T (\hat{x}_t - \bar{x}_t)$$

$$-\frac{1}{2}\sum_{t=1}^{T}\left[\frac{\partial\tilde{V}_t(\bar{x}_t, \bar{x}_{t-1}, g)}{\partial x_{t-1}}\right]^T (\hat{x}_{t-1} - \bar{x}_{t-1})$$

$$-\frac{1}{4}\sum_{t=1}^{T}\mathrm{tr}\left\{\left[\frac{\partial^2\tilde{V}_t(\bar{x}_t, \bar{x}_{t-1}, g)}{\partial x_t \partial x_t}\right](\hat{P}_t + [\hat{x}_t - \bar{x}_t][\hat{x}_t - \bar{x}_t]^T)\right\}$$

$$-\frac{1}{4}\sum_{t=1}^{T}\mathrm{tr}\left\{\left[\frac{\partial^2\tilde{V}_t(\bar{x}_t, \bar{x}_{t-1}, g)}{\partial x_{t-1} \partial x_{t-1}}\right](\hat{P}_{t-1} + [\hat{x}_{t-1} - \bar{x}_{t-1}][\hat{x}_{t-1} - \bar{x}_{t-1}]^T)\right\}$$

$$-\frac{1}{4}\sum_{t=1}^{T}\mathrm{tr}\left\{\left[\frac{\partial^2\tilde{V}_t(\bar{x}_t, \bar{x}_{t-1}, g)}{\partial x_t \partial x_{t-1}}\right](\hat{P}_{t,t-1} + [\hat{x}_t - \bar{x}_t][\hat{x}_{t-1} - \bar{x}_{t-1}]^T)\right\} - \cdots$$

$$(24)$$

where $\hat{x}_t$ and $\hat{P}_t$ are the state estimate and its covariance, respectively, $\hat{x}_0$ and $\hat{P}_0$ are their initial values, and $\hat{P}_{t,t-1}$ is the cross covariance of the state estimates at times t and $t-1$. The state estimate $\hat{x}_t$ and the covariances $\hat{P}_t$ and $\hat{P}_{t,t-1}$ are obtained using an EKF. In order to linearize the transition model for EKF at the current time step t we use the state estimates $\hat{x}_{t-1}$ from the previous time step as the point of linearization. Eq. (24) can be written in an iterative form to calculate the value of the Q function at every iteration.

The M-step involves the optimization of $Q(g, g_t)$ over g as described by the following expression:

$$\hat{g}_{t+1} = \arg\max_{g} Q(g, \hat{g}_t) \qquad (25)$$

This step could be carried out in two different ways, viz., numerical optimization or direct evaluation, as described in the following subsections.

2.2.1 Numerical Optimization of the Q Function

One way to maximize the Q function is to use the GradEM algorithm [43] where the first few iterations of Newton's algorithm are used for the M-step.

This method involves optimizing the $\mathbf{Q}$ function over $\mathbb{R}^3$. The update equation for $\hat{g}$, through the GradEM algorithm, is given by

$$\hat{g}_{k+1} = \hat{g}_k - H(\mathbf{Q})^{-1}\Delta(\mathbf{Q}) \tag{26}$$

where $\hat{g}_k$ is the estimate of g at the kth iteration of the optimization algorithm, $H(\mathbf{Q})$ and $\Delta(\mathbf{Q})$ are the Hessian and gradient of the $\mathbf{Q}$ function, respectively. Note that numerical optimization methods need to run at every time step of the E-M algorithm. In real-time implementations, the number of iterations for the optimization in Eq. (26) could be chosen based on computational capabilities. This is similar to using the first iteration of Newton's method. The Hessian of the $\mathbf{Q}$ can be numerically approximated and the analytical expression for the gradient of the $\mathbf{Q}$ function is provided in Appendix C.

2.2.2 Direct Evaluation of the Q Function
Another way to infer g is to evaluate the $\mathbf{Q}$ function for all possible g_is (the goal locations) in G and obtain $\hat{g}_{t+1}$ as described by the following expression:

$$\hat{g}_{t+1} = \arg\max_{g \in G} \mathbf{Q}(g, \hat{g}_t) \tag{27}$$

This method involving direct evaluation of the $\mathbf{Q}$ function is possible if all possible goal locations are known *a priori* and are finite. This is not an unusual case in the context of the problem scenario described. Image processing algorithms can be used to detect the objects on the workbench, along with Kinect data to extract their 3D locations.

2.3 Online Model Learning
This section describes the online learning of the NN model. The online learning of the NN weights is important to make the inference framework robust to variations in starting arm positions and various motion trajectories taken by different people. The NN weights are updated iteratively as new data become available. Using the new sensor data, the human arm motion model is updated iteratively so that the inference algorithm can adapt to the new information. The online weight update algorithm learns the weights of the NN. To this end, a state identifier is developed that computes an estimate of the state derivative based on the current state estimates obtained from the EKF and the current NN weights. The identifier state error is computed based on its estimate and measurement. The error in the state is used to update the NN weights for the next time instance. The identifier uses a robust RISE (robust integral of the sign of the error) feedback [44,45] to

ensure asymptotic convergence of the state estimates and their derivatives to the true values. The weight update equations are computed using a Lyapunov-based stability analysis. The learning algorithm is described in Algorithm 1. The mathematical details follow.

Algorithm 1 Intention Inference With Online Model Update Algorithm

Obtain demonstrations of subject reaching for different objects by recording the joint position measurements of the tracked skeleton from Microsoft Kinect for Windows;

Using a local Kalman filter, obtain position, velocity, and acceleration estimates for the demonstrations;

Label the training data based on the corresponding goal location;

Learn the NN model defined in Eq. (1) using the demonstrations;

Obtain test data from a new subject using Microsoft Kinect for Windows;

Filter the position measurements of the test data using a local Kalman filter to obtain position and velocity estimates and use them as measurements Z_T;

Initialize $\hat{x}_0$, $\hat{P}_0$, $\hat{x}_{id_0}$, and $\hat{g}_0$;

Define the parameters of the system: μ_0, P_0, Q, and R;

Define the gains for the online update algorithm: k, α, γ, β_1, Γ_W, Γ_{U_x}, and Γ_{U_g};

while *data for the current time step is present* do

Read the current measurement z_t;

    E-step:

    Using the current NN model and the previous intention estimate $\hat{g}_{t-1}$, compute $\hat{x}_t, \hat{P}_t, \hat{P}_{t,t-1}$ using the EKF;

    M-step:

    if *Numerical optimization* then

        Using the estimates obtained from the E-step, compute $\hat{g}_t$ by iteratively maximizing the **Q** function defined in Eq. (24) over $\mathbb{R}^3$ using Eq. (26);

    end

    if *Direct evaluation* then

        Using the estimates obtained from the E-step, compute $\hat{g}_t$ by maximizing the **Q** function defined in Eq. (24) over G using Eq. (27);

    end

    Online model update:

    Using the intention estimate $\hat{g}_t$ from the M-step, compute the identifier output $\dot{\hat{x}}_{id_t}$ using Eq. (28);

    Update the current NN model by changing the weights according to the adaptation laws given in Eq. (31);

end

The state identifier is given by

$$\dot{\hat{x}}_{\mathrm{id}_t} = \hat{W}_t^T \sigma\left(\hat{U}_t^T \hat{s}_t\right) + \mu_t \tag{28}$$

where $\hat{U}_t \in \mathbb{R}^{28 \times n_{\mathrm{h}}}$, $\hat{W}_t \in \mathbb{R}^{n_{\mathrm{h}} \times 24}$, $\hat{s}_t = \left[\left[\hat{x}_{\mathrm{id}_t}^T, \hat{g}_t\right]^T, 1\right]^T \in \mathbb{R}^{28}$, $\hat{g}_t \in \mathbb{R}^3$ is the current estimate of g from the E-M algorithm, $\hat{x}_{\mathrm{id}_t} \in \mathbb{R}^{24}$ is the current identifier state, and $\mu_t \in \mathbb{R}^{24}$ is the RISE feedback term defined as

$$\mu_t = k\tilde{x}_t - k\tilde{x}_0 + \nu_t \tag{29}$$

where $\tilde{x}_t = x_t - \hat{x}_{\mathrm{id}_t}$ is the state identification error at time t, and $\nu_t \in \mathbb{R}^{24}$ is the Filippov generalized solution [46] to the following differential equation:

$$\dot{\nu}_t = (k\alpha + \gamma)\tilde{x}_t + \beta_1 \mathrm{sgn}(\tilde{x}_t); \nu_0 = 0 \tag{30}$$

where $k, \alpha, \gamma, \beta_1 \in \mathbb{R}$ are positive constant control gains and $\mathrm{sgn}(\cdot)$ denotes a vector signum function. The weight update equations are given by

$$\dot{\hat{W}}_t = \mathrm{proj}\left(\Gamma_{\mathrm{w}}\,\hat{\sigma}'\hat{U}_{x_t}^T \dot{\hat{x}}_{\mathrm{id}_t}\tilde{x}_t^T\right)$$
$$\dot{\hat{U}}_{x_t} = \mathrm{proj}\left(\Gamma_{u_x}\dot{\hat{x}}_{\mathrm{id}_t}\tilde{x}_t^T \hat{W}_t^T \hat{\sigma}'\right) \tag{31}$$
$$\dot{\hat{U}}_{g_t} = \mathrm{proj}\left(\Gamma_{u_g}\dot{\hat{g}}_t\tilde{x}_t^T \hat{W}_t^T \hat{\sigma}'\right)$$

where $\mathrm{proj}(\cdot)$ is a projection operator defined in Ref. [47,48], $\hat{U}_{x_t}$ and $\hat{U}_{g_t}$ are the submatrices of $\hat{U}_t$ formed by taking the rows corresponding to $\hat{x}_{\mathrm{id}_t}$ and $\hat{g}_t$, respectively, $\hat{\sigma}'$ is the first-order derivative of the sigmoid function with respect to its inputs, and Γ_{w}, Γ_{u_x}, and Γ_{u_g} are constant matrices of appropriate dimensions. In the online learning algorithm the intention estimate $\hat{g}_t$, from the E-M algorithm, is used and, hence, for the online learning step, it is assumed to be a known parameter. The derivative of the intention estimate $\dot{\hat{g}}_t$ is computed using the finite difference method. It can be shown that the identifier defined in Eq. (28) along with the update equations defined in Eq. (31) are asymptotically stable (see Appendix D for the proof).

3. APPLICATIONS

In order to validate the proposed intention inference algorithm, we designed and carried out two case studies: (1) HRC and (2) assistive robotics. Each of the two case studies had three experiments. All experiments were conducted by using data collected from a Microsoft Kinect for Windows VI except one experiment in the HRC case study that was conducted by using Cornell's CAD-120 dataset [29]. The joint position data obtained from the subjects were preprocessed using a Kalman filter, as described in Appendix A, to obtain the velocity and acceleration estimates. After filtering, each demonstration was labeled based on the ground truth goal location. Please note that the goal location labeling was required and done only for demonstrations that were a part of the training data. The obtained measurements were processed on a standard desktop computer running an Intel i3 processor and 8 gigabytes of memory. The algorithm was coded in Matlab 2014a. The computation time for processing each frame and giving out an estimate was found to be around 0.05 seconds for all the experiments.

3.1 Human-Robot Collaboration

In all the experiments, the success/failure of a test was determined based on two criteria: (1) a test is successful if the algorithm converges to the true intention in half the time it took the subject to reach the goal location (SC1), and (2) a test is successful if the algorithm converges to the true intention before the subject's hand reaches a sphere around the goal location with radius equal to the half of the straight line distance between the start and the goal locations (SC2). In all the experiments, only data from Subject 1 was used to train the NN. The aims of each experimental study are described as follows:

- The purpose of the first experiment was to validate the ability of the proposed algorithm to infer goal locations that are different from the trained goal locations. A new set of data was collected from two new subjects.
- The purpose of the second experiment was to show that the true intention can be inferred in the presence of cluttered objects randomly placed close to each other. The training and test data had considerably different starting positions of arm motion and goal locations.
- The third experiment was conducted to validate the proposed algorithm by using Cornell's CAD-120 dataset.

3.1.1 Experiment 1

In the first experiment, a set of 32 demonstrations reaching for 4 objects (8 each) were collected from Subject 1 in order to train the NN. A set of 18 demonstrations reaching for 3 objects (6 for each) was recorded from Subject 2 and a set of 12 demonstrations reaching for 3 objects (4 for each) was recorded from Subject 3 in order to test the algorithm. The object locations were chosen to be different for Subjects 2 and 3. Note that none of the test data were used in offline training of the NN model. The starting positions of the human arm were considerably different from each other. The gains for the online learning algorithm defined in Eqs. (28) and (31) were selected to be $k = 20$, $\alpha = 5$, $\gamma = 50$, $\beta_1 = 1.25$, and the adaptation gains were chosen to be $\Gamma_W = 0.1\ I_{50 \times 50}$, $\Gamma_{U_x} = 0.2\ I_{24 \times 24}$, $\Gamma_{U_g} = 0.2\ I_{3 \times 3}$ where I denotes the identity matrix. The initial state covariance P_0, the process noise covariance Q, and the measurement noise covariance R were selected to be $0.2\ I_{24 \times 24}$, $0.1\ I_{24 \times 24}$, and $0.2\ I_{24 \times 24}$, respectively. The statistics of the number of tests conducted and successful tests are shown in Table 1. The Q function was optimized using the direct evaluation method. Fig. 3 shows a sequence of images overlaid with skeletal tracking and the inferred intentions for Subject 3.

This experiment showed that using the online algorithm the NN, trained using a small amount of training data, can be used in a generic situation where the subject is different and the end locations of the targets are different from

Table 1 Datasets for Experiment 1

	Intention g_1	Intention g_2	Intention g_3
No. of test sets	10	10	10
No. of successful tests (SC1)	9	7	7
No. of successful tests (SC2)	9	8	8

Fig. 3 Image sequence showing skeletal tracking (in *red*; *dark gray* in print version) and online inference of intention, ie, the end location (in *green*; *light gray* in print version) with online model update (the motion starts from frame 36). Data from two new subjects, with new and different goal locations, were used to test the proposed algorithm.

what the NN is trained for. The quadratic deviation was computed of each of the trajectories from corresponding straight lines between the start and goal locations. It was found that the average quadratic deviation of the trajectories in the failure cases was 25.62 while that in the successful cases was 11.94.

3.1.2 Experiment 2

In Experiment 2, four objects, randomly placed close to each other in a cluttered manner, were considered for testing the proposed algorithm. The proposed method and a Euclidean distance-based technique were compared. At any given time t, the Euclidean distance-based approach would estimate the goal location to be the object with the smallest Euclidean distance to the Subject's reaching hand. A set of 20 reaching trajectories (5 for each object) was recorded from Subject 1. The objects were reached from considerably different initial conditions and at varying speeds. The Q function was optimized using the direct evaluation method. The parameters of the approximate E-M algorithm were selected to be the same values as mentioned in Experiment 1. The statistics of number of tests conducted and successful tests are shown in Table 2. Fig. 4 shows a sequence of images overlaid with the inferred intentions.

3.1.3 Experiment 3

A set of 20 sequences (5 sequences from each of the 4 subjects) with reaching motions was randomly chosen from Cornell's CAD-120 dataset. Due to the fact that the CAD-120 dataset had only three joints (shoulder, elbow, and hand) for each arm, the state vector had to be redefined as $x_t \in \mathbb{R}^{18}$ by removing the wrist joint position and velocity from the original state definition. An NN was trained using the data collected from Subject 1 for Experiment 1 with the measurements of the wrist joint removed. However, no part of Cornell's CAD-120 dataset was used to train the NN online. The gains for the online learning algorithm defined in Eqs. (28) and (31) were selected to be $k = 20, \alpha = 5, \gamma = 50, \beta_1 = 1.25$, and the adaptation gains were

Table 2 Datasets for Experiment 2

	Intention g_1	Intention g_2	Intention g_3	Intention g_4
No. of test sets	5	5	5	5
No. of successful tests (SC1)	5	4	4	4
No. of successful tests (SC2)	5	5	4	4

Fig. 4 Image sequence showing the intention inferred by the proposed algorithm (*solid red box; solid dark gray box* in print version) and the Euclidean distance-based technique (*dashed yellow box; dashed light gray box* in print version). Four objects were placed in new locations close to each other in a cluttered manner and no new training data were used.

chosen to be $\Gamma_W = 0.1I_{35\times35}, \Gamma_{U_x} = 0.2I_{18\times18}, \Gamma_{U_g} = 0.2I_{3\times3}$. The initial state covariance P_0, the process noise covariance Q, and the measurement noise covariance R were selected to be $0.2I_{18\times18}, 0.1I_{18\times18}$, and $0.2I_{18\times18}$, respectively. The Q function was optimized using the direct evaluation method. The possible goal locations were chosen to be the objects on the table for each sequence. The proposed algorithm was able to infer the correct intention in 18 tests. In Fig. 5, a sequence of images overlaid with the inferred intentions is shown. In Fig. 6, the percentage of tests in which the intention was correctly inferred is shown as a function of time.

3.2 Assistive Robotics

A set of three experiments was conducted to showcase the use of the proposed algorithm in assistive robotics. The algorithm was used to infer underlying intention by recognizing various arm gestures performed by the user. The user was made to sit in front of a Rethink Robotics robot named Baxter and perform different gestures and the algorithm was used to recognize the gesture and perform a predefined task that will help the user.

Fig. 5 Image sequence showing the intention inferred by the proposed algorithm (*solid red box; solid dark gray box* in print version) using Cornell's CAD-120 dataset. No part of the CAD-120 dataset was used to train the NN online.

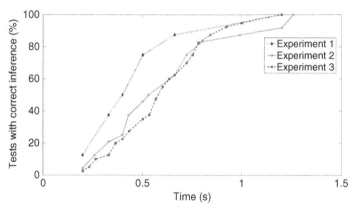

Fig. 6 Percentage of tests that correctly inferred the intention as a function of time in Experiments 1, 2, and 3.

3.2.1 Experiment 4

The first experiment of the second case study involved helping a human user by pouring water into a cup when the user requests water using a gesture that imitates drinking from a cup (see Fig. 7). The proposed algorithm was used to infer the true gesture and inform the Baxter robot that the user wants water.

Fig. 7 Image sequence showing the gesture used by the subject to request the robot to pour water into a cup he is holding.

Once the gesture is inferred, the robot can be programmed to carry out the required task—pouring water, in this case. In Fig. 8, a sequence of images, illustrating the subject performing the gesture and the robot pouring water into the subject's cup, is shown.

3.2.2 Experiment 5

In this second experiment of the second case study, the robot was programmed to hand over medicines to a human user upon request. The proposed algorithm was used for recognizing the corresponding gesture. The gesture assigned to this task involved the user stretching his/her hand forward (see Fig. 9). A sequence of images depicting the subject requesting medicine and the robot handing over the medicine is shown in Fig. 10.

3.2.3 Experiment 6

The final experiment of the assistive robotics case study involved the Baxter robot providing support to a user as he is trying to stand up. The proposed algorithm was used to infer the gesture made by the user to indicate that he is trying to stand up (see Fig. 11). Upon the inference of the gesture, the robot was programmed to provide support for the subject to get up from a chair (see Fig. 12).

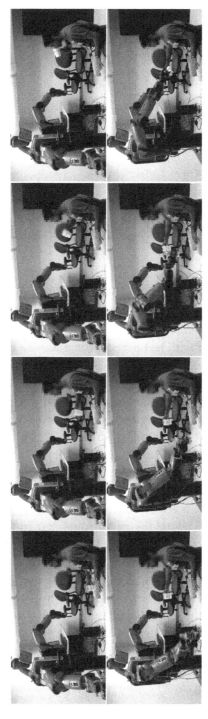

Fig. 8 Image sequence showing Baxter responding to the subject's gesture. (*Top row*: User requests for water. *Bottom row*: Baxter pours water into a cup he is holding.)

Fig. 9 Image sequence showing the gesture used by the subject to request the robot to hand over medicine.

4. DISCUSSION

Six sets of experiments were conducted to evaluate the performance of the proposed algorithm with real data collected using a Kinect sensor. The first three experiments consisted of tasks pertaining to human-robot collaboration and the second three experiments focused on assistive robotics. The first experiment, with data collected from a new subject, resulted in 26 successful tests out of 30 tests according to the distance-based success criterion (SC2). The new subject's arm motion dynamics as well as the initial conditions were considerably different from the first subject. It must be noted that the NN was trained by using data collected only from Subject 1 and was not given demonstrations everywhere in the state space. Hence, it is challenging to learn mappings that generalize to new instances and the online learning component helped in handling these challenges. Experiment 2 pointed out the need to learn the dynamics of reaching motion. A simple Euclidean distance-based approach failed in many cases where the objects were placed close to each other, with some objects placed on the way to reach other objects. We believe that the failure cases of the proposed algorithm in the experiments were due to complicated motions that are not typical arm motions. This observation was based on the fact that the average quadratic deviation of the hand trajectories from a straight line connecting the starting and end locations was found be much higher for the unsuccessful tests compared to the successful tests. The nontypical arm motions are outliers in the data representation to the NN. The third experiment, with 18 successful

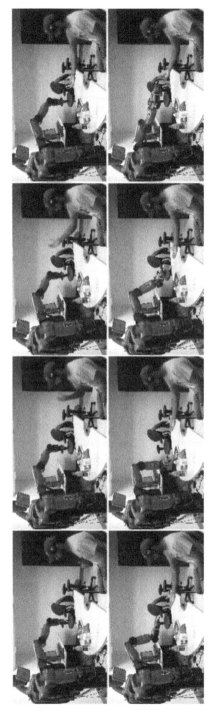

Fig. 10 Image sequence showing Baxter responding to the subject's gesture. (*Top row*: User requests for medicine. *Bottom row*: Baxter hands over medicine.)

Fig. 11 Image sequence showing the gesture used by the subject to request the robot to provide support to get up from a chair.

tests out of 20, was helpful in validating the proposed algorithm on an independent dataset, the CAD-120 dataset. Experiments 4–6 showed that the proposed algorithm can infer underlying intentions through human arm gestures.

5. CONCLUSION

In this chapter, we presented a new methodology to infer human intentions denoted by the target locations of reaching motions using an NN-based approximate E-M algorithm with online model learning. NNs were used to model the highly nonlinear human arm motion dynamics. An identifier-based online learning algorithm was developed to iteratively learn new motion dynamics as new measurements become available. A set of six experiments was carried out to validate the proposed algorithm. It was shown from Experiment 1 that the proposed algorithm, with the help of online learning, could successfully infer intentions of new human subjects with considerably different initial conditions, motion profiles, and goal locations. Experiment 2 was used to compare the proposed algorithm with a simple Euclidean distance-based approach and to show that intention could be inferred with some objects randomly placed close to each other. Experiment 3 was used to validate the proposed algorithm on Cornell's CAD-120 dataset. Experiments 4–6 showed that the proposed algorithm could be used in assistive robotics applications to infer human intentions and perform corresponding tasks.

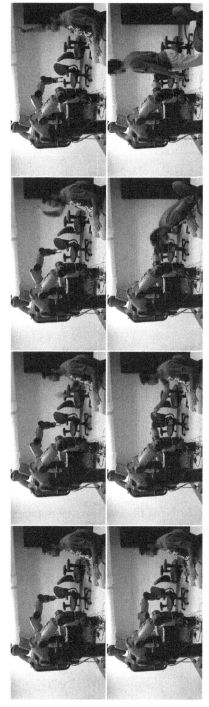

Fig. 12 Image sequence showing Baxter responding to the subject's gesture. (*Top row:* User requests for support. *Bottom row:* Baxter provides support to help the subject stand up.)

APPENDIX A KALMAN FILTER IMPLEMENTATION

The simple model of human motion from Ref. [41] is used to design and implement a standard Kalman filter. The state transition model is given by the Taylor series expansion for position, velocity, and acceleration along all three axes:

$$X_{kf}(k+1) = F_{kf}X_{kf}(k) + W_{kf}(k)$$

where $\quad X_{kf}(k) = [x_{kf}, \dot{x}_{kf}, \ddot{x}_{kf}, y_{kf}, \dot{y}_{kf}, \ddot{y}_{kf}, z_{kf}, \dot{z}_{kf}, \ddot{z}_{kf}]^T \quad$ and

$F_{kf} = \begin{bmatrix} B_1 & 0_{3\times3} & 0_{3\times3} \\ 0_{3\times3} & B_1 & 0_{3\times3} \\ 0_{3\times3} & 0_{3\times3} & B_1 \end{bmatrix}$, $B_1 = \begin{bmatrix} 1 & T_s & 0 \\ 0 & 1 & T_s \\ 0 & 0 & 1 \end{bmatrix}$, and $W_{kf}(k)$ is the Gaussian

process noise with covariance $Q_{kf} = q_{kf}\begin{bmatrix} B_2 & 0_{3\times3} & 0_{3\times3} \\ 0_{3\times3} & B_2 & 0_{3\times3} \\ 0_{3\times3} & 0_{3\times3} & B_2 \end{bmatrix}$, where

$q_{kf} = 0.02$ is the noise strength and $B_2 = \begin{bmatrix} 1 + T_s^2 & T_s & T_s^2 \\ T_s & 1 + T_s^2 & T_s \\ T_s^2 & T_s & 1 \end{bmatrix}$. The measurement model for the Microsoft Kinect sensor is given by

$$Z_{kf}(k) = H_{kf}X_{kf}(k) + V_{kf}(k)$$

where $H_{kf} = \begin{bmatrix} 1 & 0 & 0 & 0 & 0 & 0 & 0 & 0 & 0 \\ 0 & 0 & 0 & 1 & 0 & 0 & 0 & 0 & 0 \\ 0 & 0 & 0 & 0 & 0 & 0 & 1 & 0 & 0 \end{bmatrix}$, and $V_{kf}(k)$ is the Gaussian mea-

surement noise with covariance $R_{kf} = \begin{bmatrix} 0.06 & 0 & 0 \\ 0 & 0.06 & 0 \\ 0 & 0 & 0.06 \end{bmatrix}$ [41]. Given

these models, a standard Kalman filter is used to obtain the state estimates $\hat{X}_{kf}(k)$.

APPENDIX B ANALYTICAL JACOBIAN AND HESSIAN OF THE NN

The analytical Jacobian and Hessian of the NN, after training, can be derived as follows:

$$\frac{\partial f}{\partial s} = W^T \frac{\partial \sigma(U^T s)}{\partial s} = W^T \Sigma'(a) U^T \qquad (A1)$$

$$\frac{\partial^2 f}{\partial s_i \partial s_j} = W^T \left[\left(\Sigma''(a) U_i^T \right) \cdot U_j^T \right] \tag{A2}$$

where $a = U^T s$; $\Sigma'(a)$ is a diagonal matrix with elements $\frac{\partial \sigma(a_i)}{\partial a_i} = \sigma(a_i)(1 - \sigma(a_i))$; U_i and U_j are the ith and jth rows of the matrix U, respectively; $\Sigma''(a)$ is a diagonal matrix with elements $\frac{\partial^2 \sigma(a_i)}{\partial a_i^2} = \sigma(a_i)(1 - \sigma(a_i))(1 - 2\sigma(a_i))$; the product $\left(\Sigma''(a) U_i^T \right) \cdot U_j^T$ is a Hadamard product.

APPENDIX C GRADIENT OF THE Q FUNCTION

The gradient of the **Q** function used in the optimization can be derived as shown here:

$$\nabla_{gi} Q = \frac{1}{2} \sum_{t=1}^{T} \left[\left[\frac{\partial \tilde{v}}{\partial f} \right]^T \frac{\partial f}{\partial g_i} \right] + \frac{1}{2} \sum_{t=1}^{T} \left[\left[\frac{\partial f}{\partial g_i} \right]^T \left[Q^{-T} + Q^{-1} \right] [\hat{x}_t - \bar{x}_t] \right]$$

$$- \frac{1}{2} \sum_{t=1}^{T} \left[\left[\frac{\partial^2 \tilde{v}}{\partial f \partial g_i} \right]^T \frac{\partial f}{\partial x_{t-1}} + \left[\frac{\partial \tilde{v}}{\partial f} \right]^T \frac{\partial^2 f}{\partial x_{t-1} \partial g_i} \right] [\hat{x}_{t-1} - \bar{x}_{t-1}]$$

$$+ \frac{1}{4} \sum_{t=1}^{T} \mathrm{tr} \left\{ \left[\left[Q^{-T} + Q^{-1} \right] \frac{\partial^2 f}{\partial x_{t-1} \partial g_i} \right] \right.$$

$$\left[\hat{P}_{t,t-1} + [\hat{x}_t - \bar{x}_t][\hat{x}_{t-1} - \bar{x}_{t-1}]^T \right] \right\}$$

$$- \frac{1}{4} \sum_{t=1}^{T} \mathrm{tr} \left\{ \left[\left[\frac{\partial f}{\partial x_{t-1}} \right]^T \left[Q^{-T} + Q^{-1} \right] \frac{\partial^2 f}{\partial x_{t-1} \partial g_i} \right. \right.$$

$$+ \left[\frac{\partial^2 f}{\partial x_{t-1} \partial g_i} \right]^T \left[Q^{-T} + Q^{-1} \right] \frac{\partial f}{\partial x_{t-1}} + \frac{\partial^3 f}{\partial x_{t-1} \partial x_{t-1}^T \partial g_i} \frac{\partial \tilde{v}}{\partial f}$$

$$\left. \left. + \frac{\partial^2 f}{\partial x_{t-1} \partial x_{t-1}^T} \frac{\partial^2 \tilde{v}}{\partial f \partial g_i} \right] \left[\hat{P}_{t-1} + [\hat{x}_{t-1} - \bar{x}_{t-1}][\hat{x}_{t-1} - \bar{x}_{t-1}]^T \right] \right\} . \tag{A3}$$

APPENDIX D STABILITY ANALYSIS

The identification error dynamics can be described by

$$\dot{\tilde{x}} = W^T \sigma(U^T s) - \hat{U}^T \sigma(U^T s) + \epsilon(s) + d - \mu \tag{A4}$$

For brevity, explicit notation indicating the time dependence of the estimates is dropped. A filtered error is defined as

$$r \triangleq \dot{\tilde{x}} + \alpha \tilde{x} \tag{A5}$$

and its derivative with respect to time is as follows:

$$\dot{r} = W^T \sigma' U^T \dot{s} - \dot{\hat{W}}^T \sigma\left(\hat{U}^T s\right) - \dot{\hat{U}}^T \hat{s} - \hat{W}^T \hat{\sigma}' \hat{U}^T \dot{s} + \dot{\epsilon}(s) + \dot{d} - kr$$
$$- \gamma \tilde{x} - \beta_1 \mathrm{sgn}(\tilde{x}) - \alpha \dot{\tilde{x}} \tag{A6}$$

Grouping similar terms in Eq. (A6) yields

$$\dot{r} = \tilde{N} + N_{B_1} + \hat{N}_{B_2} - kr - \gamma \tilde{x} - \beta_1 \mathrm{sgn}(\tilde{x}) \tag{A7}$$

where $\tilde{N} \triangleq \alpha \dot{\tilde{x}} - \dot{\hat{W}}^T \sigma\left(\hat{U}^T s\right) - \hat{W}^T \hat{\sigma}' \hat{U}^T \dot{s} + \frac{1}{2} W^T \hat{\sigma}' \hat{U}^T \dot{\tilde{s}} + \frac{1}{2} \hat{W}^T \hat{\sigma}' \hat{U}^T \dot{\tilde{s}}$,

$N_{B_1} \triangleq W^T \sigma' U^T \dot{s} - \frac{1}{2} W^T \hat{\sigma}' \hat{U}^T \dot{s} - \frac{1}{2} \hat{W}^T \hat{\sigma}' U^T \dot{s} + \int (s) + \dot{d}$, and

$\hat{N}_{B_2} \triangleq \frac{1}{2} \tilde{W}^T \hat{\sigma}' \hat{U}^T \dot{s} + \frac{1}{2} W^T \hat{\sigma}' \tilde{U}^T \dot{s}$. To facilitate stability analysis, auxiliary term N_{B_2} is defined by replacing $\hat{s}$ in $\hat{N}_{B_2}$ by $\dot{s}$, $\tilde{N}_{B_2} \triangleq N_{B_2} - \hat{N}_{B_2}$, and $N_B \triangleq N_{B_1} + N_{B_2}$. The following bounds can be obtained [49]:

$$\left\| \tilde{N} \right\| \leq \rho_1(\|z\|)\|z\|, \left\| N_{B_1} \right\| \leq \zeta_1, \left\| N_{B_2} \right\| \leq \zeta_2,$$

$$\left\| \dot{N} \right\| \leq \zeta_3 + \zeta_4 \rho_2(\|z\|)\|z\|, \left\| \tilde{x}^T \tilde{N}_{B_2} \right\| \leq \zeta_5 \|\tilde{x}\|^2 + \zeta_6 \|\tilde{r}\|^2 \tag{A8}$$

where $z = \left[\tilde{x}^T, r^T\right]^T$, $\rho_1(\cdot)$ and $\rho_2(\cdot)$ are positive, globally invertible, and nondecreasing functions, and ζ_i, $i = 1, 2, \dots, 6$ are computable positive constants. The Lyapunov function[3] is given by

---

[3] For the stability analysis of identifier g is assumed to be known from the E-M algorithm, so it is treated as a known parameter.

$$V = \frac{1}{2} r^T r + \frac{1}{2} \gamma \tilde{x}^T \tilde{x} + P_v + Q_v \tag{A9}$$

where

$$Q_v = \frac{\alpha}{4} \left(\mathrm{tr} \left(\tilde{W}^T \Gamma_W^{-1} \tilde{W} \right) + \mathrm{tr} \left(\tilde{U}_x^T \Gamma_{U_x}^{-1} \tilde{U}_x \right) + \mathrm{tr} \left(\tilde{U}_g^T \Gamma_{U_g}^{-1} \tilde{U}_g \right) \right) \tag{A10}$$

$$\dot{P}_v = -L, P_v(0) = \beta_1 \sum_{i=1}^{n} |\tilde{x}_i(0)| - \tilde{x}^T(0) N_B(0) \tag{A11}$$

$$L \triangleq r^T (N_{B_1} - \beta_1 \mathrm{sgn}(\tilde{x})) + \dot{\tilde{x}} N_{B_2} - \beta_2 \rho_2(\|z\|) \|z\| \|\tilde{x}\| \tag{A12}$$

The Lyapunov function derivative $\dot{V}$ is given by

$$\dot{V} = r^T \dot{r} + \gamma \tilde{x}^T \dot{\tilde{x}} + \dot{P}_v + \dot{Q}_v \tag{A13}$$

where $\dot{Q}_v$ is a derivative of Eq. (A10) given by

$$\dot{Q}_v = -\frac{\alpha}{2} \left(\mathrm{tr} \left(\tilde{W}^T \Gamma_W^{-1} \dot{\hat{W}} \right) + \mathrm{tr} \left(\tilde{U}_x^T \Gamma_{U_x}^{-1} \dot{\hat{U}}_x \right) + \mathrm{tr} \left(\tilde{U}_g^T \Gamma_{U_g}^{-1} \dot{\hat{U}}_g \right) \right) \tag{A14}$$

By substituting the expressions from Eqs. (A5), (A6), (A11), and (A12), Eq. (A13) can be rewritten as follows:

$$\dot{V} = r^T \left(\tilde{N} + N_{B_1} + \hat{N}_{B_2} - kr - \gamma \hat{x} - \beta_1 \mathrm{sgn}(\tilde{x}) \right) + \gamma \tilde{x}^T (r - \alpha \tilde{x})$$
$$- r^T (N_{B_1} - \beta_1 \mathrm{sgn}(\tilde{x})) - \dot{\tilde{x}}^T N_{B_2} + \beta_2 \rho_2(\|z\|) \|z\| \|\tilde{x}\| + \dot{Q}_v \tag{A15}$$

On cancellations and simplifications, Eq. (A15) is given by

$$\dot{V} = r^T \tilde{N} + \left(\dot{\tilde{x}}^T + \alpha \tilde{x}^T \right) \hat{N}_{B_2} - k\|r\|^2 - \alpha\gamma\|\tilde{x}\|^2$$
$$- \dot{\tilde{x}}^T N_{B_2} + \beta_2 \rho_2(\|z\|) \|z\| \|\tilde{x}\| + \dot{Q}_v \tag{A16}$$

By defining $\tilde{N}_{B_2} = \hat{N}_{B_2} - N_{B_2}$, Eq. (A16) is rewritten as

$$\dot{V} = r^T \tilde{N} + \dot{\tilde{x}}^T \tilde{N}_{B_2} - k\|r\|^2$$
$$- \alpha\gamma\|\tilde{x}\|^2 + \beta_2 \rho_2(\|z\|) \|z\| \|\tilde{x}\| + \alpha\tilde{x}^T \hat{N}_{B_2} + \dot{Q}_v \tag{A17}$$

On redefining $\hat{N}_{B_2}$ as

$$\hat{N}_{B_2} = \frac{1}{2} \tilde{W}^T \hat{\sigma}' \hat{U}_x^T \dot{\hat{x}}_{\mathrm{id}} + \frac{1}{2} \tilde{W}^T \hat{\sigma}' \hat{U}_g^T \dot{g} + \frac{1}{2} \hat{W}^T \hat{\sigma}' \tilde{U}_x^T \dot{\hat{x}}_{\mathrm{id}} + \frac{1}{2} \hat{W}^T \hat{\sigma}' \tilde{U}_g^T \dot{g} \tag{A18}$$

and substituting the update equations from Eq. (31), Eq. (A17) is given by

$$
\begin{aligned}
\dot{V} = {}& r^T \widetilde{N} + \dot{\widetilde{x}}^T \widetilde{N}_{\mathrm{B}_2} - k\|r\|^2 - \alpha\gamma\|\widetilde{x}\|^2 + \beta_2\rho_2(\|z\|)\|z\|\|\widetilde{x}\| \\
& + \frac{\alpha}{2}\widetilde{x}^T \left(\widetilde{W}^T \hat{\sigma}' \hat{U}_x^T \dot{\hat{x}}_{\mathrm{id}} + \widetilde{W}^T \hat{\sigma}' \hat{U}_g^T \dot{\hat{g}} + \hat{W}^T \hat{\sigma}' \widetilde{U}_x^T \dot{\hat{x}}_{\mathrm{id}} + \hat{W}^T \hat{\sigma}' \widetilde{U}_g^T \dot{\hat{g}} \right) \\
& - \frac{\alpha}{2}\left(\mathrm{tr}\left(\widetilde{W}^T \hat{\sigma}' \hat{U}_x^T \dot{\hat{x}}_{\mathrm{id}}\widetilde{x}^T \right) + \mathrm{tr}\left(\widetilde{W}^T \hat{\sigma}' \hat{U}_g^T \dot{\hat{g}}\widetilde{x}^T \right) \right. \\
& \left. + \mathrm{tr}\left(\widetilde{U}_x^T \dot{\hat{x}}_{\mathrm{id}}\widetilde{x}^T \hat{W}^T \hat{\sigma}' \right) + \mathrm{tr}\left(\widetilde{U}_g^T \dot{\hat{g}}\widetilde{x}^T \hat{W}^T \hat{\sigma}' \right) \right)
\end{aligned}
$$

$$(A19)$$

Using the cyclic property of the trace operator and the bounds defined in Eq. (A8), Eq. (A19) is rewritten as

$$
\begin{aligned}
\dot{V} \overset{a.e.}{\leq}{}& -\alpha\gamma\|\widetilde{x}\|^2 \\
& - k\|r\|^2 + \rho_1(\|z\|)\|z\|\|r\| + \zeta_5\|\widetilde{x}\|^2 + \zeta_6\|\widetilde{r}\|^2 + \beta_2\rho_2(\|z\|)\|z\|\|\widetilde{x}\|
\end{aligned}
\quad (A20)
$$

The righthand side of Eq. (A20) is continuous almost everywhere except the Lebesgue measure zero set of times when $\widetilde{x}(t) = 0$. Substituting for $k \triangleq k_1 + k_2$ and $\gamma \triangleq \gamma_1 + \gamma_2$ and completing the squares,

$$
\begin{aligned}
\dot{V} \overset{a.e.}{\leq}{}& -(\alpha\gamma - \gamma_5)\|\widetilde{x}\|^2 \\
& - (k_1 - \gamma_6)\|r\|^2 + \frac{\rho_1(\|z\|)^2}{4k_2}\|z\|^2 + \frac{\beta_2^2\rho_2(\|z\|)^2}{4\alpha\gamma_2}\|z\|^2
\end{aligned}
\quad (A21)
$$

If the conditions $\gamma > \dfrac{\gamma_5}{\alpha}$ and $k > \zeta_6$ are met, then $\dot{V}$ can be upper bounded as follows:

$$
\dot{V} \overset{a.e.}{\leq} -\lambda\|z\|^2 + \frac{\rho(\|z\|)^2}{4\eta}
\quad (A22)
$$

where $\quad \lambda \triangleq \min\{\alpha\gamma_1 - \gamma_5, k_1 - \gamma_6\}, \qquad \eta \triangleq \min\left\{ k_2, \dfrac{\alpha\gamma_2}{\beta_2^2} \right\}, \qquad$ and

$\rho(\|z\|)^2 \triangleq \rho_1(\|z\|)^2 + \rho_2(\|z\|)^2$. A semiglobal asymptotic stability of the error dynamics in Eq. (A4) can be shown using the inequalities in Eqs. (A21) and (A22), which yields $\|\widetilde{x}(t)\| \to 0$, $\|\dot{\widetilde{x}}(t)\| \to 0$, and $\|r\| \to 0$ as $t \to \infty$ [49].

REFERENCES

[1] E. Gribovskaya, A. Kheddar, A. Billard, Motion learning and adaptive impedance for robot control during physical interaction with humans, in: IEEE International Conference on Robotics and Automation (ICRA), 2011, IEEE, 2011, pp. 4326–4332.

[2] S. Nikolaidis, K. Gu, R. Ramakrishnan, J.A. Shah, Efficient model learning for human-robot collaborative tasks, CoRR (2014). vol. abs/1405.6341.

[3] C.-S. Tsai, J.-S. Hu, M. Tomizuka, Ensuring safety in human-robot coexistence environment, in: IEEE/RSJ International Conference on Intelligent Robots and Systems (IROS), 2014, IEEE, 2014, pp. 4191–4196.

[4] C. Liu, M. Tomizuka, Modeling and controller design of cooperative robots in workspace sharing human-robot assembly teams, in: IEEE/RSJ International Conference on Intelligent Robots and Systems (IROS), 2014, IEEE, 2014, pp. 1386–1391.

[5] H. Ravichandar, A.P. Dani, Human intention inference through interacting multiple model filtering, in: IEEE Conference on Multisensor Fusion and Integration (MFI), 2015.

[6] D.A. Baldwin, J.A. Baird, Discerning intentions in dynamic human action, Trends Cogn. Sci. 5 (4) (2001) 171–178.

[7] M.A. Simon, Understanding Human Action: Social Explanation and the Vision of Social Science, SUNY Press, Albany, NY, 1982.

[8] H. Ravichandar, A. Kumar, A.P. Dani, Bayesian Human Intention Inference Through Multiple Model Filtering with Gaze-based Priors, in: International Conference on Information Fusion, 2016.

[9] H. Ravichandar, A.P. Dani, A.P. Dani, Learning Periodic Motions from Human Demonstrations using Transverse Contraction Analysis, in: IEEE American Control Conference, 2016.

[10] G.C. Goodwin, J. Aguero, Approximate E-M algorithms for parameter and state estimation in nonlinear stochastic models, in: IEEE Conference on Decision and Control, and European Control Conference, 2005, IEEE, 2005, pp. 368–373.

[11] Z. Ghahramani, S.T. Roweis, Learning nonlinear dynamical systems using an E-M algorithm, Advances in Neural Information Processing Systems, vol. 11, MIT Press, Cambridge, 1999, pp. 431–437.

[12] G. Goodwin, A. Feuer, Estimation with missing data, Math. Comput. Model. Dyn. Syst. 5 (3) (1999) 220–244.

[13] J. Preece, Y. Rogers, H. Sharp, D. Benyon, S. Holland, T. Carey, Human-Computer Interaction, Addison-Wesley Longman Ltd., Essex, 1994.

[14] M.A. Goodrich, A.C. Schultz, Human-robot interaction: a survey, Found. Trends Hum.-Comput. Interact. 1 (3) (2007) 203–275.

[15] C. Matuszek, E. Herbst, L. Zettlemoyer, D. Fox, Learning to parse natural language commands to a robot control system, Experimental Robotics, Springer, Switzerland, 2013, pp. 403–415.

[16] V.J. Traver, A.P. del Pobil, M. Perez-Francisco, Making service robots human-safe, in: IEEE/RSJ International Conference on Intelligent Robots and Systems, 2000, pp. 696–701.

[17] Y. Matsumoto, J. Heinzmann, A. Zelinsky, The essential components of human-friendly robot systems, in: International Conference on Field and Service Robotics, 1999, pp. 43–51.

[18] M.S. Bartlett, G. Littlewort, I. Fasel, J.R. Movellan, Real time face detection and facial expression recognition: development and applications to human computer interaction, IEEE Conference on Computer Vision and Pattern Recognition, vol. 5, 2003, p. 53.

[19] T. Fong, I. Nourbakhsh, K. Dautenhahn, A survey of socially interactive robots, Robot. Auton. Syst. 42 (3) (2003) 143–166.

[20] D. Kulic, E.A. Croft, Affective state estimation for human robot interaction, IEEE Trans. Robot. 23 (5) (2007) 991–1000.

[21] E. Meisner, V. Isler, J. Trinkle, Controller design for human-robot interaction, Auton. Robot. 24 (2) (2008) 123–134.

[22] D.K.E. Croft, Estimating intent for human-robot interaction, in: IEEE International Conference on Advanced Robotics, 2003, 2003, pp. 810–815.

[23] D. De Carli, E. Hohert, C.A. Parker, S. Zoghbi, S. Leonard, E. Croft, A. Bicchi, Measuring intent in human-robot cooperative manipulation, in: IEEE International Workshop on Haptic Audio Visual Environments and Games, 2009, pp. 159–163.

[24] H. Ding, G. Reißig, K. Wijaya, D. Bortot, K. Bengler, O. Stursberg, Human arm motion modeling and long-term prediction for safe and efficient human-robot-interaction, in: 2011 IEEE International Conference on Robotics and Automation, 2011, pp. 5875–5880.

[25] D. Gehrig, P. Krauthausen, L. Rybok, H. Kuehne, U.D. Hanebeck, T. Schultz, R. Stiefelhagen, Combined intention, activity, and motion recognition for a humanoid household robot, in: IEEE/RSJ International Conference on Intelligent Robots and Systems, 2011, 2011, pp. 4819–4825.

[26] O.C. Schrempf, U.D. Hanebeck, A.J. Schmid, H. Worn, A novel approach to proactive human-robot cooperation, in: IEEE International Workshop on Robot and Human Interactive Communication, 2005, 2005, pp. 555–560.

[27] O.C. Schrempf, U.D. Hanebeck, A generic model for estimating user intentions in human-robot cooperation, in: International Conference on Informatics in Control, Automation and Robotics, 2005, 2005, pp. 251–256.

[28] J. Elfring, R. Van De Molengraft, M. Steinbuch, Learning intentions for improved human motion prediction, Robot. Auton. Syst. 62 (4) (2014) 591–602.

[29] H.S. Koppula, R. Gupta, A. Saxena, Learning human activities and object-affordances from RGB-D videos, Int. J. Robot. Res. 32 (8) (2013) 951–970.

[30] H.S. Koppula, A. Saxena, Anticipating human activities using object-affordances for reactive robotic response, in: Robotics: Science and Systems, 2013, 2013.

[31] N. Hu, Z. Lou, G. Englebienne, B. Krose, Learning to recognize human activities from soft labeled data, in: Robotics: Science and Systems, Berkeley, USA, 2014, 2014.

[32] J. Mainprice, D. Berenson, Motion planning for human-robot collaborative manipulation tasks using prediction of human motion, in: Human-Robot Collaboration for Industrial Manufacturing Workshop at Robotics: Science and Systems Conference, 2014, 2014.

[33] D. Song, N. Kyriazis, I. Oikonomidis, C. Papazov, A. Argyros, D. Burschka, D. Kragic, Predicting human intention in visual observations of hand/object interactions, in: 2013 IEEE International Conference on Robotics and Automation, IEEE, 2013, pp. 1608–1615.

[34] K. Strabala, M.K. Lee, A. Dragan, J. Forlizzi, S. Srinivasa, Learning the communication of intent prior to physical collaboration, in: IEEE International Symposium on Robot and Human Interactive Communication, 2012.

[35] Z. Wang, K. Mülling, M.P. Deisenroth, H.B. Amor, D. Vogt, B. Schölkopf, J. Peters, Probabilistic movement modeling for intention inference in human robot interaction, Int. J. Robot. Res. 32 (7) (2013) 841–858.

[36] P.A. Lasota, G.F. Rossano, J.A. Shah, Toward safe close-proximity human-robot interaction with standard industrial robots, in: IEEE International Conference on Automation Science and Engineering (CASE), 2014, IEEE, 2014, pp. 339–344.

[37] R. Kelley, A. Tavakkoli, C. King, M. Nicolescu, M. Nicolescu, G. Bebis, Understanding human intentions via hidden Markov models in autonomous mobile robots, in: ACM/IEEE International Conference on Human Robot Interaction, 2008, 2008, pp. 367–374.

[38] S. Kim, A. Billard, Estimating the non-linear dynamics of free-flying objects, Robot. Auton. Syst. 60 (9) (2012) 1108–1122.

[39] S. Kim, A. Shukla, A. Billard, Catching objects in-flight, in: IEEE Transactions on Robotics, 2014, 2014.

[40] D.J. MacKay, Bayesian interpolation, Neural Comput. 4 (3) (1992) 415–447.

[41] C. Morato, K.N. Kaipa, B. Zhao, S.K. Gupta, Toward safe human robot collaboration by using multiple Kinects based real-time human tracking, J. Comput. Inf. Sci. Eng. 14 (1) (2014) 011006.

[42] A.P. Dempster, N.M. Laird, D.B. Rubin, Maximum likelihood from incomplete data via the E-M algorithm, J. R. Stat. Soc. Ser. B Methodol. 39 (1977) 1–38.

[43] K. Lange, A gradient algorithm locally equivalent to the E-M algorithm, J. R. Stat. Soc. Ser. B Methodol. 57 (1995) 425–437.

[44] B. Xian, D.M. Dawson, M.S. de Queiroz, J. Chen, A continuous asymptotic tracking control strategy for uncertain nonlinear systems, IEEE Trans. Autom. Control 49 (7) (2004) 1206–1211.

[45] P.M. Patre, W. MacKunis, K. Kaiser, W.E. Dixon, Asymptotic tracking for uncertain dynamic systems via a multilayer neural network feedforward and rise feedback control structure, IEEE Trans. Autom. Control 53 (9) (2008) 2180–2185.

[46] A.F. Filippov, Differential equations with discontinuous right-hand side, Matematicheskii sbornik 93 (1) (1960) 99–128.

[47] A.P. Dani, N. Fisher, Z. Kan, W.E. Dixon, Globally Exponentially Convergent Robust Observer for Vision-based Range Estimation, Mechatronics, Special Issue on Visual Servoing 22 (4) (2012) 381–389.

[48] A.P. Dani, N. Fischer, Z. Kan, W.E. Dixon, Globally exponentially convergent robust observer for vision-based range estimation, Mechatronics, Special Issue on Visual Servoing, 22 (4) (2012) 381–389.

[49] S. Bhasin, R. Kamalapurkar, H.T. Dinh, W.E. Dixon, Robust-identification-based state derivative estimation for nonlinear systems, IEEE Trans. Autom. Control, 58 (1) (2013) 187–192.

[50] H. Ravichandar, A. Kumar, A.P. Dani, Bayesian human intention inference through multiple model filtering with gaze-based priors, in: International Conference on Information Fusion, 2016 (to appear).

[51] H. Ravichandar, P.K. Thota, A.P. Dani, Learning periodic motions from human demonstrations using transverse contraction analysis, in: IEEE American Control Conference, 2016 (to appear).

CHAPTER NINE

Biomechanical HRI Modeling and Mechatronic Design of Exoskeletons for Assistive Applications

S. Bai, S. Christensen
Aalborg University, Aalborg, Denmark

1. INTRODUCTION

Exoskeletons are mechanical devices attached to human bodies for either power augmentation or motion assistance. Research on exoskeletons has led to many impressive solutions. Fig. 1 shows a few of these examples, with applications in either military or health care and rehabilitation. Of these, the Wilmington Robotic Exoskeleton (WREX) is a two-segment, 4-DOF (degrees of freedom) passive orthosis [1], which can be mounted on a person's wheelchair or to a body jacket. WREX uses linear elastic elements to balance the effect of gravity. The hybrid assistive limb (also known as HAL), developed by the University of Tsukuba and CYBERDYNE, Inc., is the world's first cyborg-type robot to improve, support, expand, and regenerate the physical functions of people with physical disabilities due to aging or diseases. The HAL robot detects faint bioelectric signals from the surface of the skin derived from the brain and nervous system, and translates these signals to drive its actuators to assist the patient's movements in real time [2]. HAL for medical use (lower limb type) has been commercialized for medical use in the European Union, and focuses on treatment of lower-limb paralysis. The ARMin III [3] is an arm therapy exoskeleton robot with three actuated DOFs for the shoulder and one DOF for the elbow. It was designed to improve the rehabilitation process in stroke patients. The IntelliArm [4] is a whole arm robot, which has eight actuated DOFs and two passive DOFs at the shoulder. Also, the IntelliArm has an additional DOF for hand opening and closing. A 7-DOF cable-driven exoskeleton for the upper arm was presented in [5], in which the cables were driven by electrical motors. Several other types of actuated exoskeleton

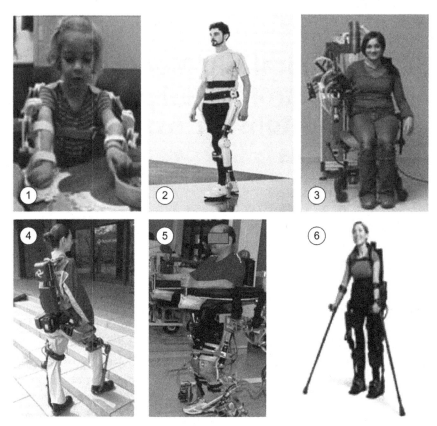

Fig. 1 Examples of exoskeletons: (1) Passive arm exoskeleton WREX, (2) HAL from Cyberdyne, Japan, (3) ARMin III rehabilitation arm, (4) ActiveLinks Powerloader Ninja exoskeleton suit, (5) NaTUre-gaits, and (6) E-leg exoskeleton from Ekso Bionics.

robots were also proposed, such as ABLE [6], CADEN-7 [7], LOPES [8], MGA [9], RehabExos [10], NaTUre-gaits [11], and Pneu-WREX [12]. For a detailed review of the state-of-the-art exoskeleton robots for upper limbs, the readers is referred to [13].

Exoskeletons can be categorized into two major groups, namely passive and active exoskeletons. Most of the preceding examples except WREX are active exoskeletons. In addition to WREX, several other passive exoskeletons have been developed recently. The NEUROExos is a variable impedance-powered elbow exoskeleton developed for rehabilitating stroke patients [14]. The exoskeleton utilizes a double shell link structure and a 4-DOF passive mechanism. NEUROExos makes use of an adaptive, passive-compliant actuator through a bio-inspired antagonistic nonlinear elastic actuation system. An upper limb exoskeleton with a 3-DOF shoulder joint and a 1-DOF elbow joint has also been designed [15]. The grounding device can increase resistance through adjustment of the spring length to train more muscle groups. A class of weight-balanced mechanisms for arm support was presented in [16]. Compared with active exoskeletons, passive exoskeleton, are inherently safe, lightweight and easy to implement, and more suitable for light-duty assistive applications.

Exoskeletons pertain to a type of dynamic system where a human is part of the motion. For both active and passive exoskeletons, the human-robot interaction (HRI) is critical for such systems, as exoskeletons work with close contact with the human body. The HRI brings challenges to all stages of exoskeleton design and development, from design simulation, mechanical design, sensing and actuation, etc. Of these, biomechanical simulation allows us a better understanding of the biomechanics of the human motion and sensory mechanisms. The mechatronics design, on the other hand, implements the physical system for a desired HRI.

This chapter introduces an approach to exoskeleton design based on biomechanical simulations of HRIs, which aims to design exoskeleton structures and motion controllers effectively in order to implement motion assistance as needed. The chapter covers topics including (1) design challenges of lightweight exoskeletons, (2) fundamentals of HRI modeling, (3) simulation modeling development, (4) mechatronics design, and also presents examples of exoskeletons.

2. CHALLENGES IN EXOSKELETON DESIGN

Exoskeleton design and development face many challenges, such as novel mechanisms and actuation, biosensing, HRI, etc. In this chapter, we briefly discuss challenges in biomechanical human-robot modeling, kinematic design, actuation, and sensing, which are major concerns in the design stage.

2.1 Biomechanical Modeling of Human-Exoskeleton System

A human-exoskeleton system consists of an exoskeleton working cooperatively with human muscles and nerve systems. The interaction between the exoskeleton and the human body determines whether the exoskeleton can implement the desired functions. To this end, biomechanical simulation plays a fundamental role in the exoskeleton design. A central issue in the modeling work is thus to simulate the response of the human body subject to external loads and forces/torques exerted by the exoskeleton.

Researchers have tried to model the HRI through musculoskeletal modeling [17]. An attempt was made to model interactions between the human and rehabilitation devices by musculoskeletal simulation [18], where parametric design of devices based on musculoskeletal performance has been conducted. Exoskeleton design through biomechanics analysis was also studied [19], with analysis of a 1-DOF exoskeleton. A human-robot model was developed for a 2-DOF assistive exoskeleton to study the influence of power assistance on the muscle activities [20].

To build a human-exoskeleton model, two submodules, namely a musculoskeletal model and an exoskeleton model, need to be developed. The musculoskeletal model deals with biomechanics analysis, and the exoskeleton model conducts the kinematics and dynamics simulation. Commercial software, such as SIMM (MusculoGraphics, Inc.), Visual 3-D (C-Motion, Inc.), AnyBody Modeling System (AnyBody Technology), and OpenSim, are available to quickly implement biomechanical models for studying the biomechanics of different biological systems. Among these software packages, AnyBody and OpenSim are two commonly used ones. The AnyBody software is mainly for inverse dynamics simulation. The software includes a well-developed comprehensive repository of all body elements, bones, muscles, etc., which allows the user to modify and define for their own model. OpenSim has embedded a forward dynamics model, which makes it possible to integrate a motion controller in the simulation.

2.2 Kinematic Structure

A main challenge with the design of the kinematic structure of power assistive upper-body exoskeleton robots is to comply with the complexity of the human body. While the human arm (from the shoulder and out) comprises a total of nine DOF, an exoskeleton that drives all DOFs is not practical. Two approaches have been considered: (1) design with redundant passive joints and (2) adopting soft coupling in exoskeletons. By using redundant passive joints, the wearer is allowed to compensate for possible misalignment with the active joints. This method is used in the SUEFUL-7 [21], where a simple passive slider mechanism automatically adjusts for the glenohumeral joint's center of rotation. On the other hand, this design is more bulky compared to exoskeletons without redundant joints and decreases the range of motion.

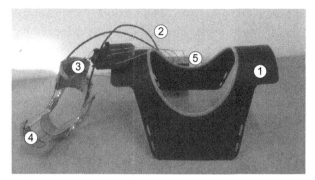

Fig. 2 Spring-loaded cable exoskeleton for gravity compensation, with parts including (1) armor, (2) cables, (3) upper arm bracket, (4) forearm bracket, and (5) spring casing wherein springs are mounted.

Adopting soft coupling in the exoskeleton refers to the fact that the segments of the exoskeletons are coupled flexibly, mainly by means of cables, rather than connected with rigid joints. An example is the spring-loaded cable-driven wearable passive exoskeleton developed at Aalborg University [22]. The arm structure of the exoskeleton is not coupled to the supporting frame rigidly, but by flexible cables instead. A total of five springs are mounted in the spring box to store the elastic potential energy and compensate for the weight of the upper arm, forearm, and external load. The soft coupling concept leads to a lighter design of the exoskeleton (Fig. 2).

2.3 Actuation

The actuation of exoskeletons must be able to drive the system safely and compliantly [23]. Back-drivability is needed to avoid the user being locked by the exoskeleton and to enhance its safety and comfort. It has to also be lightweight, compact, and preferably highly efficient.

Most active assistive exoskeletons are actuated by conventional electric motors or pneumatic muscle actuators (PMAs). Electric motors are most commonly used due to their high speed, high precision, and advanced motion control. The actuators are either placed locally at the joints or stored at the back, where the force/torque is transmitted via cables, like the design of CADEN-7. The PMA is analogous to a human muscle in that it can perform one-way actuation compliantly and flexibly. Their lightness, good power/weight ratio, and flexible structure make PMAs desirable as a portable exoskeleton. The flexible/compliant structure is an advantage over electric motors. The Robotic Upper Extremity Repetitive Trainer (RUPERT) is an active exoskeleton with four DOF that uses PMA [24]. A major problem with PMAs is their highly nonlinear behavior, which makes them more complicated and in general less accurate than electric motors. A novel actuation called the Screw Cable System (SCS) is reported for the actuation of the ABLE [6]. In this actuation, a ball screw with a cable is attached to the hollow screw. The size of the motors is kept low due to the gearing using the ball screw, which reduces the weight and overall size of the exoskeleton. In addition, the SCS has a high efficiency (at best 94%) and is back-drivable. A limitation with the design of the SCS is that it is quite bulky and weighs 13 kg, which has to be carried by the wearer.

2.4 Sensing

The exoskeleton is a robotic system for human bodies. Conventional sensors, such as positional and force sensors, are still needed. On the other hand, the robot has to be controlled collaboratively with the human body, so bio-sensors for human physiological and motion status are needed as well. A main purpose for employing the sensors is to detect the human motion intention.

From a control perspective, the user and the exoskeleton robot compose a closed-loop system. Depending on the measurement method of the HRI, the exoskeleton can be controlled either based on a cognitive human-robot interaction (cHRI) or a physical human-robot interaction (pHRI). The cHRI-based control system measures common electric signals from the central nervous system to the musculoskeletal system of humans and uses them as inputs for the exoskeleton control. The human intent is thereby identified before the occurrence of the actual motion of the wearer and the required torque/velocity/position for the motion of human joints can be predicted [25]. On the other hand, the pHRI-based control system measures the force or position changes that are the results of the motions by the human musculoskeletal system, which is then used as inputs to the exoskeleton control [26].

The SUEFUL-7 exoskeleton [21] uses a cHRI-based control system consisting of an EMG (electromyography) based impedance control. A total of 16 EMG sensors are used to predict the intention of the human, upon which a torque reference for each joint can be generated. An impedance control is used to ensure compliance of the exoskeleton. The ABLE [6] uses a pHRI-based control system that is based on a position-force feedback control principle. The torque of a given motor is a combination of gravity compensation torque and a reaction torque proportional to a position error signal. Due to the low friction and back-drivable transmission system of the ABLE, the control was conducted without any force sensors [6], which is the main drawback of pHRI-based control systems.

Of these aforementioned challenges, the biomechanics of the human body is critical for the exoskeleton design, as it plays a fundamental role in the human-exoskeleton interaction. It is essential to simulate the pHRI by means of biomechanics simulation. In this chapter, an approach to exoskeleton design based on biomechanics HRI simulation is presented.

3. BIOMECHANICAL MODELING

In a musculoskeletal model, the human body is modeled as a multibody system, in which bones and joints are treated as mechanical links and joints, while muscles exert force on the system. An illustration of the biomechanics of muscle recruitment is shown in Fig. 3, where an arm is subjected to an external load $\mathbf{d} = [\mathbf{d}_f, \mathbf{d}_m]$, which includes both force, $\mathbf{d}_f$, and moment, $\mathbf{d}_m$. The equilibrium condition of the forearm is readily obtained as

$$\sum_{i=1}^{3} \mathbf{f}_i^{(M)} + \mathbf{f}^{(R)} + \mathbf{d}_f = 0 \tag{1}$$

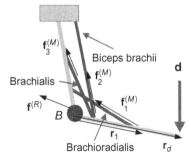

Fig. 3 A simplified biomechanics model of muscle recruitment, where **d** indicates external load which includes force and moment. $\mathbf{r}_i$, $i = 1,2,3$ are position vectors from point B to the muscle attachment points on the forearm, with only $\mathbf{r}_1$ displayed for clarity.

$$(-\mathbf{r}_d) \times \mathbf{f}^{(R)} + \sum_{i=1}^{3} (\mathbf{r}_i - \mathbf{r}_d) \times \mathbf{f}_i^{(M)} + \mathbf{d}_{\mathrm{m}} = 0 \qquad (2)$$

which can be expressed in a compact form

$$\mathbf{C}\mathbf{f} = \mathbf{d} \qquad (3)$$

where $\mathbf{f} = [\mathbf{f}^R, \mathbf{f}^{(M)}] = [\mathbf{f}^R, \mathbf{f}_1^{(M)}, \mathbf{f}_2^{(M)}, \mathbf{f}_3^{(M)}]$ is composed of a n-dimensional vector of joint reaction forces $\mathbf{f}^{(R)}$ and muscle forces $\mathbf{f}_i^{(M)}$. A muscle force is the product of a positive value $f_i^{(M)}$ and a direction vector $\hat{\mathbf{f}}_i^{(M)}$. A positive value $f_i^{(M)}$ implies that a muscle can only be subject to pulling forces, but not pushing forces. $\mathbf{C}$ is a coefficient matrix generated from the arm anatomy and muscle attachments. For the case shown in Fig. 3, $\mathbf{C}$ takes the form of

$$\mathbf{C} = \begin{bmatrix} -\mathbf{I} & -\mathbf{I} & -\mathbf{I} & -\mathbf{I} \\ \tilde{\mathbf{r}}_d & \tilde{\mathbf{r}}_d - \tilde{\mathbf{r}}_1 & \tilde{\mathbf{r}}_d - \tilde{\mathbf{r}}_2 & \tilde{\mathbf{r}}_d - \tilde{\mathbf{r}}_3 \end{bmatrix} \qquad (4)$$

where $\mathbf{I}$ is the three-dimensional identity matrix, while $\tilde{\mathbf{r}}_i, i = d, 1, 2, 3$ is the screw-symmetric matrix of vector $\mathbf{r}_i$. As the human-body biomechanics system is statically indeterminate, the muscle recruitment can be formulated as an optimization problem as

$$\begin{aligned} \min \quad & G(\mathbf{f}^{(M)}) \\ \text{s.t.} \quad & \mathbf{C}\mathbf{f} = \mathbf{d} \\ & f_i^{(M)} \geq 0, \quad i \in \{1, ..., n^{(M)}\} \end{aligned} \qquad (5)$$

The choice of the objective function $G(\mathbf{f}^{(M)})$ depends on the muscle recruitment criterion. The possible criteria include soft saturation, min/max, and polynomial muscle recruitment, etc. The polynomial criterion is adopted as

$$G(\mathbf{f}^{(M)}) = \sum_i \left(\frac{f_i^{(M)}}{N_i} \right)^p \qquad (6)$$

where N_i are normalization factors or functions, which take the form of muscle strength in this work. The power p indicates the synergy of muscles; $p = 3$ is recommended as it yields good results for most submaximal muscle efforts [27]. The ratio $f_i^{(M)}/N_i$ refers to the muscle activity.

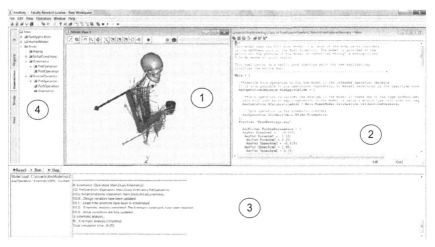

Fig. 4 User interface of the AnyBody Modeling System.

4. DEVELOPMENT OF HRI MODEL

In developing the biomechanical HRI models, we utilized the Any-Body Modeling System. The user interface of the AnyBody Modeling System is shown in Fig. 4. The features include: (1) model view window to visualize the model, (2) editor window to edit AnyScript, (3) log window, for errors, warnings and other status messages, and (4) main-tree window for study control and feature selection. Simulation results can be displayed in the AnyChart window, which pops out after a study case is simulated.

The development of the biomechanical HRI simulation model can take advantage of AnyBody's well-developed repository of biomechanics models. In the AnyBody, bones are modeled as rigid bodies. Hill-type muscle models are used, which consist of three elements, namely the contractile element (CE) that generates force and represents the muscle fibers, the passive element (PE) in parallel with CE, and a serial elastic element connected in series with the CE and PE. The software is able to be incorporated with geometric models of exoskeletons. The exoskeleton built in SolidWorks can be exported to AnyBody.

5. DESIGN EXAMPLES

We include two examples to demonstrate the biomechanical simulation and exoskeleton design.

5.1 Example I: A Spring-Loaded Exoskeleton for Gravity Compensation

The first example is taken from a design concept of spring-loaded cable-driven exoskeletons [22, 28]. This is a soft-coupling exoskeleton design, as shown in Fig. 2, where the arm structure of the exoskeleton is coupled to the supporting frame by flexible cables. In this design, springs storing elastic energy are mounted in a spring box close to the body. The spring forces are transferred via flexible cables to the distal parts of the exoskeleton. As shown in the CAD model in Fig. 5A, a group of cables are routed from the armor cuff to the elbow upper bracket. In the design presented, three cables are used for the shoulder joint, with two cables connecting the elbow upper and lower brackets. The armor of the exoskeleton is anchored on the trunk of the user. On the back of the armor, a casing holds all springs.

At the elbow joint, two anchoring nodes are designed on the elbow lower bracket. Two cables are linked to two springs from the anchoring point through the two routing points on the upper bracket. The force in the cable hereby can provide balance torque for the flexion motion. The three anchoring points on the upper bracket are the attaching nodes for the spring cables which balance the shoulder joint.

The exoskeleton is intended to assist patients with shoulder injuries. It is expected that the exoskeleton could help patients in some daily activities, such as eating or drinking with their arms. For the design concept presented, we want to confirm two design considerations: (1) whether the springs are able to provide a user necessary assistance in lifting and (2) what the proper sizes of the springs are (natural length and stiffness). These questions can be answered by biomechanical HRI simulations.

5.1.1 Human-Exoskeleton Model

The muscle recruitment in Eq. (5) then becomes

$$
\begin{aligned}
\min \quad & G(\mathbf{f}^{(M)}) \\
\text{s.t.} \quad & \mathbf{Cf} = \mathbf{d} \\
& f_i^{(M)} \geq 0, \quad i \in \{1, \dots, n^{(M)}\}
\end{aligned}
\tag{7}
$$

where the total external force $\mathbf{d}$ includes both the payload $\mathbf{d}^*$ and forces generated from springs

$$
\mathbf{d} = \mathbf{d}^* + \mathbf{Et}_c
\tag{8}
$$

where vector $\mathbf{t}_c$ contains the tensions of the incorporated springs. The coefficient matrix $\mathbf{E}$ is generated from the installation of springs in the exoskeleton and the exoskeleton attachment to the arm.

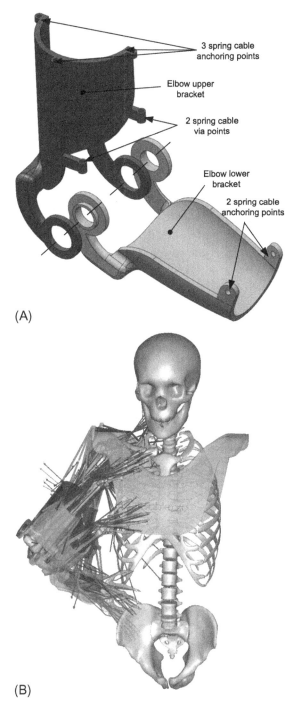

Fig. 5 Human-robot modeling: (A) CAD model of the spring-loaded cable-driven exoskeleton and (B) biomechanical model of human motion assistance in pick-up motion.

The cable tension generated by the springs is calculated by

$$\mathbf{t}_c = \mathbf{t}_{pre} + \mathbf{kl}_d \qquad (9)$$

with

$$\mathbf{k} = \mathrm{diag}\left(\left[k_{gh1}, k_{gh2}, k_{gh3}, k_{el1}, k_{el2} \right] \right) \qquad (10)$$

where $\mathbf{t}_{pre}$ denotes preload of the springs, $\mathbf{k}$ is a diagonal matrix containing the stiffness of springs for the three cables of the shoulder (glenohumeral) joint and two of the elbow joint, $\mathbf{l}_d$ is the vector of the spring elongations, with the first three diagonal entries for the springs applied to the upper arm segment and the remaining two for the springs to the forearm segment.

The preload tension of the spring is

$$\mathbf{t}_{pre} = \mathbf{kl}_{pre} \qquad (11)$$

where $\mathbf{l}_{pre}$ is an array of preload lengths of the springs.

The simulated motion is to pick up a cup, as demonstrated in Fig. 5B. The payload is 0.5 kg carried by the hand. The motion lasts for 3 s, with the joint movements realizing the hand trajectory as depicted in Fig. 6. The elbow pronation and the wrist joint rotation are not considered.

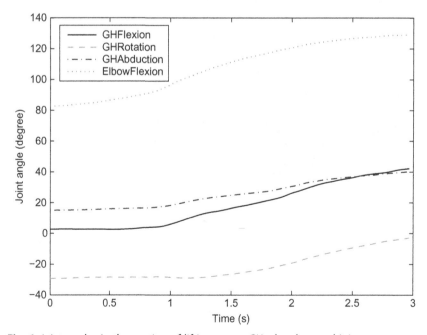

Fig. 6 Joint angles in the motion of lifting a cup. *GH*, glenohumeral joint.

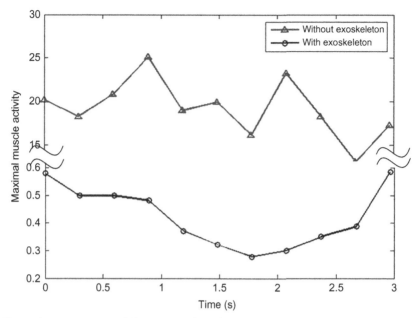

Fig. 7 The comparison of the MMA with and without assistance.

Our interest is to evaluate the maximal muscle activation (MMA) for the selected case with and without exoskeleton assistance.

Fig. 7 shows the MMAs of a simulated shoulder-injury patient with two different conditions, namely with and without exoskeleton assistance. Without assistance of the exoskeleton, normal lifting is not possible, as the calculated maximal MMA is equal to 25.1, far above the limit of MMA of 1.0. With the exoskeleton, for the selected case in the motion of picking up a cup and drinking, the simulation finds that the exoskeleton reduces the maximal MMA to 0.58, which is below the limit of MMA of 1.0, implying the possibility of picking up and drinking with the assistance of an exoskeleton.

The simulation yields the design parameters of all springs. In the simulated case, all the preload lengths of the five springs are set to 0.06 m. The sizes of springs are obtained as

$$\mathbf{x} = [k_{gh1}, k_{gh2}, k_{gh3}, k_{el1}, k_{el2}] = [1473.4, 0.3, 0.03, 102.0, 1979.0] \ \text{N/m}$$

These parameters provide a strong reference for the final selection of all components. From the results, it can be seen that it is the spring of stiffness k_{gh1} which makes a major contribution in balancing the external load in the upper arm, while the other two springs' contribution can nearly be ignored.

5.2 Example II: A 2-DOF Active Assistive Exoskeleton

This example is given for a 2-DOF exoskeleton for lifting and holding assistance. A human–robot model was developed with an exoskeleton and a musculoskeletal model, as shown in Fig. 8. The exoskeleton arm has two rotational DOF, one at the shoulder and one at the elbow. The shaft of the shoulder joint is grounded, which in practice means that it is fixed to, for instance, a back support, or a wheelchair. The musculoskeletal model is sized as a 50th percentile European male. Only the upper trunk and the right arm are included in this model.

The exoskeleton was first built in SolidWorks and then was exported to AnyBody. All joints are needed to be redefined in AnyBody. The exoskeleton is parallel to the human arms. The attachment of the exoskeleton to the human wrist is modeled as a spherical joint. The mass properties, defined with SolidWorks, are shown in Table 1.

Simulation of the pHRI was conducted for a cooperative motion of the exoskeleton and a human arm in lifting a payload of $F = 50$ N. In the simulated motion, the shoulder and the elbow joints rotate at angular velocities

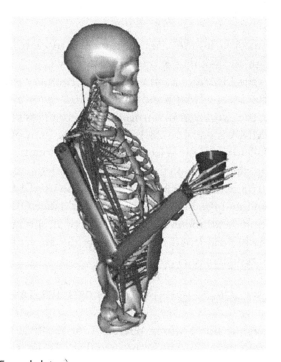

Fig. 8 A 2-DOF exoskeleton.

Table 1 Mass Properties of the Exoskeleton

	Mass (kg)	MOI $[I_{xx}, I_{yy}, I_{zz}]$ (kg· m^2)
Forearm	1.5	[0.01, 0.01, 0.15]
Upper arm	2.5	[0.02, 0.02, 0.2]

of 15 and 5 degree/s. The maximum input torques applied to the shoulder and elbow joints, namely M_s and M_e, were defined, varying in the range of [5, 30] N· m, which implies a variable assistive input power to the human body. The maximum input torques were increased without overloading the muscles. In Fig. 9, activities of three selected muscles of the upper limb are displayed. These three muscles include the biceps, brachialis, and br. radialis, which are considered to be the major extensor/flexors of arm motion. Note that each of these muscles is composed of two to six muscles, which are modeled separately in AnyBody. We evaluate the muscle activities by taking the mean activity of muscles in each group. It is seen that these muscles exert different strengths in the two selected cases of simulation with the payload of $F = 50$ N. The variations of the muscle activities are displayed in Fig. 9A and B for two different levels of assistance. The activities of all muscles in the upper limb for the case without the assistive torques are shown in Fig. 9C. It can be seen that the maximum muscle activity reduces from 0.64 for the case without assistance to 0.4 for the case with 5 N· m assistance of the exoskeleton. The muscle activity reduces further to 0.3 if the assistance increases to 10 N· m. The significance of the exoskeleton assistance is obvious.

5.2.1 Mechatronics Design

Based on the simulation results, an arm exoskeleton was designed for testing. The exoskeleton has three DOF, where one is passive (Joint 1) and two are active (Joints 2 and 3), similar to the simulation model. In the mechanical design, four attachment points are considered: the back support, upper arm strap, forearm strap and wrist strap. The lengths of the upper arms and forearms can be adjusted for different users. Similarly, the exoskeleton has a shoulder adjustment to align the axes of Joints 1 and 2 with the glenohumeral joint of the user. All adjustments are designed to fit the 5th to 95th percentile of possible users (data from ADULTDATA).

Table 2 lists the design specifications for the exoskeleton arm. Both active joints are actuated by brushless DC motors, which transfer torques to the exoskeleton via Harmonic Drives. The Harmonic Drive was selected

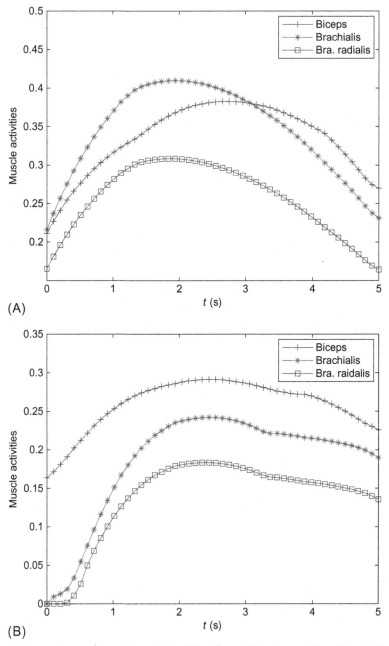

(A)

(B)

Fig. 9 Comparisons of muscle activities: (A) with assistive torque $M_s = M_e = 5$ N· m, (B) with assistive torque $M_s = M_e = 10$ N· m, and

(Continued)

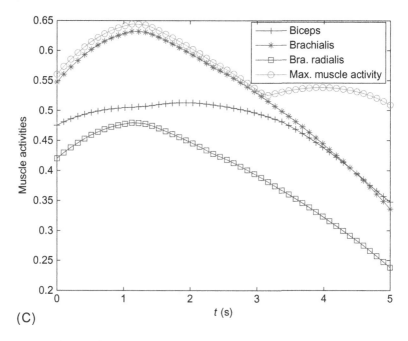

(C)

Fig. 9, Cont'd (C) without assistive torques.

Table 2 Specifications of the Exoskeleton

Range of Motion (Degree)		Motion Assistance	
Elbow rotation	Max. 135	Payload	5 kg
Shoulder flex/ext	Max. 190	Max. angular vel.	70 rpm
Shoulder abd/add	± 130	Assistive torque	11.2 N· m

for its high efficiency and back-drivability, which allows the user to move even if the motors are powered off.

The power requirement of the joints is determined from biomechanical simulations of a selection of activities of daily living. Using the power, force/torque and velocity simulation results, different compositions of the motors and gears were investigated. The output torque, τ_{out}, and velocity, ω_{out}, were calculated using the following equations:

$$\tau_{\text{out}} = \mu_T N_G \tau_{\text{in}}, \quad \omega_{\text{out}} = \frac{\omega_{\text{in}}}{N_G} \tag{12}$$

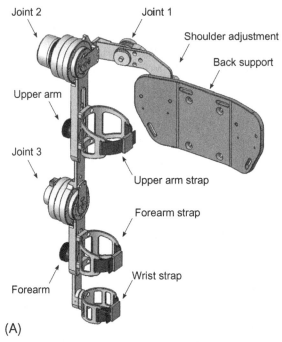

(A)

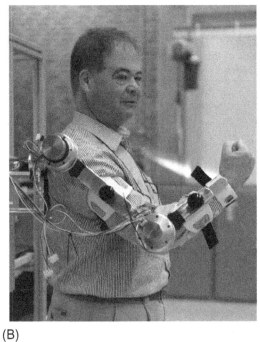

(B)

Fig. 10 A 2-DOF upper arm exoskeleton: (A) CAD model and (B) a prototype built at AAU.

Table 3 Joint Construction in the Exoskeleton

	Joint 2	Joint 3
Motor type	EC-60	EC-45
Gear type	CSD-25-50-2	CSD-25-50-2
Position sensor	Noncontact encoder	Noncontact encoder
Velocity sensor	Hall sensor in EC-60	Hall sensor in EC-45
Current sensor	EC-60 build-in sensor	EC-45 build-in sensor

where μ_T is the total efficiency of the transmission, ie, Harmonic Drive and bearings, N_G is the gear ratio of the Harmonic Drive and τ_{in} and ω_{in} are the input torque and velocity, respectively. The driving factor for selecting the motors and gears, besides the simulation results, is the total weight of the transmission system. Table 3 lists the selected motors and gears for the two joints.

Hierarchical structure with high- and low-level controls is designed for the exoskeleton, as illustrated in Fig. 11. The high-level control is the main "brain" of the exoskeleton that recognizes user intention, through the intention sensor system, and translates it into a level of assistance in terms of joint torques. The low-level control implements torque control at each joint. This control structure is achieved with a modular structure. CAN Nodes[1] are used in a master-slave configuration, where the master constitutes the high-level control and the slaves work for the low-level control at the elbow and shoulder joints. The slave CAN Node is wired up to an ESCON Module, which is a

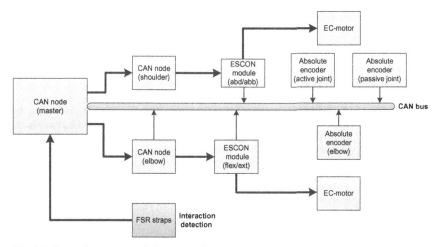

Fig. 11 Control structure of the exoskeleton.

---

[1] Microcontrollers that communicate with one another using CAN communication protocol.

dedicated microcontroller for the EC motors. The ESCON Module has built-in safety functions for both velocity and current/torque.

The intention sensor system consists of at least two strap bands placed at the upper arm and forearm. Each of the bands has eight or more force sensitive resistors (FSRs) equally spaced inside the band. The strapped band is used to detect and measure muscle activation, ie, contraction and co-contraction of the muscles. The goal is then to use the combination of the upper arm and forearm straps to detect whether the user intends to move his arm, hold a load statically or move his arm with a load applied to it.

6. CONCLUSIONS

An exoskeleton interacts directly with human bodies. The design of exoskeletons brings us new challenges in mechanics, biomechanics, and control. This chapter overviews major design problems, including the kinematic structure, human-robot modeling, mechatronics design, sensing and control, and addressing the problem of pHRIs. Examples are presented to demonstrate the biomechanical HRI modeling and mechatronics design.

The topic of this chapter is focused on the assistive exoskeletons, where the design concerns are light weight, compactness, and comfort. With the increasing interest in exoskeletons and wide areas of potential applications, the requirements may change significantly. However, the pHRI and mechatronics are still two core issues to be addressed. Novel mechanisms, sensors, and development methods are needed to advance the robotic exoskeleton for widespread and practical deployments.

ACKNOWLEDGMENTS

Parts of the results presented in this chapter are from research projects, including the patient@home (www.patientathome.dk) and EU AAL AXO-SUIT (www.axo-suit.eu) projects. The mechanical design of the upper-arm exoskeleton by MTD Precision Engineering LTD is acknowledged.

REFERENCES

[1] T. Rahman, Passive exoskeletons for assisting limb movement, J. Rehabil. Res. Dev. 43 (2006) 583–590.
[2] T. Sakurai, Y. Sankai, Development of motion instruction system with interactive robot suit HAL, in: IEEE International Conference on Robotics and Biomimetics (ROBIO), 2009, pp. 1141–1147.

[3] T. Nef, M. Guidali, R. Riener, ARMin III-arm therapy exoskeleton with an ergonomic shoulder actuation, Appl. Bionics Biomech. 6 (2) (2009) 127–142.

[4] Y. Ren, H.S. Park, L.Q. Zhang, Developing a whole-arm exoskeleton robot with hand opening and closing mechanism for upper limb stroke rehabilitation, in: IEEE International Conference on Rehabilitation Robotics (ICORR 2009), 2009, pp. 761–765.

[5] Y. Mao, S. Agrawal, Design of a cable-driven arm exoskeleton (CAREX) for neural rehabilitation, IEEE Trans. Robot. 28 (4) (2012) 922–931.

[6] P. Garrec, J.P. Friconneau, Y. Measson, Y. Perrot, ABLE, an innovative transparent exoskeleton for the upper-limb, in: IEEE/RSJ International Conference on Intelligent Robots and Systems, 2008, pp. 1483–1488.

[7] J. Perry, J. Rosen, S. Burns, Upper-limb powered exoskeleton design, IEEE/ASME Trans. Mechatronics 12 (4) (2007) 408–417.

[8] J. Veneman, R. Kruidhof, E. Hekman, R. Ekkelenkamp, E. Van Asseldonk, H. Van Der Kooij, Design and evaluation of the LOPES exoskeleton robot for interactive gait rehabilitation, IEEE Trans. Neural Syst. Rehabil. Eng. 15 (3) (2007) 379–386.

[9] C. Carignan, J. Tang, S. Roderick, Development of an exoskeleton haptic interface for virtual task training, in: IEEE/RSJ International Conference on Intelligent Robots and Systems, 2009, pp. 3697–3702.

[10] R. Vertechy, A. Frisoli, A. Dettori, M. Solazzi, M. Bergamasco, Development of a new exoskeleton for upper limb rehabilitation, in: IEEE International Conference on Rehabilitation Robotics (ICORR 2009), 2009, pp. 188–193.

[11] P. Wang, K. Low, A. McGregor, A subject-based motion generation model with adjustable walking pattern for a gait robotic trainer: NaTUre-gaits, in: 2011 IEEE/RSJ International Conference on Intelligent Robots and Systems, 2011, pp, 1743–1748.

[12] E. Wolbrecht, V. Chan, D. Reinkensmeyer, J. Bobrow, Optimizing compliant, model-based robotic assistance to promote neurorehabilitation, IEEE Trans. Neural Syst. Rehabil. Eng. 16 (3) (2008) 286–297.

[13] H. Lo, S. Xie, Exoskeleton robots for upper-limb rehabilitation: state of the art and future prospects, Med. Eng. Phys. 34 (3) (2012) 261–268.

[14] T. Lenzi, N. Vitiello, S.M.M. De Rossi, S. Roccella, F. Vecchi, M.C. Carrozza, NEUROExos: a variable impedance powered elbow exoskeleton, in: IEEE International Conference on Robotics and Automation (ICRA), 2011, pp. 1419–1426.

[15] T. Wu, S. Wang, D. Chen, Design of an exoskeleton for strengthening the upper limb muscle for overextension injury prevention, Mechanism Mach. Theory 46 (2011) 1825–1839.

[16] P. Lin, W. Shieh, D. Chen, A theoretical study of weight-balanced mechanisms for design of spring assistive mobile arm support (MAS), Mechanism Mach. Theory 61 (2013) 156–167.

[17] Y. Nakamura, K. Yamane, Y. Fujita, I. Suzuki, Somatosensory computation for man-machine interface from motion-capture data and musculoskeletal human model, IEEE Trans. Robot. 21 (1) (2005) 58–66.

[18] L.-F. Lee, M. Narayanan, S. Kannan, F. Mendel, V.N. Krovi, Case studies of musculoskeletal-simulation-based rehabilitation program evaluation, IEEE Trans. Robot. 25 (3) (2009) 634–638.

[19] P. Agarwal, M.S. Narayanan, L.F. Lee, F. Mendel, V.N. Krovi, Simulation-based design of exoskeletons using musculoskeletal analysis, in: ASME 2010 International Design Engineering Technical Conferences and Computers and Information in Engineering Conference, 2010, pp. 1357–1364.

[20] S. Bai, J. Rasmussen, Modelling of physical human-robot interaction for exoskeleton designs, in: ECCOMAS Thematic Conference: Multibody Dynamics 2011, Brussels, Belgium, 2011.

[21] R.A.R.C. Gopura, K. Kiguchi, L. Yang, SUEFUL-7: a 7DOF upper-limb exoskeleton robot with muscle-model-oriented EMG-based control, in: IEEE/RSJ International Conference on Intelligent Robots and Systems, 2009, pp. 1126–1131.

[22] L. Zhou, S. Bai, M.S. Andersen, J. Rasmussen, Design and optimization of a spring-loaded cable-driven robotic exoskeleton, in: K. Persson, J. Revstedt, G. Sandberg, M. Wallin (Eds.), Proceedings of the 25th Nordic Seminar on Computational Mechanics, Lund, Sweden, 2012, pp. 205–208.

[23] J. Veneman, R. Ekkelenkamp, R. Kruidhof, F. van der Helm, H. van der Kooij, A series elastic-and Bowden-cable-based actuation system for use as torque actuator in exoskeleton-type robots, Int. J. Robot. Res. 25 (3) (2006) 261–281.

[24] T. Sugar, Design and control of RUPERT: A device for robotic upper extremity repetitive therapy, IEEE Trans. Neural Syst. Rehabil. Eng. 15 (3) (2007) 336–346.

[25] H. Lee, W. Kim, J. Han, C. Han, The technical trend of the exoskeleton robot system for human power assistance, Int. J. Precis. Eng. Manuf. 13 (8) (2012) 1491–1497.

[26] H. Lee, B. Lee, W. Kim, M. Gil, J. Han, C. Han, Human-robot cooperative control based on pHRI (physical human-robot interaction) of exoskeleton robot for a human upper extremity, Int. J. Precis. Eng. Manuf. 13 (6) (2012) 985–992.

[27] J. Rasmussen, M. Damsgaard, M. Voigt, Muscle recruitment by the min/max criterion: a comparative numerical study, J. Biomech. 34 (3) (2001) 409–415.

[28] L. Zhou, S. Bai, M. Andersen, J. Rasmussen, Modeling and design of a spring-loaded, cable-driven, wearable exoskeleton for the upper extremity, J. Model. Ident. Contr. 3 (2015) 167–177.

CHAPTER TEN

Psychological Modeling of Humans by Assistive Robots

A. Wagner, E. Briscoe
Georgia Institute of Technology, Atlanta, GA, United States

1. INTRODUCTION

Healthcare presents a large problem space to which novel robotic solutions can be applied. An important and evolving question is how healthcare-focused robots will interact with and assist the people they serve. Assisting a patient is, in many ways, fundamentally different from interacting with a healthy adult. Being a patient implies some type of ailment that could potentially debilitate or impair the person's judgment. Patient-robot interaction may therefore require a unique perspective on how and why a robot should interact with the person. As a caregiver, situations may arise in which the robot provides companionship, protects the patient's dignity, restrains a person to prevent self-harm, or discretely observes their behavior. As researchers in this space, it is imperative that we recognize that to be effective, robots must be accommodating and flexible, with the capability to interact with a wide range of personality types, disabilities, and levels of intelligence.

The dynamic nature of patient-robot interaction presents challenges and opportunities for novel research. For some cases, the possibility of creating a "one size fits all" robot-aided therapy is unrealistic. As we move from robots as assistive tools, to robots as active collaborators in maintaining and upgrading one's physical and mental health, we must develop systems that autonomously individualize their behavior over the course of treatment with respect to the needs of the patient. Creating systems that perceive and model a person's mental state is an important problem that is currently being investigated by the human-robot interaction (HRI) and social robotics communities. This chapter presents their progress and also examines an anticipated upcoming generation of robots that will socially interact with patients, model their moods, personality, likes, and dislikes, and use this information to guide the robot's assistance related decisions. These robots will need to recognize when a person trusts them, or when a person is being deceptive; and also decide whether to act deceptively in return for a beneficial patient outcome. We review the major research in areas related to cognitive, behavioral, and psychological modeling of a patient by an assistive robot.

The chapter begins by examining the dimensions or attributes by which people are commonly characterized, especially as relevant to a person needing assistance, focusing on qualities utilized in human-human interaction, such as age and emotional state. The person's condition is another important distinguishing factor that a more general purpose assistive robot might need to consider. For example, does the person have a neurological deficit that might influence how the robot should interact with the person? A hearing deficit might make verbal commands difficult or impossible to understand for some patients. Lessons learned while developing a medication and water delivery robot for older adults suggest that the mobility of the potential user has a critical impact on their view and likelihood of using the system [1]. Emeli et al. interviewed older adults asking about their interest in a robot tasked with delivery. Although some individuals expressed interest, the person's current mobility played an important factor in their assessment of the system (Fig. 1).

Next we present methods for behavioral modeling from more traditional robotics settings, such as from the field of HRI and social robots. As will be demonstrated, the use of psychological modeling by HRI researchers is still in its infancy.

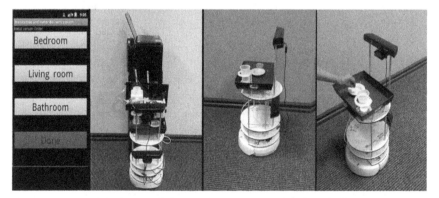

Fig. 1 A water and medication delivery robot. Users requested service from the robot by pressing a button on their smart phone indicating their location. The robot then traveled to a medication and water station for pickup of the items to be delivered, traveled to the stated location, and stopped within an arm length of the person.

We then present the dominant approaches for psychologically modeling a person. Different categories of models exist related to the fields from which they arose. Economic models, for example, focus on the decisions people make and how these decisions impact their internal state or utility. Typically, different economic models of human behavior are assessed in terms of their ability to predict people's decisions in tightly controlled laboratory experiments. These approaches, although not directly related to assistive scenarios, are important because they offer a formal, grounded, computational means for modeling a person that can be implemented in the software that controls a robot.

Cognitive models are another important category of approach for modeling people. Cognitive models attempt to approximate the different types of processes that underlie cognition. Various architectures have been advanced as different visions of how animals and people think. Often these architectures are assessed in terms of the accuracy of their predictions of human behavior, and the timing of these behaviors. Next, we discuss how human characteristics may be measured, based on explicit signals, in order to be utilized in individual-based models. The chapter concludes by summarizing the work done in the area of human modeling by assistive robots, and suggesting new avenues for future work.

2. DIMENSIONS OF HUMAN CHARACTERIZATION

Human characterization serves a critical purpose, whether conducted by other humans or machines. Several of the social sciences, such as cognitive and social psychology, share a similar goal of determining characterizations of humans that serve to explain both their commonalities and individual differences. Often investigators seek to understand how and why people differ and how those differences may be measured, wherein by defining people using common dimensions, they can better explain, understand, and predict an individual's behavior. Developing computational methods for representing these dimensions also provides a means for artificial reasoners (eg, robots) to mimic and take advantage of known human categorization techniques.

Whether explicitly or implicitly, people constantly categorize the individuals with which they interact. As cognitive processes are optimized to compensate for limited processing power by maximizing resources, humans often form impressions and infer traits based on their immediate categorization of other individuals rather than waiting to experience each specific trait firsthand [78]. Research into how people categorize other people varies, for example, by spanning both social and individuated concepts [2,3]; focusing on either a person's salient characteristics (such as height) or behavioral characteristics (such as personality); or involving top-down (category-oriented) or data-driven (attribute-oriented) inferences [4]. It is also expected that when people evaluate one another, they do so in stages, where initial categorization, confirmatory categorization, recategorization, and piecemeal processing (as described by Fiske et al. [5]) are mediated by attention and interpretation and influenced by information and motivation.

While not exhaustive, many of the dimensions used to characterize humans that have been identified through social science research, and that are especially relevant to benefiting HRI, are described in the following paragraphs.

2.1 Personality

Perhaps the most common method, either explicitly or implicitly, of describing a person is by assigning them a range of personality traits. There are several primary theories describing personality [6], including the "Big Five" [7], Eysenck's PEN [8], and Myers-Briggs Type Indicator [9]. These theories "decompose" people into multiple dimensions. For example, using the Big Five, people are characterized along five factors: openness, conscientiousness, extraversion, agreeableness, and neuroticism (see Table 1 for a

Table 1 "Big Five" Personality Dimensions [7] and Their Associated Trait Ranges

Dimension	Trait Ranges		
Conscientiousness	Organized	←——————→	Disorganized
	Careful	←——————→	Careless
	Disciplined	←——————→	Impulsive
Agreeableness	Soft-hearted	←——————→	Ruthless
	Trusting	←——————→	Suspicious
	Helpful	←——————→	Uncooperative
Neuroticism	Calm	←——————→	Anxious
	Secure	←——————→	Insecure
	Self-satisfied	←——————→	Self-pitying
Openness	Imaginative	←——————→	Practical
	Prefers variety	←——————→	Prefers routine
	Independent	←——————→	Conforming
Extraversion	Sociable	←——————→	Retiring
	Fun-loving	←——————→	Sober
	Affectionate	←——————→	Reserved

more comprehensive description of the dimensions and their associated range of traits). Despite its potential to be dynamic (there is debate as to whether people have consistent personalities [10]), personality profiling has proven valuable in predicting a number of behaviors, such as job performance in occupational settings [11].

2.2 Emotions and Moods

Understanding a person's mood or emotional state is critically important, especially with regard to potential HRI. One primary approach in the study of emotions has been to view emotions as arising from a palette of "basic emotions." These perspectives tend to perceive basic emotions as "low level" feelings such as anger, disgust, fear, joy, sadness, surprise [79]. The identification of emotion "types" based on the theory of Ortony, Clore, and Collins [12], also known as the "OCC" model, assumes that emotions develop as a consequence of certain cognitions and interpretations, which are based on one's understanding of the world in terms of agents, objects, and events (see Table 2 for the 22 emotions described in the model). How emotions might emerge is very dependent on how individuals perceive and interpret events. One can be pleased or displeased about the consequences of an event (*pleased/displeased*); one can endorse or reject the actions of an agent (*approve/disapprove*) or one can like or not like aspects of an object (*like/dislike*).

A further differentiation consists of the fact that events can have consequences for others or for oneself, and that an acting agent can have different roles. The consequences of an event for another can be divided into *desirable* and *undesirable*; the consequences for oneself as relevant or irrelevant expectations. Relevant expectations for oneself finally can be differentiated again according to whether they actually occur or not (*confirmed/disconfirmed*).

Moods, in contrast, are not usually characterized by their direction at a person or event. While someone might show an emotion (anger) toward a specific object (eg, a colleague), as the specific emotion dissipates, they may feel a general bad mood. In some models of emotion (eg, [13]), an individual's perception of a generated emotion is considered the agent's *feelings*, which are modulated by both emotion, the individual's appraisal of the current situation, and mood, which is a memory of recent emotions [14].

2.3 Intelligence

Intelligence is a well-studied individual characteristic whose measurement has been used to predict a wide variety of human behavior, such as job

Table 2 The Types and Specifications of the 22 Emotions Specified by the OCC Model [12]

Emotion Type	Specification
Joy	(Pleased about) a desirable event
Distress	(Displeased about) an undesirable event
Happy-for	(Pleased about) an event presumed to be desirable for someone else
Pity	(Displeased about) an event presumed to be undesirable for someone else
Gloating	(Pleased about) an event presumed to be undesirable for someone else
Resentment	(Displeased about) an event presumed to be desirable for someone else
Hope	(Pleased about) the prospect of a desirable event
Fear	(Displeased about) the prospect of an undesirable event
Satisfaction	(Pleased about) the confirmation of the prospect of a desirable event
Fears-confirmed	(Displeased about) the confirmation of the prospect of an undesirable event
Relief	(Pleased about) the disconfirmation of the prospect of an undesirable event
Disappointment	(Displeased about) the disconfirmation of the prospect of a desirable event
Pride	(Approving of) one's own praiseworthy action
Shame	(Disapproving of) one's own blameworthy action
Admiration	(Approving of) someone else's praiseworthy action
Reproach	(Disapproving of) someone else's blameworthy action
Gratification	(Approving of) one's own praiseworthy action and (being pleased about) the related desirable event
Remorse	(Disapproving of) one's own blameworthy action and (being displeased about) the related undesirable event
Gratitude	(Approving of) someone else's praiseworthy action and (being pleased about) the related desirable event
Anger	(Disapproving of) someone else's blameworthy action and
Love	(Liking) an appealing object
Hate	(Disliking) an unappealing object

performance [80]. Most current factor models of intelligence typically represent cognitive abilities as a three-level hierarchy, where at the highest level is a single factor referred to as g, which represents the variance common to all cognitive tasks [81]. Intelligence has also been linked to personality traits, especially openness and intellect [7], where individuals who fall higher on the openness dimension generally score higher on measures of cognitive ability [15] and are more creative [16].

2.4 Social Intelligence

In addition to traditional measures of intelligence, *social intelligence* [82] is a quality that represents a person's ability to express, recognize, and manage social signals and social behaviors [17], such as politeness and empathy. As a subset of social intelligence, *emotional intelligence* involves a person's ability to monitor his own and others' feelings and emotions and to make informed decisions based on this assessment [18]. The five key factors of emotional intelligence are: self-awareness, self-regulation, motivation (passion for work and resiliency), empathy, and social skills. In order to be useful for a robot, these dimensions must be represented within a computational framework. The following sections present some of the most common and well-known computational frameworks for modeling people.

3. CONSTRUCTING BEHAVIORAL MODELS FOR HRI

Because a robot has a physical presence in the world, the ways that people interact with these machines can, in some ways, be unique. For instance, Bethel et al. compared robot interviews to human interviews in terms of the misinformation effect [19]. The misinformation effect occurs when a person's recall is influenced by postevent information in a way that makes recall less accurate. In criminal investigations, for example, questioning by the examiner influences the eyewitnesses' memories of the event. Bethel et al. found that use of a robot interviewer resulted in greater memory recall and accuracy in spite of misinformation. In an assistive setting, use of a robot might allow a patient to more accurately state the reason for their ailments.

Human-robot research has also shown that people are likely to heed the orders of a robot. McColl and Nejat found that 87% of older adults that they tested complied with a robot's prompts to eat [20]. Salem et al. found that people typically comply with a robot, even when its instructions are in error [21]. They found that two-thirds of subjects would even pour orange juice over a plant if asked to by the robot. Most rationalized the request in some way.

It is worth noting that anthropomorphism can strongly influence the ways that a person interacts with a robot. Anthropomorphism is the tendency to attribute human characteristics to nonhuman objects. When the object being anthropomorphized is a robot, this tendency can lead to inaccurate expectations related to the robot's behavior, intentions, or communication tendencies [22]. HRI researchers have, at times, resorted to developing robotic "creatures" in order to reduce anthropomorphism [23]. Others have embraced anthropomorphism by making their robots as humanlike as possible [24]. Gratch et al. have pioneered the use of humanlike virtual characters for the purpose of training, and for psychotherapeutic effectiveness [25]. The nature of a robot's interactions with the patient will likely determine whether anthropomorphism is embraced or avoided.

The field of HRI has begun to explore the possibility of a robot modeling a person and using this information to guide its behavior. This work has typically fallen under the rubric of mental modeling and the development of shared mental models. Unlike an economic model or a cognitive model, a mental model is simply one's understanding of another person's thought process [26]. As such, the notion of a mental model is rather vague and ill-defined. Also, unlike economic and cognitive models, mental models have no natural computational underpinnings. Hence, roboticists tend to create their own. Propositions are a common format for representing an individual's beliefs [27]; probabilistic statements are also often used [28].

Belief-desire-intention models (BDI) have also been used to model the psychological state of a person. Beliefs represent the information that the person is currently aware of, or in some way knows. Desires represent the person's goals or motivations, and intentions denote a person's deliberative course of action [29]. Although useful in simulation and virtual environments, BDI methods are limited in their ability to capture the richness and variability associated with human decision making.

Partially-observable Markov decision processes (POMDPs) have also been suggested as a means for representing a model of the robot's interactive partner. POMDPs model the decision process as a chain of connected states [30]. At each time step, the robot observes information about the current state of the environment which is ultimately used to select an action during the next time step. Although the resulting behavior is provably optimal, exact POMDP solutions are computationally intractable for most nontrivial problems; approximate solutions are the norm [31].

In addition to modeling the person's mental state, it may also be useful for an assistive robot to model the risks to the patient associated with a therapy. Exoskeletons, for example, are poised to become an important therapeutic tool and perhaps even a long-term solution for some types of paralysis. It may be beneficial for these systems to model the risks associated with particular types of movements, such as climbing steps, and warn the user of the risk.

Finally, there are important ethical ramifications associated with a robot's creation and use of psychological models when interacting with a person.

Such machines may be empowered to recall an enormous amount of data associated with a person or type of person. Depending on the circumstances, this information could easily be used to manipulate or unjustly coerce a patient. Developers of robotic systems must consider the extent to which psychological modeling is appropriate and justified.

Given the limitations of current approaches, we can generate greater realism by incorporating elements of human psychology and the signals these elements produce.

4. ECONOMIC DECISION-MAKING MODELS

The earliest and most well-developed formal models of human decision making originate from economic theory. Jeremy Bentham developed the idea that economic exchange and the decision making that underlies it are driven by one's attempt to maximize pleasure and minimize pain, and can be measured as such [32]. A person's motivations and resulting behavior, Bentham argues, are a direct reflection of the utility of one's actions. At its core, utility theory claims that people use subjective assessments of the value of an action choice to make decisions. If an individual views a particular action or course of action as offering higher subjective value than some other course of action, that action is favored over others.

In contrast to many traditional psychological theories, utility theory is formalized mathematically. This mathematical grounding provides a computational framework which is implementable on an assistive robot attempting to model and predict a person's behavior. Utility theory assumes that a person receives a quantifiable pleasure, $u \in \mathbb{R}$, when making a decision. Utility functions, $U : X \rightarrow \mathbb{R}$, are used to describe an individual's preferences in relation to fixed set of choices. A preference is defined as a relation, $\preccurlyeq$, over X such that for every $x, y \in X$, $U(x) \leq U(y)$ implies $x \preccurlyeq y$. Rational behavior results when an individual selects actions which maximize their utility function. In theory, understanding a person's utility function allows one to predict the person's behavior.

Expected utility theory considers one's preferences when the utilities that result from an action choice are uncertain [33]. Uncertainty is typically modeled as a gamble resulting in outcome (x) with respect to a known probability distribution (p). The expected utility, EU, is then arrived at as the product of the outcome utilities and the probabilities. The outcome of several gambles, $x = \{x_1, x_2, ..., x_N\}$, is thus calculated as $EU[x] = \sum_i^N p_i x_i$.

John von Neumann and Oskar Morgenstern showed that an individual whose preferences satisfy certain axioms always prefer an action which maximizes their expected utility [34]. Alternatively, an individual whose preferences violate the von Neumann and Morgenstern axioms makes decisions which guarantee losses. Intentionally selecting an action which will result in losses is considered irrational. A rational individual is thus argued to be an individual who selects actions that maximize their own utility.

One of the primary criticisms of expected utility theory is that it falsely assumes that people act in a rational, utility-maximizing manner. In reality, well-known biases constantly influence human decision making. For instance, people tend to be averse to losses. A person's decisions are more strongly influenced by the desire to avoid losses (*loss aversion bias*) than the desire to seek gains. The manner in which a situation is described or framed also impacts decision making (*framing effects*). Framing a choice as a gain or a loss has been shown to bias decision making. Humans are also biased to prefer the status quo. The status quo is used as a point of reference for evaluating whether or not a choice will result in a loss or a gain. People tend to discount the value of additional gains or losses the further the gain or loss is from one's reference point. For example, the perceived difference in utility between receiving $100,000 and $100,100 is considered small, whereas the perceived difference in utility between receiving $0 and $100 is considered great, even though the value of the difference is the same. Finally, people tend to overestimate the likelihood of low-probability events and underestimate the probability of high-probability events.

Prospect theory was developed to better account for the behavior people actually exhibit when faced with choices under risk and uncertainty [35]. Prospect theory holds that the expected utility should be evaluated as $EU[x] = \sum_{i}^{N} w(p_i)v(x_i)$, where EU is the expected utility associated with making decisions x_1, x_2, ..., x_N and p_1, p_2, ..., p_N are their respective probabilities. The function v maps outcome values to utilities and the function w weighs probabilities in order to capture one's risk preferences. Fig. 2 depicts the typical shape of a prospect theory value function. The y-axis illustrates the individual's current reference point. Losses and gains are evaluated with respect to the reference point. The steeper slope on the loss portion of the graph models loss aversion. The curved tails indicate the lessening impact of additional gains or losses, a reflection of gain/loss satiation.

Fig. 3 depicts a probability weight function from prospect theory. The curved line used for the weight function mimics certainty effects by

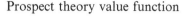

Prospect theory value function

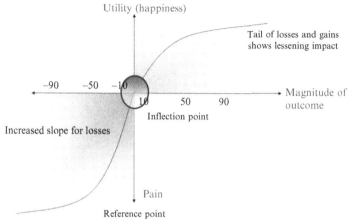

Fig. 2 An example weighted utility function from prospect theory.

Original prospect theory probability weight functions

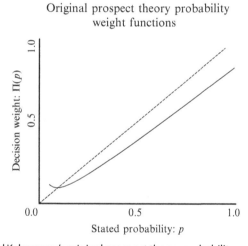

Fig. 3 Tversky and Kahneman's original prospect theory probability weight function [35].

overestimating the likelihood of low-probability events and overestimating the likelihood of high-probability events. Overall, the addition of a probability weighting and utility valuation function and the shape that these functions take allow prospect theory to better model a person's economic decision making as well as their biases.

Although prospect theory was a major advancement, a problem remained [36]. Let x and y be the outcomes that result from gambles

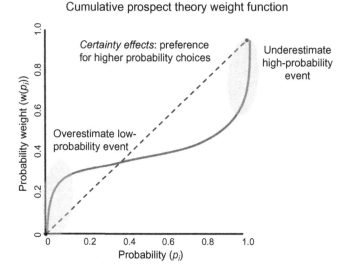

Cumulative prospect theory weight function

Fig. 4 The probability weight from function from cumulative prospect theory.

A_1 and A_2. If $P(x > t|A_1)$ is the probability that an outcome of Gamble A_1 exceeds t, then the decision (A_1) is said to "stochastically dominate" decision (A_2) if and only if $P(x_1 > t|A_1) \geq P(x_2 > t|A_2)$ for all t, assuming $A_1 \neq A_2$. Unfortunately, because of the shape of the weighted probability function, prospect theory predicts situations in which A_2 is chosen over A_1 in spite of the fact that A_1 is stochastically dominated by A_2. *Cumulative prospect theory* is a theoretical and practical refinement of prospect theory which does not violate stochastic dominance [37]. Fig. 4 depicts a weighted probability function that maintains stochastic dominance.

Many extensions and refinements of these behavioral decision theories have been proposed. In contrast to utility theory, regret theory focuses an individual's motivation to minimize the negative feelings associated with regret [38–40]. Regret theory holds that the preference over choices A_1 can be represented formally as:

$$A_1 \succeq A_2 \Leftrightarrow \sum_i^N p_i(v(A_1(x_i)) - v(A_2(x_i)))$$

where v is the utility function. Regret theory explains certain decisions better than prospect theory. For a more formal and detailed treatment of regret theory, see Ref. [40].

The preceding theories are based on rational models of human behavior. Decision field theory, on the other hand, is a dynamic theory that does not assume a fixed, rational set of preferences [41]. Decision field theory includes a model of preference evolution that allows it to generate a better account of decision regularity and make predictions about decision time under certain conditions. For more detailed information related to decision field theory, Busemeyer and Diederich provide a survey of the field [42].

For robots trying to model a person, decision theories offer a computational starting point. They are easily implemented on a robot and share some common conceptual traits with more traditional robotics frameworks such as reinforcement learning. Yet decision theories also tend to assume rational or semirational behavior and underestimate the idiosyncratic, emotional, and automatic nature of human behavior. Moreover, some argue that a single scalar utility value is too impoverished a model to represent a person's state. Nevertheless, this approach has been employed successfully for more than 60 years by economists and others. With respect to robotics, Wagner showed that a robot could learn categories of models related to different types of people [43]. These stereotyped models could then be used to make predictions about interactions with newly encountered people. These techniques could presumably be used by an assistive robot as a strategy for bootstrapping its early interactions with a new patient. Later, with time and experience, the system could tailor its interactions to needs of the particular patient.

4.1 Neuroeconomics

Given the limitations of utility-based economic models of human decision making, researchers began to investigate methods that augment traditional economic models with results and data from neuroscience. They called this new direction *neuroeconomics*. Neuroeconomics is an interdisciplinary field which seeks to build from utility-based theories while also including evidence from neuroscience. As a field, neuroeconomics makes a concerted effort to include automatic and emotional processing in their models of human behavior [44]. There is a great deal of evidence that human decision making is strongly influenced by automatic, unconscious processes [45,46] as well as emotion [47]. Emotion, in particular, has a tendency to hijack a person's decision-making faculties and generate behavior which is typically viewed as irrational. Emotion also impacts learning which, in turn, influences decision making [13]. Significant evidence suggests that memories are imbued with affective information, which upon recall, generates similar emotions to the time the memory was created.

Neuroeconomics embraces the use of brain imaging as a means to better understand the mind's processing while making decisions. Typically, brain images are used to compare brain activity while people are engaged in an experimental or control task. Often the tasks used involve economic games such as the prisoner's dilemma or the investment game. King-Casas et al., for example, used a hyperscan functional Magnetic Resonance Imager (fMRI) to monitor participant's neural responses while playing a two-person investment game [48]. In this game one person assumes the role of investor and the other of trustee. The investor is given $20 and may invest any portion of the money with the trustee. Money that is invested with the trustee then appreciates (is multiplied by 3). The trustee must then decide how much to return to the investor. King-Casas et al. monitored subjects over 10 rounds of play. They found neural correlates which indicated a person's intention to trust their partner.

These types of studies provide insight into the areas of the brain that are activated in specific decision-making situations as well as the temporal process underlying that decision. For instance, it has been shown that situations involving distrust activate the parts of the brain associated with the fear emotion [49], whereas situations involving trust do not activate any specific area of the brain [50].

Significant work in this area has also shown that decisions are made when the brain integrates inputs from multiple systems to generate a utility-like value. This multiple systems model claims that information from these different systems is processed in a qualitatively individualized manner and weighted accordingly [51]. For example, the emotional or affective system quickly and unconsciously processes information to generate a reflexive response. The analytic system, on the other hand, is a slower, conscious effort which influences behavioral decisions. The behavior that results is a combination of signals from these different sources (Fig. 5).

The multiple systems model of cognition can be used to explain several aspects of human behavior. For instance, Shiv and Fedorikhin [52] showed that self-regulation decreases with increased cognitive load by asking people to choose between cake or fruit salad while remembering either a 2-digit or 7-digit number. Significantly more people choose cake when asked to remember the 7-digit number. Presumably remembering the larger number marshals resources away from the executive functions of the mind, allowing one's lower functions to play a larger role in decision making.

Multiple systems model

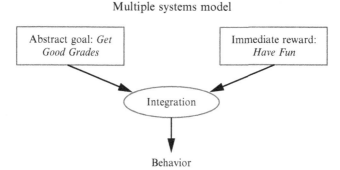

Fig. 5 An example of the multiple system's model.

Integration of these signals is not necessarily limited to behavior. Perceptual processes, for example, are intimately influenced by motor activity. Jessica Witt has found that motor experience derived from practice actually increases a person's perceived size of the target [53]. For instance, better baseball batters perceive pitched baseballs as larger. The same has been demonstrated with field goal kickers and golfers. Witt even found that, when compared with nonthreatening objects, people perceived spiders as moving faster [54].

This interplay of different, competing systems influences people's decisions during interactions with a robot. Robinette et al. examined a situation in which a robot guided subjects to a meeting room [55]. Later, an emergency evacuation occurred and the same robot offered to guide them out of the building. In virtual environments subjects that were initially led by a robot that navigated poorly generally choose not to use the robot. In a real instantiation of the experiment, subjects universally chose to follow the robot regardless of how poorly it had navigated earlier. The multiple systems hypothesis provides a possible explanation for this behavior. In the virtual environment, people's high executive functions consider the robot's reputation when making a decision. In the real environment, when alarms are sounding, their lower more automatic cognitive functions dominate and the robot's reputation is no longer considered.

4.2 Cognitive Architectures

The challenge of evaluating potential interaction outcomes may be addressed by representing features as components in a human cognitive architecture, such as those proposed by Anderson and Lebiere [83], Revelle et al. [15], or Newell [84], which consists of computational components that, in coordination, are argued to produce human-level intelligence. These architectures allow a robot to "understand" and predict human behavior on some level [56]. Several of these architectures, such as H-CogAff, ACT-R, and Soar, operate at a high level of comprehensiveness, where they include key psychological moderators, such as personality.

By utilizing the cues provided by both humans and contextual information, cognitive architecture(s) may greatly enhance a robot's ability to interact with an individual. Trafton et al. [56] specifically discuss an expansion of ACT-R, ACT-R/E to provide robots with theory of mind (knowledge of others' cognition) so as to improve their functionality. With an understanding of how people might perform in different situations, the robot can better achieve its own goals. For example, reasoning about how people are likely to react during emergencies (such as a fire) in order to protect themselves may impart a robot with knowledge that will aid it to locate victims. Likewise, the ability to detect and reason about emotion may be a critical component of robotic caregivers [57]. These types of models can be used as a source of information by the robot for decision making.

In regard to assistive robots, one should expect that the person's mental state will strongly influence their decision making relative to the robot. For instance, emotional patients, younger and older patients, and those with traumatic brain injuries may not consider the longer-term benefits of the robot. It is important that an assistive robot respond to the person it is trying to help. Failing to monitor and react to changes in the person's mood or behavior is likely to lead to little desire by the person to use the robot and may even result in patients injuring themselves to prevent assistance.

Overall, economic models of human decision making may have a large role to play with respect to assistive robotics technologies. These models offer a formal, computational means for the systems to predict and possibly understand the person's behavior. While these models have limitations, there are currently few well-developed alternatives. Perhaps the most well-developed alternatives are the cognitive models presented in the next section.

5. INFERRING PSYCHOLOGICAL MODELS

Implementing social science research allows an artificially intelligent machine to better represent, understand, and ultimately interact with a human. Better machine understanding of human behavior may result in more natural and beneficial HRI (eg, [58]). In order to take advantage of the knowledge that has been provided by social science research, machines must be able to perceive and interpret information about the human with which it is interacting. Efforts in social signal processing [59,60] have determined that social signals commonly identified by psychologists and sociologists can be recognized and captured by machinery, such as microphones and cameras, and processed into intelligence through algorithmic techniques, such as machine learning. The ability to utilize these signals allows a machine (robot) to create "mental" models of people in a way very similar to humans. These models then allow for reasoning about human behavior, which can be optimized towards tailoring successful machine interactions.

Nonverbal behavior is a continuous source of signals that convey information about the traits of people, such as their emotions and personality [85]. Ekman and Friesen [61] notably categorized communication types that result from nonverbal cues. These include: affective/attitudinal/cognitive states (eg, fear, joy, stress, disagreement, ambivalence, and inattention), emblems (culture-specific interactive signals like wink or thumbs up), manipulators (eg, touching objects in the environment), illustrators (eg, finger pointing and raised eyebrows), and regulators (eg, head nods and smiles).

Using nonverbal behaviors is not necessarily sufficient, as they carry a great deal of ambiguity [85]. For example, an awkward posture may indicate aggression or an injury. Culture is also a contributing factor to individual differences [86], and must be considered during nonverbal cue interpretation. To overcome these complications, it is advisable to marry features across modalities, though attention to the context in which the behavior arises is also critically important.

In order to model the behavior of a person, it is necessary to discover the subset of features relevant to a specific signal [17]. Often, the best approach is to select the most relevant features from all available data; however, this may result in only selecting features that are not relevant to any specific individual but only to an average model.

5.1 Detecting and Modeling Psychological Characteristics
5.1.1 Personality
Humans exhibit a number of cues that can be used to infer personality (eg, [62,63]). These cues may take many forms, such as the language that a person uses [64], a person's environment [65], or the tone in which they speak [66].

Personality cues may be used to create profiles to understand and, in some cases, predict an individual's behavior. For example, Walters et al. [87]

investigated whether personality, as characterized by certain personality traits, could be used to predict the likely approach distance preferred by the human subjects in robot interaction experiments. They found that their variable "proactiveness," primarily made up of creativity and impulsiveness, was positively correlated with the preferred human-to-robot approach distance. A person's personality may also be utilized to help optimize robotic interactions, where extroverted individuals may prefer language that emphasizes friendliness and warmth, and introverted individuals may prefer slow movements and more silence.

5.1.2 Emotion and Mood

The previously described OCC model, which defines 22 emotions (described in Table 1), is extremely useful for modeling emotional agents as the authors explicitly constructed the model to allow for "reasoning about emotion" [12], by assuming that individuals perform a subjective assessment of their relationship to the environment. Objects, events, and actions are evaluated in an appraisal process based on specific criteria, and result in multiple emotions of different intensities. As a person interacts with objects or agents, they evaluate the benefits or potential harm that they may cause, based on concerns such as goals, standards, or tastes. If those concerns are satisfied, which may be detected through recognition of cues such as facial recognition [67], the individual experiences a positive emotion (eg, admiration, joy); otherwise, a negative emotion (eg, frustration) is elicited. Past research has found many cues to emotion detection, much of it arising from research in *affective computing* [68]. Past research has also determined other methods for automatic emotion detection, such as facial expression analysis using artificial machine vision [69], voice analysis to detect the emotion of the interlocutor [88], or multimodal analysis [70]. The detection efficacy of modalities differs across emotions, where some emotions are better identified by voice, such as sadness and fear, but others are better detected through facial expression analysis, such as happiness and anger [89]. By representing the generation of emotions, and, over a longer temporal scale, moods, artificial agents can utilize small cues, such as changes in facial expressions, to provide additional meaning to communications and detect subtle changes in an individual.

5.1.3 Measuring General Intelligence

Though intelligence is traditionally measured through standardized testing (eg, [71]), other methods infer intelligence through its relation to other observable traits, such as personality and expressed interests (eg, [72,73, 90]) and even physical characteristics (eg, [74]). Understanding a person's intelligence may also aid in HRIs. For example, when interacting with highly intelligent persons, interactions may benefit from presenting ideas in more technical depth, using words that are more difficult and asking challenging questions.

5.1.4 Measuring Social Intelligence

Inferring social intelligence is a bit more straightforward than general intelligence, as relevant behaviors may be expressed through various cues in reaction to interactions, such as gestures and facial expressions. The use and incorporation of human social signals is often referred to as *socially aware computing* [59]. Vinciarelli et al. [17] describe the primary social cues as falling into the following classes: physical appearance, gestures and posture, face and eyes behavior, vocal behavior, and space and environment. Relevant for social intelligence, Salovey and Mayer [18] proposed a model that identified four different factors of emotional intelligence: the perception of emotion, the ability to reason using emotions, the ability to understand emotion, and the ability to manage emotions. An individual's place at each of these dimensions may be used to determine optimal interaction methods [75].

5.2 Utilizing Context

Modeling behavior is extremely dependent on the situation and context in which that behavior is exhibited. Human behavior is highly variable, changing and adapting according to the situation. What may be construed in one situation (eg, a smile upon greeting another, indicating happiness) can be representative of something completely different in another (eg, a smile after seeing another get hurt, indicating malice). In order to utilize a behavioral cue in one context, it is necessary to understand that context—potentially by determining the 5 W+ questions (who, what, when, where, why, and how) [17].

Similar to categorizing individuals, categorizing the situation may provide meaningful knowledge for optimizing a robot's response. For example, in situations where a decision must be made immediately, such as in emergencies, individuals may be forced to rely on instinct- or experience-based processes, which may be viewed as irrational [76]. Decision making in these situations differs from nonemergency reasoning, where there are higher stakes, higher uncertainty, and increased time pressure [77]. These considerations support the need for a comprehensive means to understand and represent human behavior, especially as it is likely to change in different contexts and situations.

6. CONCLUSIONS

Robots are becoming an important facet of physical and rehabilitative therapy. Exoskeletons and autonomous social robots may soon assist people with daily tasks. We argue that as robots become prevalent, it will be important that they model the people and patients that they interact with in a way that includes the many different psychological facets that make humans human. We have explored many of the major research directions by which these models are realized and developed. The inclusion of methods for behavioral and psychological modeling as a part of a robot's decision making is a new and exciting area of research. As this avenue of research grows, one should expect to see robots that are less aligned with our traditional notion of robots. These systems will appear more human and interactive. Although these models will likely be beneficial for a number of tasks, researchers and the community at large must give serious consideration to the ethical implications of creating robots which psychologically model humans.

REFERENCES

[1] V. Emeli, A.R. Wagner, C.C. Kemp, A robotic system for autonomous medication and water delivery, Technical report GT-IC-12-01, College of Computing, Georgia Institute of Technology, 2012.

[2] M.B. Brewer, A dual process model of impression formation, in: Advances in Social Cognition, Lawrence Erlbaum Associates, Inc, Hillsdale, NJ, vol. 1, 1988.

[3] S.T. Fiske, S.L. Neuberg, A continuum model of impression formation, from category-based to individuating processes: influences of information and motivation on attention and interpretation, Adv. Exp. Soc. Psychol. 23 (1990) 1–74.

[4] T.M. Ostrom, Between-theory and within-theory conflict in explaining context effects in impression formation, J. Exp. Soc. Psychol. 13 (5) (1977) 492–503.

[5] S.T. Fiske, M. Lin, S.L. Neuberg, The continuum model: ten years later, in: Dual-Process Theories in Social Psychology, Guilford Press, New York, NY, 1999, pp. 231–254.

[6] R. Ewen, Personality: A Topical Approach, Lawrence Erlbaum Associates, London, 1998.

[7] P.T. Costa, R.R. McCrae, NEO PI-R Professional Manual, Psychological Assessment Resources, Odessa, FL, 1992.

[8] H.J. Eysenck, The Biological Basis of Personality, Charles C. Thomas, Springfield, IL, 1967.

[9] I.B. Myers, The Myers-Briggs Type Indicator: Manual, Consulting Psychologists Press, Palo Alto, CA, 1962.

[10] A. Bandura, Social cognitive theory of personality, in: Handbook of Personality: Theory and Research, Guilford Press, New York, 1999, pp. 154–196.

[11] M.R. Barrick, M.K. Mount, The big five personality dimensions and job performance: a meta-analysis, Pers. Psychol. 44 (1) (1991) 1–26.

[12] A. Ortony, G.L. Clore, A. Collins, The Cognitive Structure of Emotions, Cambridge University Press, Cambridge, 1990.

[13] A. Damasio, Descartes' Error: Emotion, Reason, and the Human Brain, Avon Books, New York, 1994.

[14] R.P. Marinier, J.E. Laird, Computational modeling of mood and feeling from emotion, Power 1 (1) (2007) 461–466.

[15] W. Revelle, J. Wilt, A. Rosenthal, Personality and cognition: the personality-cognition link, in: A. Gruszka, G. Matthews, B. Szymura (Eds.), Handbook of Individual Differences in Cognition: Attention, Memory and Executive Control, Springer, New York, 2010, pp. 27–49.

[16] R.R. McCrae, Creativity, divergent thinking, and openness to experience, J. Pers. Soc. Psychol. 52 (6) (1987) 1258–1265.

[17] A. Vinciarelli, M. Pantic, H. Bourlard, Social signal processing: survey of an emerging domain, Image Vis. Comput. 27 (12) (2009) 1743–1759.

[18] P. Salovey, J. Mayer, Emotional intelligence, Imagin. Cognit. Personal. 9 (3) (1990) 185–211.

[19] C.L. Bethel, D. Eakin, S. Anreddy, J.K. Stuart, D. Carruth, Eyewitnesses are misled by human but not robot interviewers, in: 8th ACM/IEEE International Conference on Human-Robot Interaction (HRI), IEEE, 2013.

[20] D. McColl, G. Nejat, A socially assistive robot that can monitor affect of the elderly during meal-time assistance, J. Med. Devices 8 (3) (2014) 030941.

[21] M. Salem, G. Lakatos, F. Amirabdollahian, K. Dautenhahn, Would you trust a (faulty) robot? Effects of error, task type, and personality on human-robot cooperation and trust, in: 10th ACM/IEEE International Conference on Human-Robot Interaction, 2015.

[22] B.R. Duffy, Anthropomorphism and the social robot, Robot. Auton. Syst. 42 (3) (2003) 177–190.

[23] C. Breazeal, Designing Sociable Robots, MIT Press, Cambridge, MA, 2002.

[24] H. Ishiguro, S. Nishio, Building artificial humans to understand humans, J. Artif. Organs 10 (3) (2007) 133–142.

[25] G. Stratou, S. Scherer, J. Gratch, L.-P. Morency, Automatic nonverbal behavior indicators of depression and PTSD: exploring gender differences, in: International Conferences on Affective Computing and Intelligent Interaction, Geneva, Switzerland, 2013.

[26] P.N. Johnson-Laird, Mental Models: Towards a Cognitive Science of Language, Inference, and Consciousness, Cambridge University Press, Cambridge, 1983.

[27] P. Felli, T. Miller, C. Muise, A.R. Pearce, L. Sonenberg, Artificial social reasoning: computational mechanisms for reasoning about others, in: The International Conference on Social Robotics, {ICSR}, 2014.

[28] M.A. Goodrich, D. Yi, Toward task-based mental models of human-robot teaming: a Bayesian approach, in: Proceedings of HCI International, Las Vegas, Nevada, 2013.

[29] X. Zhao, J. Venkateswaran, Y.J. Son, Modeling human operator decision making in manufacturing systems using BDI agent paradigm, in: Proceedings of Annual Industrial Engineering Research Conference, 2005, pp. 14–18.

[30] S. Nikolaidis, J. Shah, Human-robot teaming using shared mental models, in: ACM/IEEE HRI, 2012.

[31] M. Hauskrecht, Value function approximations for partially observable Markov decision processes. J. Artif. Intell. Res. 13 (2000) 33–94, http://dx.doi.org/10.1613/jair.678.

[32] J. Bentham, The Principles of Morals and Legislation, Clarendon Press, Oxford, 1789. p. 1 (Chapter I).

[33] D. Bernoulli, Exposition of a new theory on the measurement of risk. Econometrica 22 (1) (1954) 22–36, http://dx.doi.org/10.2307/190982.

[34] J. von Neumann, O. Morgenstern, Theory of Games and Economic Behavior, Princeton University Press, Princeton, NJ, 1953.

[35] D. Kahneman, A. Tversky, Prospect theory: an analysis of decision under risk, Econometrica 0012-9682, 47 (2) (1979) 263, http://dx.doi.org/10.2307/1914185.

[36] M.H. Birnbaum, J.B. Navarrete, Testing descriptive utility theories: violations of stochastic dominance and cumulative independence, J. Risk Uncertain. 17 (1998) 49–78.

[37] A. Tversky, D. Kahneman, Advances in prospect theory: cumulative representation of uncertainty, J. Risk Uncertain. 5 (4) (1992) 297–323, http://dx.doi.org/10.1007/BF00122574.

[38] D.E. Bell, Regret in decision making under uncertainty, Oper. Res. 30 (5) (1982) 961–981.

[39] P.C. Fishburn, The Foundations of Expected Utility, Theory & Decision Library, Springer, New York, 1982.

[40] G. Loomes, R. Sugden, Regret theory: an alternative theory of rational choice under uncertainty, Econ. J. 92 (4) (1982) 805–824.

[41] J.R. Busemeyer, J.T. Townsend, Decision field theory: a dynamic cognition approach to decision making, Psychol. Rev. 100 (1993) 432–459.

[42] J.R. Busemeyer, A. Diederich, Survey of decision field theory, Math. Soc. Sci. 43 (3) (2002) 345–370.

[43] A.R. Wagner, A representation for interaction, in: Proceedings of the ICRA 2008 Workshop: Social Interaction with Intelligent Indoor Robots (SI3R), Pasadena, CA, USA, 2008.

[44] C. Camerer, G. Loewenstein, D. Prelec, Neuroeconomics: how neuroscience can inform economics, J. Econ. Lit. XLIII (2005) 9–64.

[45] J.A. Bargh, T.L. Chartrand, The unbearable automaticity of being, Am. Psychol. 54 (7) (1999) 462–479.

[46] R.M. Shiffrin, W. Schneider, Controlled and automatic human information processing: II. Perceptual learning, automatic attending and a general theory, Psychol. Rev. 84 (2) (1977) 127–190.

[47] J.E. LeDoux, The Emotional Brain: The Mysterious Underpinnings of Emotional Life, Simon and Schuster, New York, 1996.

[48] B. King-Casas, D. Tomlin, C. Anen, C.F. Camerer, S.R. Quartz, P.R. Montague, Getting to know you: reputation and trust in two-person economic exchange, Science 308 (2005) 78–83.

[49] J.S. Winston, B.A. Strange, J. O'Doherty, R.J. Dolan, Automatic and intentional brain responses during evaluation of trustworthiness of faces, Nat. Neurosci. 5 (2002) 277–283.

[50] F. Krueger, K. McCabe, J. Moll, N. Kriegeskorte, R. Zahn, Neural correlates of trust, Proc. Natl. Acad. Sci. 104 (2007) 20084–20089.

[51] D. Laibson, Neuroeconomics and the Multiple Systems Hypothesis, Mannheim Summer School, Mannheim, 2009.

[52] B. Shiv, A. Fedorikhin, Heart and mind in conflict: the interplay of affect and cognition in consumer decision making, J. Consum. Res. 26 (3) (1999) 278–292.

[53] J.K. Witt, S.A. Linkenauger, J.Z. Bakdash, D.R. Proffitt, Putting to a bigger hole: golf performance relates to perceived size, Psychon. Bull. Rev. 15 (2008) 581–585.

[54] J.K. Witt, M. Sugovic, Spiders appear to move faster than non-threatening objects regardless of one's ability to block them, Acta Psychol. 143 (2013) 284–291.

[55] P. Robinette, A. Wagner, A.M. Howard, The effect of robot performance on human robot trust, IEEE Trans. Hum. Mach. Syst., forthcoming.

[56] G. Trafton, L. Hiatt, A. Harrison, F. Tamborello, S. Khemlani, A. Schultz, Act-r/e: an embodied cognitive architecture for human-robot interaction, J. Hum.-Robot Interact. 2 (1) (2013) 30–55.

[57] M. Swangnetr, D.B. Kaber, Emotional state classification in patient–robot interaction using wavelet analysis and statistics-based feature selection, IEEE Trans. Hum. Mach. Syst. 43 (1) (2013) 63–75.

[58] L. Hiatt, A. Harrison, J. Trafton, Accommodating human variability in human-robot teams through theory of mind, in: Proceedings of the International Joint Conference on Artificial Intelligence, 2011.

[59] A. Pentland, Socially aware computation and communication, IEEE Comput. 38 (3) (2005) 33–40.

[60] A. Pentland, Social dynamics: signals and behavior, in: International Conference on Developmental Learning, 2004.

[61] P. Ekman, W.V. Friesen, The repertoire of nonverbal behavior: categories, origins, usage, and coding, Semiotica 1 (1) (1969) 49–98.

[62] B. Lepri, N. Mana, A. Cappelletti, F. Pianesi, M. Zancanaro, Modeling the personality of participants during group interactions, in: User Modeling, Adaptation, and Personalization, Springer, Berlin, 2009, pp. 114–125.

[63] J.S. Uleman, Spontaneous versus intentional inferences in impression formation, in: S. Chaiken, Y. Trope (Eds.), Dual Process Theories in Social Psychology, Guilford Press, New York, 1999.

[64] Y.R. Tausczik, J.W. Pennebaker, The psychological meaning of words: LIWC and computerized text analysis methods, J. Lang. Soc. Psychol. 29 (1) (2010) 24–54.

[65] S.D. Gosling, S.J. Ko, T. Mannarelli, M.E. Morris, A room with a cue: personality judgments based on offices and bedrooms, J. Pers. Soc. Psychol. 82 (3) (2002) 379.

[66] D.A. Kenny, C. Horner, D.A. Kashy, L.C. Chu, Consensus at zero acquaintance: replication, behavioral cues, and stability, J. Pers. Soc. Psychol. 62 (1) (1992) 88.

[67] P. Ekman, E.L. Rosenberg, What the Face Reveals: Basic and Applied Studies of Spontaneous Expression Using the Facial Action Coding System (FACS), Oxford University Press, Oxford, 1997.

[68] R.W. Picard, Affective computing for HCI, in: HCI (1), 1999, pp. 829–833.

[69] A. Chakraborty, A. Konar, U.K. Chakraborty, A. Chatterjee, Emotion recognition from facial expressions and its control using fuzzy logic, IEEE Trans. Syst. Man Cybern. Syst. Hum. 39 (4) (2009) 726–743.

[70] T. Bänziger, D. Grandjean, K.R. Scherer, Emotion recognition from expressions in face, voice, and body: the Multimodal Emotion Recognition Test (MERT), Emotion 9 (5) (2009) 691–704.

[71] R.J. Sternberg, M.K. Gardner, A componential interpretation of the general factor in human intelligence, in: H.J. Eysenck (Ed.), A Model for Intelligence, Springer-Verlag, New York, 1982, pp. 231–254.

[72] P.L. Ackerman, E.D. Heggestad, Intelligence, personality, and interests: evidence for overlapping traits, Psychol. Bull. 121 (2) (1997) 219.

[73] K. Kleisner, V. Chvátalová, J. Flegr, Perceived intelligence is associated with measured intelligence in men but not women, PLoS One 9 (3) (2014) e81237.

[74] M.D. Prokosch, R.A. Yeo, G.F. Miller, Intelligence tests with higher g-loadings show higher correlations with body symmetry: evidence for a general fitness factor mediated by developmental stability, Intelligence 33 (2) (2005) 203–213.

[75] J. Chan, G. Nejat, Social intelligence for a robot engaging people in cognitive training activities, Int. J. Adv. Robot. Syst. 9 (2012) 1–13.

[76] X. Pan, C.S. Han, K. Dauber, K.H. Law, A multi-agent based framework for the simulation of human and social behaviors during emergency evacuations, AI Soc. 22 (2) (2007) 113–132.

[77] G. Proulx, Understanding human behaviour in stressful situations, in: Workshop to Identify Innovative Research Needs to Foster Improved Fire Safety in the United States, Delegate Binder Section 7, April 15–16, National Academy of Sciences, Washington, DC, 2002, pp. 1–5.

[78] S.T. Fiske, S.E. Taylor, Social Cognition, Addison-Wesley Pub. Co., Reading, MA, 1984.

[79] P. Ekman, W.V. Friesen, P. Ellsworth, What emotion categories or dimensions can observers judge from facial behaviour? in: P. Ekman (Ed.), Emotion in the Human Face, second ed., Cambridge University Press, Cambridge, 1982, pp. 98–110.

[80] I.J. Deary, M.C. Whiteman, J.M. Starr, L.J. Whalley, H.C. Fox, The impact of child-hood intelligence on later life: following up the Scottish mental surveys of 1932 and 1947, J. Pers. Soc. Psychol. 86 (1) (2004) 130.

[81] A.R. Jensen, The G Factor: The Science of Mental Ability, Praeger, Wesport, 1998.

[82] M. Argyle, The Psychology of Interpersonal Behaviour, Penguin, Harmondsworth, 1967.

[83] J.R. Anderson, C. Lebiere, The Atomic Components of Thought, Erlbaum, Mahwah, NJ, 1998.

[84] A. Newell, Unified Theories of Cognition, Harvard University Press, Cambridge, MA, 1994.

[85] V.P. Richmond, J.C. McCroskey, S.K. Payne, Nonverbal Behavior in Interpersonal Relations, Prentice Hall, Englewood Cliffs, NJ, 1991.

[86] U.C.E. Segerstråle, P.E. Molnár, Nonverbal Communication: Where Nature Meets Culture, Lawrence Erlbaum Associates, Inc., Mahwah, NJ, 1997.

[87] M.L. Walters, K. Dautenhahn, K.L. Koay, C. Kaouri, S. N. Woods, C.L. Nehaniv, I. Werry, The influence of subjects' personality traits on predicting comfortable human-robot approach distances', in: Proceedings of Cog Sci 2005 Workshop: Toward Social Mechanisms of Android Science, 2005, pp. 29–37.

[88] S. Dar, Z. Khaki, Emotion recognition based on audio speech, IOSR J. Comput. Eng. 1 (2013) 46–50.

[89] L.C. De Silva, T. Miyasato, R. Nakatsu, Facial emotion recognition using multi-modal information, in: Information, Communications and Signal Processing, 1997, in: Proceedings of 1997 International Conference on ICICS, Vol. 1, IEEE, 1997, pp. 397–401.

[90] H.G. Gough, A nonintellectual intelligence test, J. Consult. Psychol. 17 (4) (1953) 242.

CHAPTER ELEVEN

Adaptive Human-Robot Physical Interaction for Robot Coworkers

J. Ueda*, W. Gallagher[†], A. Moualeu*, M. Shinohara*, K. Feigh*
*Georgia Institute of Technology, Atlanta, GA, United States
[†]NASA's Goddard Space Flight Center, Greenbelt, MD, United States

1. INTRODUCTION

As robotic technology advances, the area of human-robot interaction (HRI) is expanding rapidly. Robots can no longer be designed in isolation from their operators. HRI systems must consider that the human is an integral part of the system. This applies especially to physical HRI (pHRI) systems, where the operator is in physical contact with the robot. Recent trends in robotic research and applications have particularly emphasized a push for the development of true robotic coworkers, acting as direct human partners. Industrial settings, for instance, are increasingly utilizing robotics and automation to streamline difficult jobs. Yet, some situations render automation difficult due to the usually strict tolerances required for repetitive tasks. For example, on vehicle assembly lines, the placement of a vehicle component, such as a door, must be done within tolerances, but the location of the vehicle itself may vary slightly. It is often more efficient to have a human accomplish certain tasks, especially those that require force sensitivity, which is best accomplished through human touch. Load requirements in most industrial settings usually exceed a worker's physical capability. Hence, assistive robotic devices can be useful in aiding the completion of such heavy-duty tasks through force amplification. Teleoperated systems could be used, but can introduce possible errors and require slower motion, due to remote sensing. A system that allows direct interaction with an operator is preferred. Haptics is a popular control method, since tactile sensation constitutes a very intuitive tool for controlling a robotic device. Force feedback and haptic controllers are common in applications ranging from gaming to industrial machines.

Physical contact established in pHRI creates a coupled system with static and dynamic properties that significantly differ from those of each isolated system prior to interaction [1]. Without an appropriate controller, this coupling can result in reduced stability, making the robot harder to operate. Instability increases task completion time, decreases performance, and creates an opportunity for injury. Humans naturally increase muscle stiffness to control instability or oscillation. Unfortunately, this creates a stiffer coupled system with more instability. Generic robot controllers cannot directly measure or adjust to operator stiffness. A system with access to information about the operator's motion could adjust accordingly, and thereby increase stability, bolster safety, and make operation easier. This would enable increased operator performance, which in industrial settings can result in increased efficiency.

This research seeks to develop methods that allow a haptic robot controller to adjust to changes in pHRI, by expanding the information about the human operator available to the controller. Fig. 1 shows a conceptual illustration of such a system. The specific aims are to understand the relationship between system stability and performance characteristics and neuromuscular adaptations in pHRI, and to develop novel control methods that account for classification and prediction of operator physical and/or cognitive states, in order to adaptively tune controller parameters for increased pHRI stability and performance.

2. HAPTIC STABILITY

Haptic interaction, which requires physical human-robot contact, creates a coupled and bilateral system in which the device responds to the force applied by the operator and the operator adjusts the applied force based on the device's motion. Some haptic systems attempt to resist the motion of the operator so as to provide a virtual environment for them to feel [2–5], while many attempt to amplify the motion of the operator so as to increase human capability [6–8]. Force-assisting devices are the primary concern for this research, but some studies relative to other types of haptic devices may also be applicable. In all cases, the device is controlled based on the measured force applied by the operator.

Devices using force control have been shown to become unstable under contact with stiff environments or the presence of a time delay, both

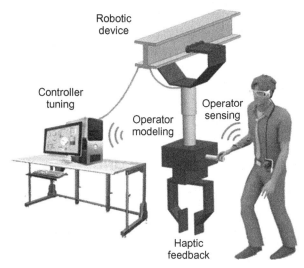

Fig. 1 Conceptual drawing of a haptically controlled robotic device with a controller that adjusts based on an estimated operator model.

of which are often present under contact with a human operator [9–11]. Human reaction times can be orders of magnitude larger than the typical period of a single control loop, and the demonstrated human reaction under instability is to increase contact stiffness. In addition, sensor time delays constitute further sources of instability in the system. Introducing compliance into the robotic system can mitigate this issue, but this inherently decreases the performance of the system, a trade-off which has been well documented [9–12]. Since the goal of this research is to increase performance, added compliance to the robotic system would not be beneficial.

Stability of haptic systems during HRI has been analyzed using both root-locus methods [11] and Lyapunov theory [10]. These studies have provided useful stability bounds, which are highly dependent on the stiffness of the human operator. However, these bounds do not account for deliberate stiffening of the human operator's arm, and therefore are not sufficient for the current study, which requires further stability analysis. In the case of teleoperated systems, stability is often viewed from the perspective of passivity, and this view has been extended to haptic devices as well [2,9,13–16]. While such a perspective could provide a useful condition for the stability of a system, force-assisting devices are by nature not strictly passive [8]. Several more recent studies have explored combined methods in a way that could be applicable to analyzing haptic force assistive systems [3,4]. Other studies have explored the design of robust controllers for interacting systems and teleoperation [12,16–19], but require a priori knowledge of the range of the system parameters.

3. HUMAN OPERATOR MODELING

3.1 Human Arm Stiffness

Stiffness, the static component of impedance, is defined as the change in force over a change in distance from a given neutral point. While the applied force is readily measurable with sensors, the change in distance is not, because as the operator moves, the neutral point moves as well. In addition, there are stiffness distinctions related to the motion of the human arm: (a) muscle stiffness, the resistance of a single muscle to changes in length; (b) joint stiffness, the resistance of a joint to changes in joint angle; and (c) endpoint stiffness, the resistance of the entire arm to changes in endpoint location. Endpoint stiffness is of most interest for the design of a robotic controller, but it is affected by both individual muscle stiffness and joint stiffness. Hence, in this chapter, most mentions of "stiffness" refer to endpoint stiffness, unless specified otherwise.

Much of the current understanding of human muscles came from Hill's work, which models muscles primarily as springs with a force-generating component [20], as well as Bernstein's discussion of human motor control [21]. Because muscles can only generate contractile forces, joints in the body normally have two or more antagonistic muscles, which act on opposite sides of the same joint. Increases in joint stiffness have been linked to simultaneous activation (ie, cocontraction) of these antagonistic muscles [22,23]. The stiffness measured at the endpoint emerges from the combined effects of the elastic properties of muscles and joints, as well as the neural control circuits that act on the limb [24]. For instance, an important consideration is that the moment arm of a particular muscle on a given joint changes as the arm posture changes, so the stiffness of a joint will vary based on posture [25], affecting the endpoint stiffness of the entire limb [26]. Several studies have shown that endpoint stiffness cannot be controlled independently of force and position [14,27,28], which implies that an estimate of endpoint stiffness is indicative of either involuntary reactions to the environment or intended voluntary applied force and motion. In general, these studies found a roughly linear increase in endpoint stiffness with voluntary force.

Antagonistic cocontraction is the vehicle for stiffness modulation in the musculoskeletal system. It has been demonstrated that the brain attempts to correct for an inability to maintain a desired target by increasing arm stiffness, which is a result of increased cocontraction [29–35], and has a similar response when trying to resist movement [36,37]. Also, the reverse has been demonstrated during smooth movements or when not trying to resist motion, which results in lower stiffness with less cocontraction [29,31]. Therefore, for the purpose of designing a system that detects the body's reaction to unstable situations, it should be possible to measure the level of cocontraction in the operator's arm and use it as an indication of stiffness level.

3.2 Musculoskeletal Modeling

Ueda et al. have developed an integrated model (see Fig. 2A) that can characterize the coupling between the human musculoskeletal system and robot dynamics at the level of individual muscles [38–43]. This upper-right musculoskeletal model is a 5-rigid link model with 12 joints from the waist to the wrist and 51 arm muscles. For a human model with M joints and N muscles,

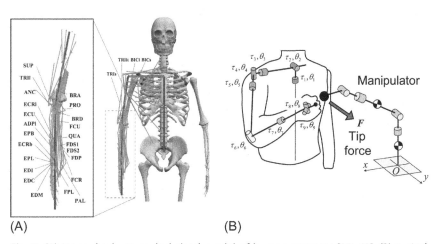

(A)   (B)

Fig. 2 (A) Upper-body musculoskeletal model of human operator [38–43]; (B) typical pHRI system.

the relationship between joint torques and end-point force in a typical pHRI system (see Fig. 2B) is given by Eq. (1), under static assumptions and neglected dynamics of both robot and human body.

$$\tau_h = J^T(\theta)F \tag{1}$$

Here, $\tau_h \in \mathfrak{R}^M$ is a vector of human joint torques, $\theta = [\theta_1, ..., \theta_M]^T \in \mathfrak{R}^M$ is a vector of joint angles and $F = [F_x, F_y, F_z]^T$ is the tip force. $J(\theta)$ is the Jacobian whose transpose maps end-point force to joint torques. The relations from joint torques (and hence tip force) to muscle forces are given by Eq. (2).

$$\tau_h = M_A(\theta)f = J^T(\theta)F \tag{2}$$

Here, $M_A \in \mathfrak{R}^{M \times N}$ is the moment arm matrix of muscles, and $f = [f_1, ..., f_N]^T \in \mathfrak{R}^N$ is the human operator's muscle forces vector. The latter is subject to boundary constraints given by Eq. (3).

$$0 \leq f_j \leq f_{\max j} \quad (j = 1, ..., N) \tag{3}$$

The values of $f_{\max j}$ are defined in Appendix A. Using an optimization method (eg, Crowinshield's optimality principle, see Appendix A), a baseline of muscle activity f_o can be determined if zero muscle coactivation is assumed. As previously mentioned, muscle cocontraction is invisible in the joint space, a phenomenon characterized by Eq. (4):

$$f = f_o + \left(I - M_A^+ M_A\right)\beta(t) \tag{4}$$

Here, M_A^+ represents the pseudo-inverse of M_A, $\beta(t) \in \mathfrak{R}^N$ represents the nullspace of the mapping between muscle forces and joint torques, ie, muscle coactivation. Current commercial and open-source biomechanical modeling software (eg, OpenSim, The Anybody Modeling System) assume that $\left(I - M_A^+ M_A\right)\beta(t)$ is zero in their analysis. However, $\beta(t)$ can take any value; a change in $\beta(t)$ affects human arm stiffness, and therefore the stability and performance of the whole system. This research study hypothesizes that $\beta(t)$ is a random variable.

4. HAPTIC ASSIST CONTROL

4.1 Impedance Masking Approach

For the case of force-assisting devices, the simplest control method is pure force amplification, where the force generated by a device is simply a gain multiplied by the applied force [44]. However, such systems can be difficult and unnatural to control given large gains. An impedance controller, alternatively, can be used in such situations.

Impedance-based control methods have shown to be effective in increasing system stability [45]. Compensating systems have been demonstrated to be capable of increasing stability without sacrificing performance [46]. Many haptic systems use impedance control, which provides a masking of actual system dynamics and allows the operator to feel a system with desired mass m_d, damping b_d, and stiffness k_d [44]. Modeling the operator as a system with mass m_o, damping b_o, and stiffness k_o, applying a force f_o, and the haptic device as having mass m_h, damping b_h, and stiffness k_h, and capable of measuring the applied force f_m, then Eq. (5) is the equation of motion of the contact point of the operator with the device and Eq. (6) is the equation of motion of the device itself [47,48]. The control force, f_h, is determined based on the desired parameters, as given by Eq. (7). From these follows the derivation of the equation of motion in Eq. (8), which demonstrates how the controller can mask the device's dynamics. This makes the load on the user lighter and allows easier operation, which, when combined with haptic feedback, can give a very natural feeling to operating the robot. The choice of desired stiffness k_d is usually zero and the damping gain b_d is the adjusted system parameter in the tuning scheme. Figs. 3 and 4, respectively, show the overall and detailed block diagrams for such a system, which yields the transfer function for the outer block as given by Eq. (9), with X_d giving the desired position in the s domain. The impedance-masking control

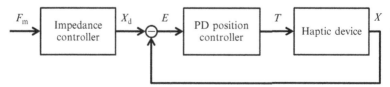

Fig. 3 Block diagram of a haptic system incorporating an impedance controller with an inner PD position controller.

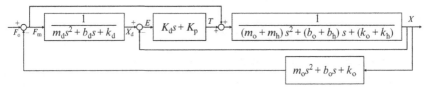

Fig. 4 The s domain block diagram of a haptic control system, including controller, device and operator characteristics.

scheme often incorporates a position controller that drives the error e between current and desired endpoint positions to zero (Fig. 4). The most basic approach for position control would be to use a proportional-derivative (PD) controller with fixed proportional gain K_p and derivative gain K_d, yielding a transfer function as shown in Eq. (10), where T is the motor torque and E gives the error in position from X_d in the s domain.

$$f_o - f_m = m_o \ddot{x} + b_o \dot{x} + k_o x \tag{5}$$

$$f_h + f_m = m_h \ddot{x} + b_h \dot{x} + k_h x \tag{6}$$

$$f_h = (m_d - m_h)\ddot{x} + (b_d - b_h)\dot{x} + (k_d - k_h)x \tag{7}$$

$$f_o = (m_o + m_d)\ddot{x} + (b_o + b_d)\dot{x} + (k_o + k_d)x \tag{8}$$

$$\frac{X_d(s)}{F_m(s)} = \frac{1}{m_d s^2 + b_d s + k_d} = \frac{1}{m_d s^2 + b_d s} \tag{9}$$

$$\frac{T(s)}{E(s)} = K_d s + K_p \tag{10}$$

4.2 Linear Quadratic Gaussian Control

The combined human and robot dynamics can be modeled using the equation of motion of a typical mass-spring-damper system with a state-space representation given by Eq. (11) in the continuous time domain, where the system input u is the applied torque.

$$\dot{x} = Ax + Bu, \, y = Cx + Du \tag{11}$$

where

$$A = \begin{bmatrix} 0 & 1 \\ -\dfrac{k}{m} & -\dfrac{b}{m} \end{bmatrix}, B = \begin{vmatrix} 0 \\ \dfrac{1}{m} \end{vmatrix}, C = [1 \ 0], D = 0$$

If the stiffness k, and in the most general situation, the damping b, is changing, then the A matrix is not constant. When its value with time is known, a control law can be designed around it. However, in the case of the coupled human-robot system, this value is changing stochastically based on the response of the human. Therefore, it is desirable to utilize a controller designed for such a case of uncertain system parameters. This new controller can be substituted for an inner position controller in the impedance control scheme. This will ensure that, even as the actual system parameters vary, the device can still maintain the motion of the desired system model.

Assuming that the sampling rate is high enough to prevent significant loss of information, the discrete version of the system is given by Eq. (12).

$$x_{i+1} = Ax_i + Bu_i, \quad y_i = Cx_i + Du_i \tag{12}$$

Using state feedback, a standard linear quadratic regulator (LQR) optimal controller can be designed to stabilize a system with minimum control effort [49], relying on the assumption that the system is time–invariant. It can be adapted to control stochastic systems by modeling variation as additive or multiplicative noise [50]. When combined with Kalman's filter to reduce uncertainty and noise [51], the linear quadratic Gaussian (LQG) control method is obtained [52], which has become one of the fundamental techniques for control under uncertainty. Fig. 5 shows a diagram of the LQG controller. With time-invariant system matrices A and B in Eq. (12), the LQR control input can be obtained using Eq. (13). In the equation, L is derived by minimizing the cost function J in Eq. (14) for all t. Q_1 and Q_2 are weighting matrices designed to emphasize the minimization of error and control effort, respectively, and are positive semidefinite.

$$u_t = -Lx_t \tag{13}$$

$$J = \sum_{t=0}^{\infty} x_t^T Q_1 x_t + u_t^T Q_2 u_t \tag{14}$$

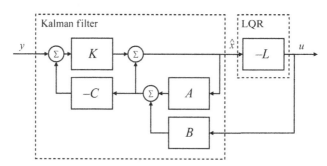

Fig. 5 Block diagram of an LQG controller.

This problem can be solved through dynamic programming. The optimal control law can be found from the solution of the Riccati equation shown in Eq. (15), and is given by Eq. (16).

$$\Phi = Q_1 + A^{\mathrm{T}}\Phi A - A^{\mathrm{T}}\Phi B\left(Q_2 + B^{\mathrm{T}}\Phi B\right)^{-1}B^{\mathrm{T}}\Phi A \qquad (15)$$

$$L = \left(Q_2 + B^{\mathrm{T}}\Phi B\right)^{-1}B^{\mathrm{T}}\Phi A \qquad (16)$$

This solution is optimal and minimizes control effort. However, it relies on the accuracy of the system model and the assumption of time invariance. Consider, instead, the system with sensor output given by Eq. (17), where v_t is process noise and w_t is sensor noise.

$$x_{t+1} = Ax_t + Bu_t + v_t y_t, \quad y_t = Cx_t + w_t \qquad (17)$$

Both process and sensor noises represent unknowable variables a priori and are not directly measurable. Using the Kalman filter in Eq. (18) to perform state estimation can provide more accurate values for the control law, given in Eq. (19).

$$\hat{x}_{t+1} = A\hat{x}_t + Bu_t + K\left(y_{t+1} - C(A\hat{x}_t + Bu_t)\right) \qquad (18)$$

$$u_t = -L\hat{x}_t \qquad (19)$$

The K matrix found from the Riccati equation in Eq. (20) and control law solution in Eq. (21) is combined with the L found previously. The diagonals of W and V give the expected range of the elements of w_t and v_t, respectively.

$$\Gamma = V + A\Gamma A^{\mathrm{T}} - A\Gamma C^{\mathrm{T}}\left(W + CPC^{\mathrm{T}}\right)^{-1}C\Gamma A^{\mathrm{T}} \qquad (20)$$

$$K = \Gamma C^{\mathrm{T}}(W + C\Gamma C)^{-1} \qquad (21)$$

4.3 Control of a System With Stochastically Varying Parameters

While the LQG controller holds promise for stabilizing uncertain systems, it is not specifically designed for systems with stochastically varying parameters, but rather for time-invariant systems disturbed by white noise. It has been demonstrated that the standard LQG controller, while optimal for a known linear system, is not robust to large variations in system parameters due to its reliance on the standard Kalman filter [53,54]. Previous work has shown that the human operator arm stiffness can be approximated using a series of Gaussian distributions [47,55], so it should be possible to determine a priori an expected value for A and an estimated variance. Therefore, the ideal controller would be designed to accept a distribution for the system parameters and calculate an optimal control law that will be well conditioned over most of that distribution. Numerous advances have been made in the control of stochastic systems and have provided a variety of robust and optimal controllers that are effective [56–66]. Ultimately, the most promising of these approaches for this research incorporated an assumed distribution for the system parameters directly into the derivation of the optimal control law, ensuring that the controller would be optimal for their expected values, but still perform well as they varied [67], with Fujimoto providing such a system that was designed to minimize the variance of the control signal [68,69].

Consider again the system of Eq. (12), with A and B now stochastically varying parameters given by some distribution that does not change with time with expected values $E[A]$ and $E[B]$. In this case, it is desirable to not only minimize the control effort, but also the variance of the system's tracking error, so the cost function is modified as in Eq. (22), with the additional weight parameter, S, which must also be positive semidefinite.

$$J = E\left[\sum_{t=0}^{\infty}\left(x_t^T Q_1 x_t + u_t^T Q_2 u_t + tr[Q_3 cov[x_{t+1}, x_t]]\right)\right] \tag{22}$$

Using the methodology of Fujimoto [68,69], the optimal control gain is found by solving Eq. (23) numerically, with the parameter Π given by Eq. (24) as an initial guess and the covariances Σ_{XY} given by Eq. (25).

$$L = \left(E[B^T \Pi B] + \Sigma_{BB} + Q_2\right)^{-1}\left(E[B^T \Pi A] + \Sigma_{BA}\right) \tag{23}$$

$$\begin{aligned}\Pi = \ & Q_1 + \Sigma_{AA} + E[A]^T \Pi E[A] - \left(E[A]^T \Pi E[B] + \Sigma_{AB}\right) \\ & \left(E[B]^T \Pi E[B] + \Sigma_{BB} + Q_2\right)^{-1}\left(E[B]^T \Pi E[A] + \Sigma_{BA}\right)\end{aligned} \tag{24}$$

$$\Sigma_{XY} = E[X^T Q_3 Y] - E[X]^T Q_3 E[Y] \tag{25}$$

This controller, called the stochastic linear quadratic regulator (SLQR), as shown by Fig. 6, is an optimal controller designed from the expected value and covariance of A and B to minimize the control effort and variance of the

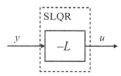

Fig. 6 Block diagram of an SLQR controller.

system. The emphasis on each of these can be varied by choosing appropriate values for the Q's. The state feedback diagram is drastically simpler, as the state estimator is removed, since the calculation of L already incorporates an estimate of the stochasticity of the system.

5. SYSTEM INTEGRATION

Literature has demonstrated that adjusting the impedance of a system is an effective method for adapting to the motion of an operator [45,46,70]. Integrating a stiffness estimating classifier with the impedance controller utilizing one of the inner position controllers yields the complete system shown in the block diagram in Fig. 7. In the final system, the position of the device is controlled solely by the force applied to the device as with a standard impedance controller. However, additional data related to arm stiffness are used to adjust the way in which the controller performs this task. For this reason, the stiffness classifier model would ideally need to be updated at a rate faster than that of the control loop [71].

5.1 Human Arm Stiffness Estimation

There is a long history of human stiffness estimation, particularly through movement perturbations applied with a robotic device [72–75,27]. Limitations of current methods include some one-joint motion restrictions, systematic errors and the potential of movement disruption. The challenge in directly measuring operator arm stiffness in most dynamic control situations has led to proposed estimation techniques based on physiological measures known to correlate [76,77], notably electromyogram (EMG) signals, used as a measure of muscle activity.

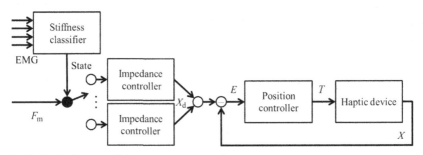

Fig. 7 Block diagram of the complete control system.

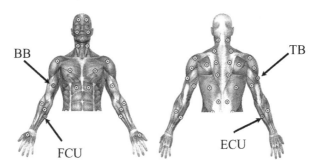

Fig. 8 EMG placement for muscles including biceps brachii (BB), triceps brachii (TB), flexor carpi ulnaris (FCU), and extensor carpi ulnaris (ECU).

Given that antagonistic muscle cocontraction is the vehicle of stiffness modulation in the musculoskeletal system [22], careful selection of the best pairs of antagonistic muscles known to affect the wrist and elbow joints is important. The degree of antagonism between two muscles forming an antagonistic pair can be evaluated using the angle between their moment arm vectors [78,79]. Two completely antagonistic muscles should act along opposite directions, resulting in an angle of antagonism close to 180 degrees. Based on this principle, two muscle pairs can be chosen out of several great candidates [47,48]. The biceps brachii and triceps brachii at the elbow constitute the first pair, whereas the flexor carpi ulnaris and extensor carpi ulnaris at the wrist constitute the second pair (see Fig. 8).

For each antagonistic muscle pair (i, j), processed EMG values (f_i, f_j) are scaled with respect to maximum recorded EMG values (f_i^{max}, f_j^{max}) to obtain normalized muscle activity $(f_i^{\%}, f_j^{\%})$ values between 0 and 1. Cocontraction, $cc_{i,j}$, for the muscle pair, can be determined by taking the minimum normalized muscle activity at each sample time [47,48,80,81], as shown in Eq. (26).

$$cc_{i,j} = \min \left(f_i^{\%}, f_j^{\%} \right) \tag{26}$$

5.2 Stiffness Classification
5.2.1 Threshold-Based Classifier
Using cocontraction levels, $C_i(t)$, a simple threshold-based classifier can be developed and used to determine if the operator arm stiffness level, Z, falls into one of H discrete levels, $z_1, z_2, \ldots, z_H$ [47]. This is done using $H-1$

adjustable thresholds, $l_{i1}, l_{i2}, \ldots, l_{i(H-1)}$, for each pair of muscles, $i = E, W$, as shown in Eq. (27).

$$Z = \begin{cases} z_K & \text{if } C_W(t) \geq l_{W(H-1)} \text{ or } C_E(t) \geq l_{E(H-1)} \\ z_{K-1} & \text{if } C_W(t) \geq l_{W(H-2)} \text{ or } C_E(t) \geq l_{E(H-2)} \\ \vdots & \\ z_2 & \text{if } C_W(t) \geq l_{W1} \text{ or } C_E(t) \geq l_{E1} \\ z_1 & \text{if } C_W(t) < l_{W1} \text{ and } C_E(t) \geq l_{E1} \end{cases} \qquad (27)$$

5.2.2 Hidden Markov Model Classifier

There is a long history of using a Bayesian approach to estimate the intent of an operator interacting with a device [82–84]. As a probabilistic approach, it allows for reasoning under uncertainty. Since the operator arm stiffness, k_o, in this study is not directly measurable, hidden Markov models (HMMs) seem to be the natural choice for a predictive model. These models have proven to be easy to train and to use for inference. Other predictive models, such as dynamic Bayesian networks, artificial neural networks, random forests, and support vector machines (SVMs) could also be explored but may incur more training data and computation time.

An HMM is a doubly Markov (stochastic) process with an underlying stochastic process that is *hidden* and can only be observed through another set of stochastic processes [85]. A Markov process, as illustrated in Eq. (28), describes the probability of being in a current state (at time t) as a function of the preceding state only, where q_t is the actual state at time t and S is the discrete state of the Markov chain.

$$P\big[q_t = S_j \big| q_{t-1} = S_i, q_{t-2} = S_k, \ldots\big] = P\big[q_t = S_j \big| q_{t-1} = S_i\big] \qquad (28)$$

An HMM uses a transition model $U = P(X_t|X_{t-1})$ and a sensor model $O = P(Y_t|Z_t)$ to perform estimation of a current, hidden-state X_t from a list of N possible discrete states $S_1, \ldots, S_N$, based on the preceding state X_{t-1} and a currently observed metric Y_t. U is an $N \times N$ matrix of probabilities where each row represents probabilities for a single previous state and each column a current state. O is an $N \times D$ matrix, where D represents the number of possible discrete observations. Fig. 9 displays a representation of an HMM for the case of the current study. Given the two models, the currently observed measurements, and the previous state, the probability of the system

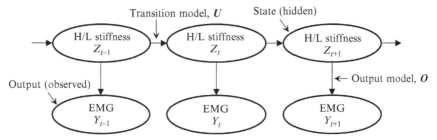

Fig. 9 Representation of a stiffness prediction model using an HMM.

being in a current state can be found using Eq. (29), where k is a normalizing constant.

$$P(X_t|Y_1, \ldots, Y_t) = kP(Y_t|X_t) \sum_{S_k} P(X_t|X_{t-1} = S_k)$$

$$P[X_{t-1} = S_k| X_1, \ldots, X_{t-1}]$$

(29)

Putting this equation in matrix form, as shown in Eq. (30), we can obtain a state probability vector, P_t, with O_t being an $N \times N$ matrix with diagonal elements giving the likelihood of each state given the current sensor measurement, and P_{t-1} being the previous likelihood vector. This can be applied recursively to find the current state given the starting state, P_o, and a sequence of sensor measurements, O_t's. Using this approach, called filtering, the most likely current state is the one with the highest probability [86].

$$P_t = \kappa O_t U P_{t-1}$$

(30)

5.2.3 SVM Classifier

The use of a deterministic classifier, a SVM, as a decision tool for gain tuning of the impedance controller was also investigated. SVM classifiers have been used in myoelectric control applications, with high accuracy results in EMG pattern recognition of healthy subjects [87]. Given a set of linearly separable training data samples $(x_i, y_i)_{1 \leq i \leq n}$, where $x_i \in \mathfrak{R}^d$, $y_i \in \{-1, 1\}$, the general form of the classification function is $g(x) = w \cdot x + b$, which represents a hyperplane. Identifying the optimal separating hyperplane (OSH) through maximization of the margin $1/\|w\|$ between the hyperplane and support vectors, the closest data points, corresponds to minimizing the objective function in Eq. (31). Classification of nonseparable data introduces slack variables $\xi_i \geq 0$ and a penalty factor $C > 0$, leading to Eq. (32).

$$\min_{w,b} \frac{1}{2} \|w\|^2$$ (31)

$$\text{Subject to}: y_i(w \cdot x_i + b) \geq 1, \quad i = 1, \ldots, n$$

$$\min_{w,b} \frac{1}{2} \|w\|^2 + C\sum_{i=1}^{n} \xi_i$$ (32)

$$\text{Subject to}: y_i(w \cdot x_i + b) \geq 1 - \xi_i, \quad i = 1, \ldots, n$$

In the nonlinear case, a mapping from input space to a feature space where corresponding features are linearly separable occurs, and the OSH can be constructed. This is equivalent to replacing every dot product in Eqs. (31) or (32) with a nonlinear kernel function. The most common function used is the (Gaussian) radial basis function (RBF) kernel, shown in Eq. (33). Although SVM is fundamentally a two-state classifier, multiclass SVM classification can be achieved using different approaches, among which the most common is the *one-versus-all* approach. This approach constructs separate SVMs, each corresponding to a class where the characteristic training data points are considered positive and the rest negative. There are various SVM formulations capable of multiclassification. This paper uses the ν-SVM formulation with an RBF kernel. A number of parameters, including $\nu \epsilon (0, 1]$, are employed to control the number of support vectors, the shape of the separating hyperplanes, among other factors. More details about binary and multiclass SVM methods, formulations and parameters can be found in [88–93].

$$K(x, x') = \exp\left(-\frac{\|x - x'\|_2^2}{2\sigma^2}\right)$$ (33)

SVM classification of EMG data requires a priori knowledge of model parameters. Consensus in the literature over the most appropriate model selection approach is missing. In this study, statistical criteria were explored as possible decision tools in the model selection process: the Akaike information criterion (AIC), final prediction error, and the minimum description length (MDL) [94]. Criticism of any of the criteria can be found in the literature. For instance, concerns about the AIC of inconsistency and choice of too complex models have been expressed in previous literature [95]. In general, a high-order model would fit a given dataset better, from a prediction error standpoint, due to the increase in degrees of freedom. Yet, greater memory size and computation time is required to achieve such high performance. Hence, choosing the smallest model order that is able to sufficiently

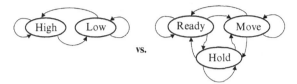

Fig. 10 Stiffness level versus operator intent model approaches.

characterize the given dataset with reduced risk of data under fitting is ideal. Such a goal is achievable by using the MDL criterion in a recursive fashion. Computation of the MDL can be done through Eq. (34), where V_n is an index related to the prediction error or the residual sum of squares, p is the number of model parameters, and N is the number of data points [96]. The optimal model order choice is the one that minimizes Eq. (34) and therefore allows the shortest description of the given dataset.

$$MDL = V_n\left(1 + \frac{p \ln N}{N}\right) \tag{34}$$

5.2.4 Stiffness Level Approach Versus Operator Intent Approach

HMMs are known to work best when classifying unique states rather than trying to identify whether a value is above or below a threshold. A more useful approach to using an HMM as a human operator model is to focus on an intended subtask instead of stiffness level. While stiffness level is indicative of the operator's motion, estimating the phase of motion can help eliminate issues linked to the stochasticity and wide variability of human arm stiffness. Therefore, refining the training approach of both the previous HMM and SVM models can be done, in order to be able to estimate which subtask of a given task an operator is performing. For example, a pick-and-place task can be subdivided into three distinct subtasks: holding steady, moving, and relaxed. Fig. 10 shows a graphical comparison between two transition models, one using a stiffness level approach and the other using an intent-based approach.

6. SYSTEM VALIDATION AND EXPERIMENTAL EVALUATION

6.1 Haptic Device

The testing device used in this study is a 1 degree-of-freedom (1-DOF) haptic paddle device shown in Fig. 11. The device can produce a maximum

Fig. 11 1-DOF haptic paddle device.

force of 100 N at the handle and was determined to have a maximum frequency response of 10 Hz. The haptic system set-up includes a multiaxis force/torque sensor (ATI Industrial Automation, Apex, NC, USA) with stand-alone controller, a brushless motor (Anaheim Automation, Inc., Anaheim, CA, USA) linked to an incremental optical encoder (US Digital, Vancouver, WA, USA) and controlled by a servo motor drive (Pacific Scientific, CA, USA). The haptic robot is run using a CompactRio real-time controller programmable and deployable through a user interface in LabView software (National Instruments, Austin, TX, USA). EMG signals in this experiment were initially acquired through an 8-channel Myopac Junior system (RUN Technologies, Mission Viejo, CA, USA) and later using an 8-channel EMG wireless system (Cometa Srl, Milan, Italy) with true differential electrodes (common-mode rejection ratio >120 dB) without reference electrode.

6.2 Stability Analysis

Using the 1-DOF haptic interface as a testbed, conditions under which a haptic system grows unstable as the operator arm stiffness increased were reproduced. Fig. 12 plots the magnitude of device oscillation while an operator attempts to hold the device steady. The time delay for the force feedback and the stiffness of the operator's arm were independently varied to characterize the stability of the device. As either variable increased, the magnitude of the uncontrollable oscillation grew. With no time delay

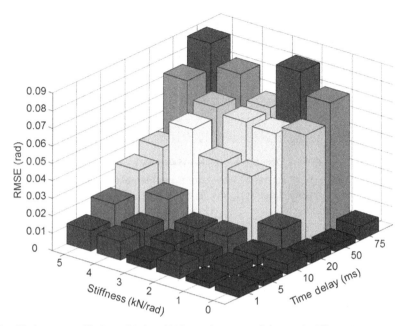

Fig. 12 Larger oscillations (higher RMS error) as time delay and stiffness increase.

and minimal stiffness, the device was much more stable. This shows that the increased stiffness and time delay combine to prevent the system from remaining stable.

6.3 Stiffness Distribution

Analysis of the stiffness distribution obtained from testing showed that having more than two stiffness levels provided no statistically significant advantage, as represented in Fig. 13. Further, the data showed that the stiffness level generated by operators was inconsistent and not necessarily just high enough for the applied force. The stiffness, k_o, dataset collected was poorly fit by several common distributions, but proved to be well fit by two separate Gaussians, shown in Fig. 14, supporting a two-level classification scheme.

6.4 System Validation: Contact With Rigid Surface

To validate the theoretical stability analysis, it was necessary to test the stability of the haptic paddle device under real-world conditions. Therefore, an experiment was designed to create a situation which is usually unstable for

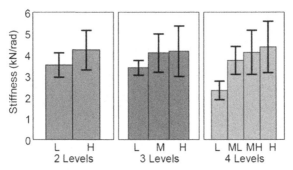

Fig. 13 Average of classified stiffness levels for experimental data; error bars show standard deviation of classified points: *L*, low; *M*, medium; *H*, high.

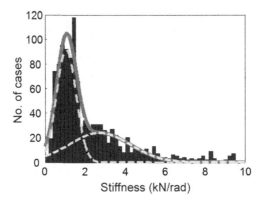

Fig. 14 Stiffness, k_o, data fit with two normal distributions $\left(R^2 = 0.917\right)$.

haptic devices and evaluate how well the compensating system could stabilize the device as compared to a low damping fixed gain system.

Haptic devices are difficult to hold against a rigid surface due to the reaction force of contact between the two [97]. When an operator attempts to do so, the device repeatedly bounces off the surface and becomes unstable. It was expected that the operator would stiffen their arm to hold the device against the surface, so the damping coefficient would increase when the compensation was on, stabilizing the system when needed. Participants were asked to hold the device against a fixed rigid surface with and without the compensating controller, and the device position was recorded over time. To measure the stability of the system, the root-mean-square error (RMSE) of the distance to the surface was calculated for each attempt, which was expected to be minimal for a goal of maintaining contact with the surface. Each participant was oriented with the EMG measurement system and haptic device, and then had

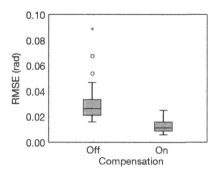

Fig. 15 Mean and variance of RMSE data for compensation on and off. Participants show a significant decrease with compensation on (* represents an "extreme outlier," a data-point located at least three times the interquartile range below the first quartile or above the third quartile respectively).

the EMG measurement system connected. After using the device for a short time to minimize learning effects, participants were asked to place the handle of the device against a rigid surface and hold it in contact for 5 s. This was repeated several times with the compensation both on and off. The results were analyzed by the analysis of variance (ANOVA) method to find statistically significant differences between controller states.

The number of participants was chosen based on the desired power, $1 - \beta$, of the resulting statistical analysis, which indicates the chance of statistical errors, β, with a typical target power $1 - \beta = 0.95$. This required a minimum of 16 participants to obtain statistically significant results. The experiment included 20 participants (12 males; 8 females; age range 19–37), resulting in $1 - \beta = 0.965$ and a required critical $F = 1.29$ for statistical significance. This was performed followed an approved institutional review board (IRB) protocol. As demonstrated by Fig. 15, participants were able to reduce the average RMSE with the compensation on. The ANOVA analysis showed a statistically significant result ($F = 55.72$, $p \leq 0.05$).

6.5 Validation Experiment: Contact With Rigid Surface

The effect of the five stiffness classifiers (fixed high, fixed low, threshold, HMM, SVM) and three position controllers (PD, LQG, SLQR) on system performance was evaluated in comparison at different stages of the study. Performance trials in this study consisted of speed and accuracy trials. In each task, the participant was given a series of targets to reach on a computer screen in front of the device. Fig. 16 shows the visual display of the experimental simulation. For speed trials, subjects were instructed to move the

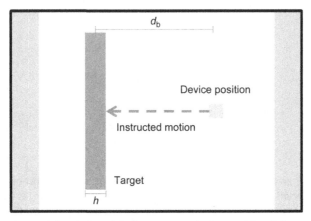

Fig. 16 Experimental simulation; right box, current device position; left gray box, target position.

device to the target as fast as possible and told that overshooting was allowed, though slightly penalized. In the accuracy trials, the instructions provided were to move the device as accurately as possible while avoiding overshooting. The participants were allowed to practice each type of trial until they felt comfortable with each task. The order of trials was randomized using a Latin Square design of experiments, in order to eliminate any effects from confounding variables. For each of the trials, scores for each move were calculated and displayed on screen for the subject. A total score was computed at the end of each trial. Calculation of scores was done according to equations provided in Appendix B, with a higher performance corresponding to a lower score. A typical experimental run is shown in Fig. 17.

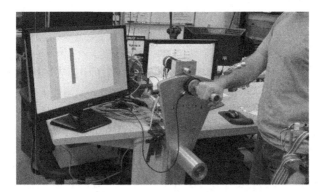

Fig. 17 Haptic performance experimental run.

6.6 Threshold Versus HMM Classification Performance

Results from performance experiments performed on 24 participants (16 male, 8 female, age range 19–42) are shown in Figs. 18 and 19. Fig. 18 shows the distribution of the data based on accuracy, separated by classifier in Fig. 18B and by controller in Fig. 18C, while Fig. 19 shows the results of the speed tasks in a similar fashion. All statistics are presented with a corrected F value using Pillai's trace. Also, unless otherwise stated, $p < 0.001$ may be assumed.

For the variation of classifier, the results demonstrated that the change of the classifier had a statistically significant effect on both tests ($F = 162.0$ and $PV = 0.112$). For the variation of the controller, the results were also statistically significant ($F = 273.9$ and $PV = 0.125$). Also, testing for interactions between the two independent variables showed that the combination of

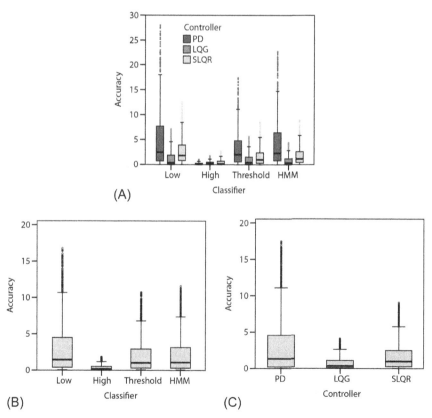

(A)

(B)

(C)

Fig. 18 Accuracy results for HMM performance experiment (lower score = better performance). (A) All cases, (B) separated by classifier, and (C) separated by controller.

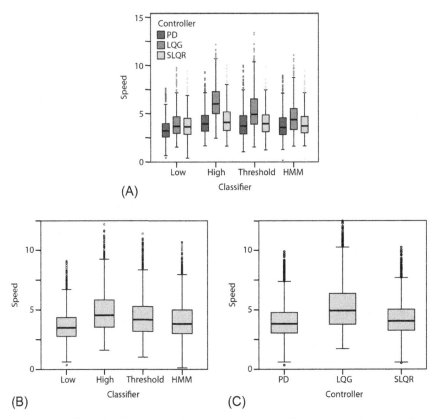

Fig. 19 Speed results for HMM performance experiment (lower score = better performance). (A) All cases, (B) separated by classifier, and (C) separated by controller.

classifier and controller variation has a statistically significant effect on the results ($F = 33.52$ and $PV = 0.48$). Pairwise comparisons between each of the classifiers showed statistically significant differences with $p < 0.001$, except for between the HMM and threshold, which did not demonstrate a statistically significant difference with $p = 0.116$. Between the controllers, all pairs were significantly different with $p < 0.001$, except for between the LQG and SLQR, which showed significance with $p = 0.017$.

Pairwise comparisons between each of the classifiers showed statistically significant differences ($p < 0.001$), except between the HMM and threshold classifiers, which did not demonstrate a statistically significant difference ($p = 0.116$). Between the controllers, all pairs were significantly different ($p < 0.001$), except between the LQG and SLQR controllers ($p = 0.017$).

6.7 Threshold Versus SVM Classification Performance

Three participants (all male; ages 23–28) were asked to resist the displacement of the haptic device, which generated a randomized set of force levels and directions at the handle. This testing was also used to generate the participant's distribution of endpoint stiffness, using force and position recorded signals. The number of classes for this approach during model selection was determined by the MDL criterion approach. Using an MDL algorithm [98] recursively, an optimal model order that can no longer be minimized is obtained [99].

Two approaches to SVM classification were employed: a value-based approach that directly related EMG activation levels to stiffness levels, and an intent-based approach that mapped EMG levels to predetermined cognitive states. Fig. 10 illustrates an example of this distinction using a two-value state model versus a three-intent state model. For the value-based approach, training muscle cocontraction data for an SVM classifier were generated using EMG measurements obtained during a series of static (hold) trials.

Multiclass SVM training using a LIBSVM LabView library tool [93] was carried (ν-SVC, RBF kernel). In addition to training the classifier, optimal control gains of the SLQR controller previously described were calculated for each class (state) in each approach, ultimately leading to three sets of gain matrices in the intent-based approach, and an MDL-based number of gain matrices in the value-based approach. The complete calibration procedure is summarized in [100].

Figs. 20 and 21 summarize the performance results obtained using the three different classifiers (threshold, SVM-value, and SVM-intent), each of which was tested in combination with one of the three position controllers (PD, LQR, SLQR). No statistically significant difference ($p > 0.05$) was found between the three classifiers when compared based on the accuracy performance trials. For the speed trials, the only statistical difference found was between the threshold and value-based SVM, when both are combined with the SLQR position controller, with the Threshold classifier increasing speed performance by about 50% on average. When looking at the difference between classifiers, irrespective of the position controller that was used, we found a statistically significant difference ($p < 0.05$) between the Threshold classifier and each of the two SVMs, but not between the two SVMs. In terms of accuracy, a significant difference in performance is only observable between using PD control and LQG control, whereas for speed trials, the

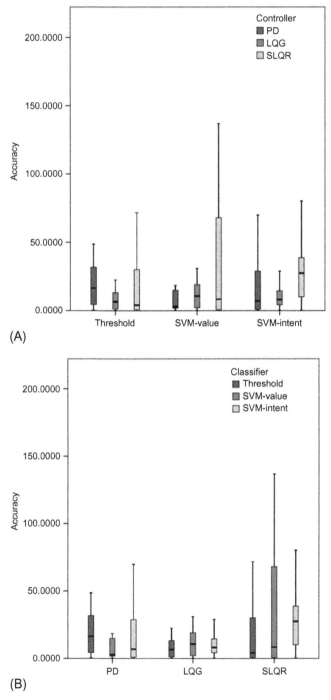

Fig. 20 Accuracy results for SVM performance experiment (lower score = better performance). (A) Separated by classifier and (B) separated by controller.

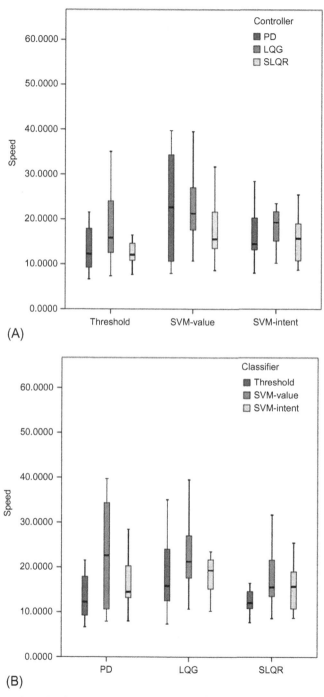

Fig. 21 Speed results for SVM performance experiment (lower score = better performance). (A) Separated by classifier and (B) separated by controller.

SLQR was observed to be significantly faster than the LQG and slightly faster, though not significantly, than PD (p-value between 0.055 and 0.059).

7. LIMITATIONS AND SOLUTION APPROACHES

The results of the experiments in this research study very strongly suggest a statistically significant difference between the noncompensating system and the compensating one. Also, there is a trade-off between the speed and accuracy of the device. This trade-off is clearly demonstrated since, for all controllers, a high damping setting consistently yielded higher speed scores and lower accuracy scores, respectively indicating that the target was reached more slowly yet more accurately. All classifiers (threshold, HMM, and SVM) seemed to achieve a balance between these two performance metrics, speed and accuracy. Overall, the results also indicate that, contrary to expectations, the threshold classifier mostly performed better than HMM and SVM, with varied level of statistical significance. For all classifiers, the accuracy of the model was found to be below 90% on average (51–87% range). Hence, achieving higher classification accuracy is critical to improving the performance of the proposed control approach. Such improvement can be achieved by identifying the optimal training parameters such as training data size and training time. Another important factor to consider for improved HMM or SVM classification is the need for better mapping from cocontraction levels to operator physical or cognitive state. Proper tuning of impedance gains is dependent on the accuracy of this mapping. Analysis of training data used for intent-based classification reveals an overlapping region between cocontraction levels recorded in each state. Hence, in the level-based approach, for instance, it is likely that sorting of the training classes in ascending order of vector norm from origin to class distribution mean does not translate in ascending endpoint stiffness levels. In this sense, an intent-based approach may show more promise. However, prior knowledge of cognitive states is a necessary condition.

For controllers, little difference was found between SLQR and PD controllers during speed trials, and both were significantly better than the LQG controller. On the other hand, the PD controller was the worst performing controller of the three during accuracy trials, with no statistically significant difference between LQG and SLQR. In general, use of the LQG controller resulted in a more "sluggish" response from the system and the effect of changing controllers during accuracy trials became insignificant as damping was increased. The slower response from the LQG controller currently makes it the least desirable of the three controllers. Out of the remaining two, the SLQR controller shows greater promise, given the presence of a trade-off between stability and performance in the system. Despite their comparable potential, PD controllers can fail to provide the best performance and a guarantee of stability under a wide range of operating conditions, due to a fixed gain approach that discounts changes in dynamic system parameters in stochastic systems. It can be shown, on the other hand, that LQR-based controllers have a gain margin within $[0.5, \infty)$ and a phase margin of at least 60 degrees [101]. In addition, mean and mean square stochastic stability are guaranteed with the use of the SLQR controller [68,69].

In addition, the current control system relies only on detecting changes in stiffness through EMG measurement. The quality of this correlation could likely be enhanced with less noisy muscle activity measurements. Further analysis of the collected data indicated that the operator's strategy for choosing the appropriate stiffness level for a given situation was not straightforward. It would be expected for a person to choose a stiffness level that is just high enough for the applied force, but the data showed that the stiffness level for a given applied force was inconsistent. While the strategy that a human uses to choose the appropriate stiffness level is unknown, it is clearly more complicated than balancing the applied force with minimum effort. The use of musculoskeletal models, such as that of Ueda et al. [38–43], can provide insights into stiffness modulation at the endpoint level and even be used for stiffness prediction when adapted for real-time computation.

8. CONCLUSION

Instability of haptic force feedback devices under human contact can lead to undesired oscillations in the combined human and machine system. This chapter discusses research towards a novel controller design for force amplifying systems that can successfully increase the system's stability on demand, allowing for higher performance than similar systems with low and fixed damping gains, yet while retaining stability when necessary. This increase in stability was accomplished by estimating changes in arm endpoint stiffness based on cocontraction levels measured using EMGs. Justification for the use of EMGs is based on experimental validation and is consistent with previous literature. Typical haptic controllers, which combine an impedance controller with an inner position controller, constitute the basis of the proposed novel control approach. In this research, the gains of the impedance controller were adjusted based on stiffness or state classification using a stiffness classifier. Three classifiers were investigated: a threshold classifier, a HMM classifier, and a SVM classifier. Under low stiffness situations, the parameters were chosen to maintain low damping and allow for fast movement. However, when stiffness increased, the gains were adjusted to increase damping and allow the device to easily be held steady. Experiments showed that the system demonstrated improved stability in stiff situations and improved performance with regards to both speed and accuracy of task completion. Due to the inherent stochastic nature of the stiffness levels generated by users, the impedance controller's inner position controller was changed from a standard PD controller to one of two stochastically tolerant controllers. An LQG showed improvement and better modeled the system's parameters, but still required some performance sacrifices with regards to speed. An LQR specifically adapted to systems with stochastically varying parameters (SLQR) provided better results than the standard PD controller without the sacrifices of the LQG controller.

Ultimately, the design allowed a haptic controller to be more intelligent, by gathering more information about the operator, therefore enabling it to adjust, decreasing the workload on the operator and increasing overall performance. By allowing such a system to estimate the intentional state of the operator, it can overcome some of the issues with fixed gain systems. This more intelligent controller can simplify the operator's job and allow for performance increase, which in industrial settings could translate into higher productivity. By combining the compensation system with an advanced controller capable of predicting, and even controlling, the muscle activity levels of the user, it would be possible to design robots that interact very closely with human operators. Success in this research would benefit communities interested in adaptive shared control approaches, notably automobile, aerospace, construction, and medical industries.

APPENDIX A CROWNINSHIELD'S OPTIMALITY PRINCIPLE

Nominal human muscle force f_o can be determined through minimization of a physiologically based optimality criterion $u(f)$, under Crowninshield's method [102], as shown in Eq. (A1).

$$u(f) = \sum_{j=1}^{N} \left(\frac{f_j}{PCSA_j} \right)^r \to \min$$

$$\text{subject to} \begin{cases} \tau_h = Af \\ f_{\min_j} \le f_j \le f_{\max_j} \quad (j = 1, \ldots, N = 51) \end{cases} \tag{A1}$$

Here, τ_h is the human joint torque vector, A is the moment arm matrix of muscles, $PCSA_j$ is the physiological cross-sectional area (PCSA), and $f_{\max_j} = \varepsilon \cdot PCSA_j$ is the maximum muscle force of the jth muscle. In this paper, $\varepsilon = 0.7 \times 10^6 \, (\mathrm{N/m^2})$ is given according to [103]. $PCSA_j$'s are given according to [104]. $f_{\min_j} = 0$, $\forall j$ and $r = 2$ are used. See [102] for the choice of r.

APPENDIX B PERFORMANCE SCORES CALCULATION

Calculation of performance scores are based on equations found in [47]. For the speed-based tasks, the performance score, j_{speed}, is the amount of time the user took to move to the target, t_b, normalized by the distance to the target, d_b. A penalty for overshooting the target is added using the overshoot time, t_o, in order to differentiate fast and accurate motions from motions that are only fast. The resulting equation is given by Eq. (A2), with a lower score signifying a better performance.

$$j_{\text{speed}} = \frac{t_b}{d_b} + c_s t_s \tag{A2}$$

For the accuracy task, the score, j_{accuracy}, was based on the point of closest approach to the center of the target, $d_{\min}$, or the maximum overshoot, d_s, if the participant overshoots the target. This is normalized with respect to the width of the target, h, then squared to more strongly penalize poor accuracy. An additional penalizing factor is added for overshooting the target. Also, a small factor, $\dot{x}_z$, that accounts for the velocity to reach the target, was added as well to differentiate between simply accurate motions and both fast and accurate motions. The final equation for the score is given by Eq. (A3).

$$j_{\text{accuracy}} = \begin{cases} \left(\dfrac{d_s}{h}\right)^2 + \dfrac{d_s}{\omega} + \dfrac{\dot{x}_b}{\dot{x}_{\max}} & \text{if overshoot} \\[3mm] \left(\dfrac{d_{\min}}{h}\right)^2 - \dfrac{\dot{x}_b}{\dot{x}_{\max}} & \text{if no overshoot} \end{cases} \tag{A3}$$

ACKNOWLEDGMENTS

The authors would like to acknowledge General Motors and the National Science Foundation for their sponsorship (National Robotics Initiative Grant Numbers 1142438 and 1317718, and Computing Research Infrastructure Grant Number 1059362). The authors would also like to thank Timothy McPherson of the Biorobotics and Human Modeling Lab, J.D. Huggins of the Intelligent Machine Dynamics Lab, the Center for Robotics and Intelligent Machines, the Applied Physiology Department, and the Cognitive Engineering Center at Georgia Tech for their assistance and advice. Opinions expressed in this document can be attributed to the authors alone and do not necessarily reflect the views of any institution.

REFERENCES

[1] S. P. Buerger, N. Hogan, Complementary stability and loop shaping for improved human-robot interaction, IEEE Trans. Robot. 23 (2) (2007) 232–244.

[2] J. E. Colgate, G. G. Schenkel, Passivity of a class of sampled-data systems: application to haptic interfaces, J. Robot. Syst. 14 (1) (1997) 37–47.

[3] J. J. Gil, E. Sánchez, T. Hulin, C. Preusche, G. Hirzinger, Stability boundary for haptic rendering: influence of damping and delay, J. Comput. Inf. Sci. Eng. 9 (1) (2009) 011005.

[4] T. Hulin, C. Preusche, G. Hirzinger, Stability boundary for haptic rendering: influence of human operator, in: IEEE/RSJ International Conference on Intelligent Robots and Systems, IROS 2008, IEEE, 2008, pp. 3483–3488.

[5] H. Kazerooni, T. J. Snyder, Case study on haptic devices: human-induced instability in powered hand controllers, J. Guid. Control. Dyn. 18 (1) (1995) 108–113.

[6] H. Kazerooni, A. Chu, R. Steger, That which does not stabilize, will only make us stronger, Int. J. Robot. Res. 26 (1) (2007) 75–89.

[7] H. Kazerooni, J. Guo, Human extenders, J. Dyn. Syst. Meas. Control. 115 (2B) (1993) 281–290.

[8] P. Y. Li, Design and control of a hydraulic human power amplifier, in: ASME 2004 International Mechanical Engineering Congress and Exposition, American Society of Mechanical Engineers, 2004, pp. 385–393.

[9] E. Colgate, N. Hogan, An analysis of contact instability in terms of passive physical equivalents, in: Proceedings of the 1989 IEEE International Conference on Robotics and Automation, IEEE, 1989, pp. 404–409.

[10] V. Duchaine, C. M. Gosselin, Investigation of human-robot interaction stability using Lyapunov theory, in: IEEE International Conference on Robotics and Automation, ICRA 2008, IEEE, 2008, pp. 2189–2194.

[11] T. Tsumugiwa, R. Yokogawa, K. Yoshida, Stability analysis for impedance control of robot for human-robot cooperative task system, in: Proceedings of the 2004 IEEE/RSJ International Conference on Intelligent Robots and Systems, IROS 2004, vol. 4, IEEE, 2004, pp. 3883–3888.

[12] J. E. Colgate, N. Hogan, Robust control of dynamically interacting systems, Int. J. Control. 48 (1) (1988) 65–88.

[13] B. Hannaford, A design framework for teleoperators with kinesthetic feedback, IEEE Trans. Robot. Autom. 5 (4) (1989) 426–434.

[14] D. Lawrence, Stability and transparency in bilateral teleoperation, IEEE Trans. Robot. Autom. 9 (5) (1993) 624–637.

[15] P. Y. Li, Passive control of bilateral teleoperated manipulators, in: Proceedings of the American Control Conference 1998, vol. 6, IEEE, 1998, pp. 3838–3842.

[16] Y. Yokokohji, T. Yoshikawa, Bilateral control of master-slave manipulators for ideal kinesthetic coupling-formulation and experiment, IEEE Trans. Robot. Autom. 10 (5) (1994) 605–620.

[17] R. J. Anderson, M. W. Spong, Bilateral control of teleoperators with time delay, IEEE Trans. Autom. Control 34 (5) (1989) 494–501.

[18] J. E. Colgate, Robust impedance shaping telemanipulation, IEEE Trans. Robot. Autom. 9 (4) (1993) 374–384.

[19] D. Lee, P. Y. Li, Toward robust passivity: a passive control implementation structure for mechanical teleoperators, in: Proceedings of the 11th Symposium on Haptic Interfaces for Virtual Environment and Teleoperator Systems, HAPTICS 2003, IEEE, 2003, pp. 132–139.

[20] A. V. Hill, The heat of shortening and the dynamic constants of muscle, Proc. R. Soc. Lond. B Biol. Sci. 126 (843) (1938) 136–195.

[21] N. A. Bernstein, The Co-ordination and Regulation of Movements, Pergamon Press, Oxford, 1967.

[22] N. Hogan, Adaptive control of mechanical impedance by coactivation of antagonist muscles, IEEE Trans. Autom. Control 29 (8) (1984) 681–690.

[23] A. M. Smith, The coactivation of antagonist muscles, Can. J. Physiol. Pharmacol. 59 (7) (1981) 733–747.

[24] J. McIntyre, F. A. Mussa-Ivaldi, E. Bizzi, The control of stable postures in the multijoint arm, Exp. Brain Res. 110 (2) (1996) 248–264.

[25] W. M. Murray, S. L. Delp, T. S. Buchanan, Variation of muscle moment arms with elbow and forearm position, J. Biomech. 28 (5) (1995) 513–525.

[26] T. Flash, F. Mussa-Ivaldi, Human arm stiffness characteristics during the maintenance of posture, Exp. Brain Res. 82 (2) (1990) 315–326.

[27] E. J. Perreault, R. F. Kirsch, P. E. Crago, Voluntary control of static endpoint stiffness during force regulation tasks, J. Neurophysiol. 87 (6) (2002) 2808–2816.

[28] E. J. Perreault, R. F. Kirsch, P. E. Crago, Multijoint dynamics and postural stability of the human arm, Exp. Brain Res. 157 (4) (2004) 507–517.

[29] S. J. De Serres, T. E. Milner, Wrist muscle activation patterns and stiffness associated with stable and unstable mechanical loads, Exp. Brain Res. 86 (2) (1991) 451–458.

[30] T. E. Milner, C. Cloutier, A. B. Leger, D. W. Franklin, Inability to activate muscles maximally during cocontraction and the effect on joint stiffness, Exp. Brain Res. 107 (2) (1995) 293–305.

[31] D. J. Bennett, J. M. Hollerbach, Y. Xu, I. W. Hunter, Time-varying stiffness of human elbow joint during cyclic voluntary movement, Exp. Brain Res. 88 (2) (1992) 433–442.

[32] E. Burdet, R. Osu, D. W. Franklin, T. E. Milner, M. Kawato, The central nervous system stabilizes unstable dynamics by learning optimal impedance, Nature 414 (6862) (2001) 446–449.

[33] D. W. Franklin, E. Burdet, R. Osu, M. Kawato, T. E. Milner, Functional significance of stiffness in adaptation of multijoint arm movements to stable and unstable dynamics, Exp. Brain Res. 151 (2) (2003) 145–157.

[34] D. W. Franklin, T. E. Milner, Adaptive control of stiffness to stabilize hand position with large loads, Exp. Brain Res. 152 (2) (2003) 211–220.

[35] T. E. Milner, C. Cloutier, Compensation for mechanically unstable loading in voluntary wrist movement, Exp. Brain Res. 94 (3) (1993) 522–532.

[36] F. Lacquaniti, F. Licata, J. F. Soechting, The mechanical behavior of the human forearm in response to transient perturbations, Biol. Cybern. 44 (1) (1982) 35–46.

[37] T. Tsuji, P. G. Morasso, K. Goto, K. Ito, Human hand impedance characteristics during maintained posture, Biol. Cybern. 72 (6) (1995) 475–485.

[38] M. Ding, Y. Kurita, J. Ueda, T. Ogasawara, Pinpointed muscle force control taking into account the control DOF of power-assisting device, in: ASME 2010 Dynamic Systems and Control Conference, American Society of Mechanical Engineers, 2010, pp. 341–348.

[39] M. Ding, J. Ueda, T. Ogasawara, Pinpointed muscle force control using a power-assisting device: system configuration and experiment, in: 2nd IEEE RAS & EMBS International Conference on Biomedical Robotics and Biomechatronics, BioRob 2008, IEEE, 2008, pp. 181–186.

[40] J. Ueda, M. Ding, M. Matsugashita, R. Oya, T. Ogasawara, Pinpointed control of muscles by using power-assisting device, in: 2007 IEEE International Conference on Robotics and Automation, IEEE, 2007, pp. 3621–3626.

[41] J. Ueda, M. Hyderabadwala, V. Krishnamoorthy, M. Shinohara, Motor task planning for neuromuscular function tests using an individual muscle control technique, in: IEEE International Conference on Rehabilitation Robotics, ICORR 2009, IEEE, 2009, pp. 133–138.

[42] J. Ueda, D. Ming, V. Krishnamoorthy, M. Shinohara, T. Ogasawara, Individual muscle control using an exoskeleton robot for muscle function testing, IEEE Trans. Neural Syst. Rehabil. Eng. 18 (4) (2010) 339–350.

[43] J. Ueda, M. Matsugashita, R. Oya, T. Ogasawara, Control of muscle force during exercise using a musculoskeletal-exoskeletal integrated human model, in: Experimental Robotics, Springer, Berlin, Heidelberg, 2008, pp. 143–152.

[44] B. Siciliano, L. Villani, Robot Force Control, vol. 540, Springer Science & Business Media, New York, 2012.

[45] N. Hogan, Controlling impedance at the man/machine interface, in: Proceedings of the 1989 IEEE International Conference on Robotics and Automation, IEEE, 1989, pp. 1626–1631.

[46] F. Mobasser, K. Hashtrudi-Zaad, Adaptive teleoperation control using online estimate of operator's arm damping, in: 2006 45th IEEE Conference on Decision and Control, IEEE, 2006, pp. 2032–2038.

[47] W. J. Gallagher, Modeling of Operator Action for Intelligent Control of Haptic Human-Robot Interfaces, Robotics PhD Thesis, Georgia Institute of Technology, 2013.

[48] W. Gallagher, D. Gao, J. Ueda, Improved stability of haptic human-robot interfaces using measurement of human arm stiffness, Adv. Robot. 28 (13) (2014) 869–882.

[49] K. Zhou, J. C. Doyle, K. Glover, Robust and Optimal Control, vol. 40, Prentice Hall, Upper Saddle River, NJ, 1996, p. 146.

[50] W. L. De Koning, Infinite horizon optimal control of linear discrete time systems with stochastic parameters, Automatica 18 (4) (1982) 443–453.

[51] R. E. Kalman, A new approach to linear filtering and prediction problems, J. Fluids Eng. 82 (1) (1960) 35–45.

[52] M. Athans, The role and use of the stochastic linear-quadratic-Gaussian problem in control system design, IEEE Trans. Autom. Control 16 (6) (1971) 529–552.

[53] B. D. Anderson, J. B. Moore, Optimal Filtering, Prentice Hall, Englewood Cliffs, NJ, 1979.

[54] J. C. Doyle, Guaranteed margins for LQG regulators, IEEE Trans. Autom. Control 23 (4) (1978) 756–757.

[55] W. Gallagher, J. Ueda, A haptic human-robot interface accounting for human parameter stochasticity, in: 2014 IEEE International Conference on Robotics and Automation (ICRA), IEEE, 2014, pp. 4256–4261.

[56] I. R. Petersen, M. R. James, P. Dupuis, Minimax optimal control of stochastic uncertain systems with relative entropy constraints, IEEE Trans. Autom. Control 45 (3) (2000) 398–412.

[57] E. Yaz, Minimax control of discrete nonlinear stochastic systems with noise uncertainty, in: Proceedings of the 30th IEEE Conference on Decision and Control, 1991, IEEE, 1991, pp. 1815–1816.

[58] B. Bernhardsson, Robust performance optimization of open loop type problems using models from standard identification, Syst. Control Lett. 25 (2) (1995) 79–87.

[59] M. Sternad, A. Ahlén, Robust filtering and feedforward control based on probabilistic descriptions of model errors, Automatica 29 (3) (1993) 661–679.

[60] M. Kárný, Towards fully probabilistic control design, Automatica 32 (12) (1996) 1719–1722.

[61] M. Kárný, T. V. Guy, Fully probabilistic control design, Syst. Control Lett. 55 (4) (2006) 259–265.

[62] E. Gershon, D. J. Limebeer, U. Shaked, I. Yaesh, Robust H∞ filtering of stationary continuous-time linear systems with stochastic uncertainties, IEEE Trans. Autom. Control 46 (11) (2001) 1788–1793.

[63] E. Gershon, U. Shaked, I. Yaesh, Robust H∞ estimation of stationary discrete-time linear processes with stochastic uncertainties, Syst. Control Lett. 45 (4) (2002) 257–269.

[64] Y. Wang, L. Xie, C. E. de Souza, Robust control of a class of uncertain nonlinear systems, Syst. Control Lett. 19 (2) (1992) 139–149.

[65] C. E. Wong, A. M. Okamura, The snaptic paddle: a modular haptic device, in: First Joint Eurohaptics Conference, 2005 and Symposium on Haptic Interfaces for Virtual Environment and Teleoperator Systems, 2005. World Haptics 2005, IEEE, 2005, pp. 537–538.

[66] S. S. Chang, T. Peng, Adaptive guaranteed cost control of systems with uncertain parameters, IEEE Trans. Autom. Control 17 (4) (1972) 474–483.

[67] F. Farokhi, K. H. Johansson, Limited model information control design for linear discrete-time systems with stochastic parameters, in: 2012 IEEE 51st Annual Conference on Decision and Control (CDC), IEEE, 2012, pp. 855–861.

[68] K. Fujimoto, S. Ogawa, Y. Ota, M. Nakayama, Optimal control of linear systems with stochastic parameters for variance suppression: the finite time horizon case, in: Proceedings of the 18th IFAC World Congress, 2011, pp. 12605–12610.

[69] K. Fujimoto, Y. Ota, M. Nakayama, Optimal control of linear systems with stochastic parameters for variance suppression, in: 2011 50th IEEE Conference on Decision and Control and European Control Conference (CDC-ECC), IEEE, 2011, pp. 1424–1429.

[70] J. K. Salisbury, Active stiffness control of a manipulator in Cartesian coordinates, in: 1980 19th IEEE Conference on Decision and Control Including the Symposium on Adaptive Processes, IEEE, 1980, pp. 95–100.

[71] G. Ellis, Observers in Control Systems: A Practical Guide, Academic Press, London, 2002.

[72] F. A. Mussa-Ivaldi, N. Hogan, E. Bizzi, Neural, mechanical, and geometric factors subserving arm posture in humans, J. Neurosci. 5 (10) (1985) 2732–2743.

[73] T. Tsuji, M. Kaneko, Estimation and modeling of human hand impedance during isometric muscle contraction, in: Proceedings of the ASME Dynamic Systems and Control Division (DSC), Atlanta, GA, 1996, pp. 575–582.

[74] E. Burdet, R. Osu, D. Franklin, T. Yoshioka, T. Milner, M. Kawato, A method for measuring endpoint stiffness during multi-joint arm movements, J. Biomech. 33 (12) (2000) 1705–1709.

[75] E. J. Perreault, R. F. Kirsch, P. E. Crago, Effects of voluntary force generation on the elastic components of endpoint stiffness, Exp. Brain Res. 141 (3) (2001) 312–323.

[76] R. Osu, H. Gomi, Multijoint muscle regulation mechanisms examined by measured human arm stiffness and EMG signals, J. Neurophysiol. 81 (4) (1999) 1458–1468.

[77] R. Osu, D. W. Franklin, H. Kato, H. Gomi, K. Domen, T. Yoshioka, M. Kawato, Short-and long-term changes in joint co-contraction associated with motor learning as revealed from surface EMG, J. Neurophysiol. 88 (2) (2002) 991–1004.

[78] J. N. A. L. Leijnse, The controllability of the unloaded human finger with superficial or deep flexor, J. Biomech. 30 (11) (1997) 1087–1093.

[79] A. Jinha, R. Ait-Haddou, P. Binding, W. Herzog, Antagonistic activity of one-joint muscles in three-dimensions using non-linear optimisation, Math. Biosci. 202 (1) (2006) 57–70.

[80] P. L. Gribble, L. I. Mullin, N. Cothros, A. Mattar, Role of cocontraction in arm movement accuracy, J. Neurophysiol. 89 (5) (2003) 2396–2405.

[81] K. A. Thoroughman, R. Shadmehr, Electromyographic correlates of learning an internal model of reaching movements, J. Neurosci. 19 (19) (1999) 8573–8588.

[82] Y. Yamada, Y. Umetani, H. Daitoh, T. Sakai, Construction of a human/robot coexistence system based on a model of human will-intention and desire. in: Proceedings of the IEEE International Conference on Robotics and Automation, 1999, pp. 2861–2867, http://dx.doi.org/10.1109/ROBOT.1999.774031.

[83] A. J. Schmid, O. Weede, H. Worn, Proactive robot task selection given a human intention estimate. in: Proceedings of the 16th IEEE Conference on Robot and Human Interactive Communication (RO-MAN), 2007, pp. 726–731, http://dx.doi.org/10.1109/ROMAN.2007.4415181.

[84] T. Takeda, Y. Hirata, K. Kosuge, Dance step estimation method based on HMM for dance partner robot, IEEE Trans. Ind. Electron. 54 (2) (2007) 699–706.

[85] L. Rabiner, B.-H. Juang, An introduction to hidden Markov models, ASSP Mag. 3 (1) (1986) 4–16. New York: IEEE.

[86] S. Russell, P. Norvig, Artificial Intelligence: A Modern Approach, Prentice-Hall, Englewood Cliffs, NJ, 1995. p. 25.

[87] B. Cesqui, P. Tropea, S. Micera, H. I. Krebs, EMG-based pattern recognition approach in post stroke robot-aided rehabilitation: a feasibility study, J. Neuroeng. Rehabil. 10 (2013) 75.

[88] K. P. Burnham, D. R. Anderson, Model Selection and Multimodel Inference: A Practical Information-Theoretic Approach, Springer Science & Business Media, New York, 2002.

[89] T. Hastie, R. Tibshirani, J. Friedman, Linear methods for regression, in: The Elements of Statistical Learning, Springer, New York, 2009, pp. 43–99.

[90] R. E. Kass, A. E. Raftery, Bayes factors, J. Am. Stat. Assoc. 90 (430) (1995) 773–795.

[91] J. Schoukens, Y. Rolain, R. Pintelon, Modified AIC rule for model selection in combination with prior estimated noise models, Automatica 38 (5) (2002) 903–906.

[92] E. J. Wagenmakers, S. Farrell, AIC model selection using Akaike weights, Psychon. Bull. Rev. 11 (1) (2004) 192–196.

[93] C. C. Chang, C. J. Lin, LIBSVM: a library for support vector machines, ACM Trans. Intell. Syst. Technol. 2 (3) (2011) 27. Software available at: http://www.csie.ntu.edu.tw/~cjlin/libsvm.

[94] D. M. Wilkes, G. Liang, J. A. Cadzow, ARMA model order determination and MDL: a new perspective, in: 1992 IEEE International Conference on Acoustics, Speech, and Signal Processing, ICASSP-92, vol. 5, IEEE, 1992, pp. 525–528.

[95] T. Roos, J. Rissanen, On sequentially normalized maximum likelihood models, Compare 27 (31) (2008) 256.

[96] LabVIEW™. System identification toolkit user manual. http://www.ni.com/pdf/manuals/371001c.pdf.

[97] N. Hogan, E. Colgate, Stability problems in contact tasks, Robot. Rev. 1 (1989) 339–348.

[98] C.A. Bouman, M. Shapiro, G.W. Cook, C.B. Atkins, H. Cheng, Cluster: an unsupervised algorithm for modeling Gaussian mixtures, 2005, 7, http://www.ece.purdue.edu/~bouman.

[99] A. Moualeu, W. Gallagher, J. Ueda, Support vector machine classification of muscle cocontraction to improve physical human-robot interaction, in: 2014 IEEE/RSJ International Conference on Intelligent Robots and Systems (IROS 2014), IEEE, 2014, pp. 2154–2159.

[100] A. Moualeu, J. Ueda, Haptic control in physical human-robot interaction based on support vector machine classification of muscle activity: a preliminary study, in: ASME 2014 Dynamic Systems and Control Conference, American Society of Mechanical Engineers, 2014. V003T42A005.

[101] M.G. Hollars, Experiments on the end-point control of a two-link robot with elastic drives, 1986, http://dx.doi.org/10.2514/6.1986-1977.

[102] R. D. Crowninshield, R. A. Brand, A physiologically based criterion of muscle force prediction in locomotion, J. Biomech. 14 (11) (1981) 793–801.

[103] D. Karlsson, B. Peterson, Towards a model for force predictions in the human shoulder, J. Biomech. 25 (2) (1992) 189–199.

[104] MotCo Project, http://motco.info/data/pcsa.html.

INDEX

Note: Page numbers followed by *f* indicate figures, *t* indicate tables, and *b* indicate boxes.

Printed in the United States
By Bookmasters